Instructor's Manual
Test Item File

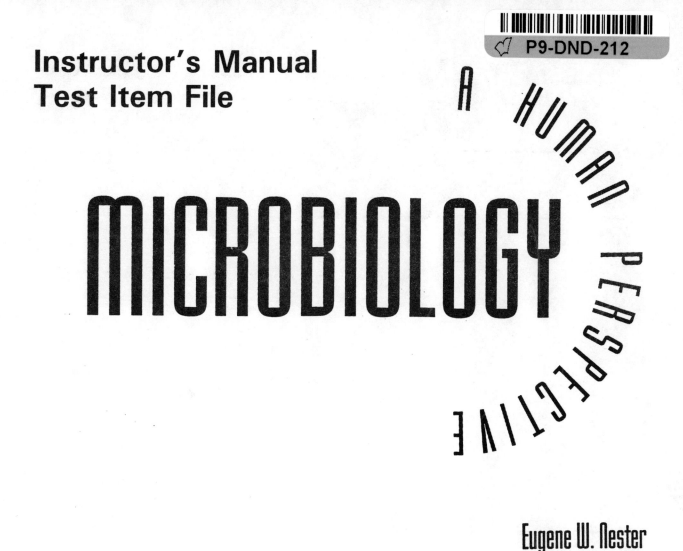

MICROBIOLOGY

A HUMAN PERSPECTIVE

Eugene W. Nester
University of Washington

C. Evans Roberts
University of Washington

Martha T. Nester

Prepared by
Clementine A. deAngelis

Wm. C. Brown Publishers

Dubuque, IA Bogota Boston Buenos Aires Caracas Chicago
Guilford, CT London Madrid Mexico City Sydney Toronto

Copyright © 1995 by Wm. C. Brown Communications, Inc. All rights reserved

A Times Mirror Company

ISBN 0-697-12761-3

Printed in the United States of America by Wm. C. Brown Communications, Inc., 2460 Kerper Boulevard, Dubuque, Iowa, 52001

10 9 8 7 6 5 4 3 2 1

Contents

Preface

This *Instructor's Manual* is designed to accompany the first edition of *Microbiology: A Human Perspective* by Nester, Roberts, and Nester. The chapters in this manual correspond to those in the textbook. The material provided herein is intended to assist instructors in the preparation of classroom lectures and the development of exams.

This manual is divided into two sections. The first section consists of listings of the transparencies and 35mm slides that also accompany *Microbiology: A Human Perspective,* a compilation of audiovisual material sources, and instructional material for each of the 33 chapters of the text. The second section, the *Test Item File,* contains objective questions correlated to each chapter in the text.

In the first section, the instructional material for each chapter opens with an introduction that could be used as an overview of the chapter content. Also included are chapter objectives, important terms, essay questions written to encourage independent thought accompanied by logical responses, and the answers to the section review and end-of-chapter critical thinking questions that appear within the text.

The second section of the manual, the *Test Item File,* includes 60–100 objective questions pertinent to the material found in each chapter. Multiple choice, true/false and matching questions are included. This *Test Item File* is intended as an aid to instructors in preparing exams. The questions can be used as written, modified to more closely represent the instructor's course material, or used as a basis for developing essay or other types of questions. Since the questions are based solely on the text, the instructor may want to devise additional questions that incorporate lecture material or to revise some of the questions to cover additional material.

The *Test Item File* in this manual is also available on WCB's computerized testing software (MicroTest III). This computerized test generator enables you to make customized exams quickly and easily. No programming experience is needed. Tests of up to 1,000 questions can be created.

Printing an exam yourself requires access to a personal computer—an IBM (DOS or WINDOWS) that uses 3.5" or 5.25" diskettes or a Macintosh. Diskettes are available through your local William C. Brown sales representative or by phoning Educational Services at 800–228–0458. The package you receive will contain complete instructions for making up an exam.

William C. Brown Publishers also provides support services, via mail (Judi David, Wm. C. Brown Publishers, 2460 Kerper Blvd., Dubuque, IA 52001) or by phone (800–258–2385), to assist you in the use of the test generator software, as well as in the printing of tests if you do not have access to a suitable computer.

Finally, please let us know about any inaccuracies (or suggestions for improvement) that you discover in this *Instructor's Manual.* Please contact us through the Biology II Editor, Wm. C. Brown Publishers.

Transparency List

35 mm Slide Set
Microbiology: A Human Perspective
Nester/Roberts/Nester

1. *E. coli* attaching to epithelial cells in the small intestine of a pig.
   Figure 3.46   (Courtesy Harley W. Moon, USDA)

2. Wet mount of vaginal secretions containing *Trichomonas*.
   Figure 10.8(b)   (Courtesy of Dr. Thomas R. Fritsche, M.D., Ph.D., Clinical Microbiology Division, University of Washington, Seattle)

3. Wet mount of a stool sample containing roundworm eggs.
   Figure 10.8(c)   (Courtesy of Dr. Thomas R. Fritsche, M.D., Ph.D., Clinical Microbiology Division, University of Washington, Seattle)

4. Phase contrast photomicrograph of a very long free-living spirochete. Note the size relative to the cocci.
   Figure 11.3   (Courtesy J.T. Staley and J.P. Dalmasso)

5. Cultures from different strains of *Pseudomonas aeruginosa* from infected patients.
   Figure 11.6   (Evans Roberts)

6. *Simonsiella species* (gliding bacterium). Harmless residents of the human mouth.
   Figure 11.13   (Courtesy of J.T. Staley and J.P. Dalmasso)

7. Polymorphism in a protozoan, *Naegleria* amebaform. This protozoan exists in this form in human tissues.
   Figure   12.20(a)   (Courtesy of Fritz Schoenknecht)

8. Polymorphism in a protozoan, *Naegleria* flagellate form.
   Figure   12.20(b)   (Courtesy of Fritz Schoenknecht)

9. Polymorphism in a protozoan, *Naegleria* cyst form.Figure   12.20(c)   (Courtesy of Fritz Schoenknecht)

10. *Giardia lamblia* (trophozoite).   Can cause giardiasis.
    Box 12.4 Figure 1(a)   (Courtesy of Mary F. Lampe)

11. *Giardia lamblia* (cyst).
    Box 12.4 Figure 1(b)   (Courtesy of Mary F. Lampe)

12. Two strains of *E. coli* inoculated onto a sorbitol indicator medium.
    Figure 18.5   (Evans Roberts)

13. *Staphylococcus aureus* in pus, Gram-stained smear.
    Box 19.1 Figure 1(a)   (Evans Roberts)

14. *Streptococcus pyogenes* growing on blood agar.
    Figure 19.7   (Evans Roberts)

15. *Rickettsia rickettsii* growing within a cell of a rodent.
    Figure 19.9 (Rocky Mountain Laboratory, Hamilton, Montana, Centers for Disease Control)

16. *Candida albicans* in pus (Gram-stained smear).
    Figure 19.25(b)   (Evans Roberts)

17. Bacterial flora of the nasopharynx.
    Figure 20.3   (Evans Roberts)

18. *Streptococcus pyogenes*. Chain formation as revealed by dark-field examination of a fluid culture.
    Figure 20.5   (Evans Roberts)

19. Bacitracin susceptibility of *Streptococcus pyogenes*.
    Figure 20.6   (Evans Roberts)

20. *Corynebacterium diphtheriae*.   The cause of diphtheria.
    Figure 20.12   (Evans Roberts)

21. *Streptococcus pneumoniae*. Gram stain of sputum from a patient with pneumococcal pneumonia.
    Figure 21.2   (Evans Roberts)

22. *Streptococcus pneumoniae* capsules.
    Box 21.1 Figure 1(a)   (Courtesy of J.T. Staley and J.P. Dalmasso

23. *Streptococcus pneumoniae*. Inhibition of growth by optochin.
    Box 21.1 Figure 1(c)   (Evans Roberts)

24. *Klebsiella pneumoniae* in infected sputum. Pneumonia caused by *Klebsiella* species can result in permanent lung damage.
    Figure 21.4(a)   (Evans Roberts)

25. *Histoplasma capsulatum*. Yeast phase organisms in macrophage of an infected mouse.
    Figure 21.21   (Evans Roberts)

26. *Histoplasma capsulatum*. Characteristic tuberculate macroconidia in mold phase culture.
    Figure 21.22   (Evans Roberts)

27. Acute necrotizing ulcerative gingivitis. Gram stain of exudate.
Figure 22.7(b) (Evans Roberts)

28. *Campylobacter jejuni* growth on medium under microaerophilic conditions.
Figure 23.5(b) (Evans Roberts)

29. *Clostridium botulinum*. Note spores near the ends of the organisms.
Figure 23.8 (Evans Roberts)

30. Appearance of *E. coli* on blood and MacConkey agars.
Figure 24.4 (Evans Roberts)

31. Appearance of *Treponema pallidum* under dark-field illumination.
Figure 24.14 (Courtesy of F. Schoenknecht and P. Perine)

32. *Candida albicans*. When incubated in serum, blastospores develop tubelike projections called germ tubes.
Box 24.6 Figure 2 (Evans Roberts)

33. Plaques produced by poliomyelitis virus in a tissue cell culture monolayer.
Figure 25.10 (Evans Roberts)

34. Smear of brain tissue from a rabid dog (Sellers stain) showing Negri bodies.
Figure 25.13 (Evans Roberts)

35. Swarming growth of *Clostridium tetani* on agar after a 4-hour incubation.
Figure 26.7(b) (Evans Roberts)

36. Mold form of *Sporothrix schenckii*. *Sporothrix schenckii* cause sporotrichosis.
Box 26.3 Figure 1 (Evans Roberts)

37. Yeast phase of *Sporothrix schenckii* as seen in a stained smear of infected tissue.
Box 26.3 Figure 2 (Evans Roberts)

38. *Toxoplasma gondii* cyst. Cause of toxoplasmosis, the most common opportunistic brain infection in AIDS patients.
Box 28.4 Figure 2 (Courtesy of Dr. Thomas R. Fritsche, M.D., Ph.D., Clinical Microbiology Division, University of Washington, Seattle)

39. *Pediculus humanus* (body louse) is the vector for *Rickettsia prowazekii*, the cause of typhus.
Appendix Figure IX-3 (Courtesy of S. Eng and F. Schoenknecht)

40. *Phthirus pubis* (crab louse) is commonly sexually transmitted.
Appendix Figure IX-4 (Courtesy of S. Eng and F. Schoenknecht)

41. *Sarcoptes scabiei* (scabies mite) is readily identified by its microscopic appearance in scrapings of a scabies burrow.
Appendix Figure IX-6 (Courtesy of S. Eng and F. Schoenknecht)

42. Body of an adult *Enterobius vermicularis* (pinworm) surrounded by thousands of its ova.
Appendix Figure IX-8(a) (Courtesy of S. Eng and F. Schoenknecht)

43. *Enterobius vermicularis* (pinworm) ova.
Appendix Figure IX-8(b) (Courtesy of S. Eng and F. Schoenknecht)

44. Hookworm larva. Species identification of the worm causing hookworm disease is made by microscopic examination of the larvae.
Appendix Figure IX-11 (Courtesy of S. Eng and F. Schoenknecht)

45. *Strongyloides stercoralis* in feces. The ova of this intestinal roundworm hatch in the intestine.
Appendix Figure IX-13 (Courtesy of S. Eng and F. Schoenknecht)

46. *Ascaris lumbricoides*, the largest of the roundworms infecting the human intestine.
Appendix Figure IX-15 (Courtesy of S. Eng and F. Schoenknecht)

47. *Schistosoma mansoni* ovum. Schistosomiasis is a chronic disease contracted by swimming in fluke infested water.
Appendix Figure IX-21 (Courtesy of S. Eng and F. Schoenknecht)

48. *Schistosoma mansoni* cercaria, the form that penetrates the skin and initiates infestation.
Appendix Figure IX-22 (Courtesy of S. Eng and F. Schoenknecht)

49. *Diphyllobothrium latum* (fish tapeworm) ovum. Diagnosis of intestinal tapeworms can be made by identifying ova in the feces.
Appendix Figure IX-24 (Courtesy of S. Eng and F. Schoenknecht)

50. *Taenia solium* (tapeworm) scolex. Acquired from eating inadequately cooked pork.
Appendix Figure IX-25 (Courtesy of S. Eng and F. Schoenknecht)

Sources of Audiovisual Materials

Abbot Laboratories
Film Service Department
North Chicago, IL 60605

American Chemical Society
1155 16th Street, N.W.
Washington, D.C. 20036

American Edwards Laboratories
17221 Red Hill Avenue
Irvine, CA 92715

American Hospital Association
840 North Lake Shore Dr.
Chicago, IL 60611

American Medical Association
535 N. Dearborn Street
Chicago, IL 60610

American Nurses Association
2420 Pershing Road
Kansas City, MO 64108

American Society for Microbiology
1913 "I" Street, N.W.
Washington, D.C. 20006

American Society of Clinical Pathologists
2100 W. Harrison
Washington, D.C. 20006

Armed Forces Institute of Pathology
Audio Visual Support Center
Washington, D.C. 20305

Armstrong Industries
3660 Commercial Avenue
Northbrook, IL 60062

Association Films/Geigy Pharmaceutical
7838 San Fernando Road
Sun Valley, CA 91352

Audio Visual Medical Marketing, Inc.
850 Third Avenue
New York, NY 10022

Ayerst Laboratories, Association-Sterling Films
600 Garden Ave.
Ridgefield, NJ 07657

Barr Films
P.O. Box 5667
Pasadena, CA 91107

Batten, Batten, Hudson, and Swab
820 Keo Way
Des Moines, IA 50309

Bausch & Lomb
635 St. Paul Street
Rochester, NY 14602

BBC TV/Open University/The Media Guild
11526 Sorrento Valley Rd. Suite J
San Diego, CA 92121

BFA Educational Media
2211 Michigan Avenue
Santa Monica, CA 90406

Bio Service Corporation
500 S. Racine Avenue, Rm 302
Chicago, IL 60607

Biology Media
2437 Durant Ave. Suite 206
Berkeley, CA 94704

Bluestone Video Makers
4018-22nd Street
San Francisco, CA 94114

BMA Audio Cassettes
200 Park Avenue
New York, NY 10003

BNA Communications, Inc.
9401 Decoverly Hall Rd.
Rockville, MD 20850

Carolina Biological Supply Co.
2700 York Road
Burlington, NC 27215

Carousel Films and Video, Inc.
1501 Broadway
New York, NY 10036

Centron Films
1621 W. 9th Street
Lawrence, KS 66044

Charles Pfizer and Co.
630 Flushing Avenue
Brooklyn, NY 10006

Churchill Films
662 N. Robertson Boulevard
Los Angeles, CA 90069

CIBA Pharmaceutical Co.
Audio-Visual Department
Summit, NJ 07901

Clay Adams, Division of Becton-Dickinson and Co.
299 Webro Road
Parsippany, NJ 07540

Communications Skill Corporation
50 Sanford Street
Fairfield, CT 06430

Communications World
2316-2nd Avenue
Seattle, WA 98121

Coronet Films
65 East S. Water St.
Chicago, IL 60601

Davis and Geck Film Library, American Cyanamid Co.
1 Casper Street
Danbury, CT 06810

Dayton Lab Services
3235 Dayton Avenue
Loraine, OH 44055

Eastman Kodak Company
343 State Street
Rochester, NY 13650

Educational Graphic Aids, Inc.
1315 Norwood Avenue
Boulder, CO 80302

Eli Lilly and Co., Educational Resources Library
P.O. Box 100B
Indianapolis, IN 46206

Elliot Scientific
185 E. 85th St.
New York, NY 10028

Encyclopedia Britannica Educational
Corporation
2494 Teagarden Street
San Leandro, CA 94577

Filmakers Library, Inc.
133 E. 558th Street, Suite 703A
New York, NY 10022

Films Incorporated
733 Green Bay Road
Wilmette, IL 60091

Fisher Scientific Company
Educational Materials Division
4901 W. Le Moyne Ave.
Chicago, IL 60651

Florida State University
Instructional Media Center
Tallahassee, FL 32306

Harper and Row Audio-Visuals
2350 Virginia Ave.
Hagerstown, MD 21740

Harwyn Medical Photographers
4814 Larchwood Ave.
Philadelphia, PA 19143

Human Relation Media
175 Tompkins Avenue
Pleasantville, NY 10570

IBIS Media
P.O. Box 308
Pleasantville, NY 10570

Indiana University
Audio-Visual Center
Bloomington, IN 47401

International Film Bureau, Inc.
332 South Michigan Avenue
Chicago, IL 60604

J. B. Lippincott Company
East Washington Square
Philadelphia, PA 19105

John Wiley & Sons, Inc.
605 Third Avenue
New York, NY 10016

Learning Corporation of America
1350 Avenue of the Americas
New York, NY 10019

Lederle Laboratories
Film Library
Pearl River, NY 10053

Life Sciences Associates
1 Fenimore Road
Bayport, NY 11705

McGraw-Hill Book Co.
330 W. 42nd Street
New York, NY 10036

McGraw-Hill/CRM Films
P.O. Box 641
Del Mar, CA 92014

Meced, Inc.
12815-120th Avenue, N.E.
Kirkland, WA 98033

Medcom Products, Inc.
1633 Broadway
New York, NY 10019

Media Visuals
4 Midland Avenue
Hicksville, NY 11801

MEP/McGraw-Hill Publications
4530 W. 77th Street
Edina, MN 55435

Millipore Corporation
80 Asby Road
Bedford, MD 01730

Milner-Fenwick, Inc.
3800 Liberty Heights Ave.
Baltimore, MD 21215

Multi Media Publishing, Inc.
1393 S. Inca Street
Denver, CO 80223

National Audiovisual Center
General Services Administration
Washington, D.C. 20409

National Educational Media, Inc.
21601 Devonshire Street # 300
Chatsworth, CA 91311

National Health Films
P.O. Box 13973, Station K
Atlanta, GA 30324

National Medical Audiovisual Center
8600 Rockville Pike
Bethesda, MD 20209

Oklahoma State University
Audiovisual Center
Stillwater, OK 74078

PBS Video
475 L'Enfant Plaza S.W.
Washington, D.C. 20024

Perennial Education, Inc.
P.O. Box 855 Rivinia
Highland Park, IL 60034

Pfizer Medical Film Library
470 Park Avenue, South
New York, NY 10016

Photography Division
Office of Information
U.S. Department of Agriculture
Washington, D.C. 20250

Prentice-Hall Media
150 White Plains Road
Tarrytown, NY 10591

Science Software Systems, Inc.
11899 W. Pico Blvd.
West Los Angeles, CA 90064

Scientificom
706 North Dearborn Street
Chicago, IL 60610

Shell Film Library
1433 Sadlier Circle W. Drive
Indianapolis, IN 46239

Teaching Films, Inc.
P.O. Box 66824
Houston, TX 77006

The National Foundation, Supply Division
P.O. Box 2000
White Plains, NY 10602

Time-Life Video
100 Eisenhower Drive
Paramus, NJ 07652

**University of California,
Extension Media Center**
2223 Fulton Street
Berkeley, CA 94720

University Park Press
300 North Charles Street
Baltimore, MD 21202

Upjohn Professional Film Library
7170 Portage Road
Kalamazoo, MI 49001

W.B. Saunders
West Washington Square
Philadelphia, PA 19105

Wm. C. Brown Publishers
2460 Kerper Blvd.
Dubuque, IA 52001

CHAPTER 1

Introduction

Microbiology is the study of life forms too small to be seen with the naked eye. The speculation of unseen life began when humans first questioned the forces of nature, health, and disease. Pasteur and his colleagues experimented with and theorized about this invisible world and finally proved that it did indeed exist. Schleiden and Schwann proved that all cells came from previous cells and resembled the cells from which they came.

In the two hundred and fifty years since the discovery era of microbiology, advancements have been made that allow us to speculate on the molecules of life itself. More powerful microscopy now allows for visualization of the tiniest of infectious agents. Biotechnology has led us to the discovery of the molecules of heredity and life itself.

Objectives

After completing this chapter, the student should be able to:

1. Understand the significance of microorganisms to history.
2. List some important contributors to the development of microbiology.
3. Discuss the theory of spontaneous generation and its downfall.
4. Trace the development of the medical aspects of microbiology.
5. Project the future challenges to microbiologists.
6. Describe the beneficial aspects of microbiology.
7. Understand the cell theory.
8. Discuss the basic cell types.
9. Describe the similarity among cells.
10. Describe the members of the microbial world.
11. Discuss how organisms are named.

Important Terms

microorganisms

viruses

dark-field illumination

endospores

bioremediation

cells

cell theory

enzymes

deoxyribonucleic acid (DNA)

adenosine triphosphate (ATP)

coenzyme

amino acids

organelles	viroids
mitochondria	infectious agents
chloroplasts	prions
prokaryotes	binomial nomenclature
eukaryotes	genus
eubacteria	species
algae	strains
fungi	varieties
protozoa	

Essay

1. Describe the reasons behind the discrepancies in the experimental results of Tyndall and Pasteur when they attempted to refute the theory of spontaneous generation.

    The experimental conditions were not identical. The steps of the scientific method specify that all phases of the experimental conditions be as similar as possible with each repetition of the trial. Tyndall attempted to repeat Pasteur's success with his swan-necked flask refutation of spontaneous generation but neglected to use the same media. Pasteur used infusions prepared from sugar or yeast. Tyndall used hay infusions. The former would not likely contain spores, whereas the latter probably would. Spores are much more difficult to destroy and therefore the infusion much more difficult to sterilize.

2. List some recent applications of microorganisms.

    Some of the most important are:
    (1) source of protein for feeding animals
    (2) make plants resistant to insect attacks
    (3) produce vaccines against rabies, herpes, malaria, etc.
    (4) by genetic engineering produce insulin, interferon, etc.

3. What features are used to separate the major groups of bacteria?

    The features used to separate these organisms into groups include:
    (1) appearance of cells
    (2) means by which they move
    (3) nutritional characteristics
    (4) presence or absence of a rigid cell wall
    (5) reproduction in the absence of living cells

4. Describe some of the future challenges of microbiology.

    The future challenges include the following:
    (1) Solving the mystery of new diseases like AIDS, Lyme disease and toxic shock syndrome.

(2) Why are diseases that were almost eliminated now increasing in number like measles, mumps, and whooping cough?

(3) Discovering solutions to the puzzle of increasing microbial resistance to antibiotics.

5. Describe the basic cell types.

   Most cells can be described as being either prokaryotic or eukaryotic. Prokaryotic cells are simple, without membrane-bound nuclei or membrane-bound organelles. Eukaryotic cells are larger and more complex, and have a membrane-bound nucleus and membrane-bound organelles. A cell type has been identified that does not fit into either of these categories. It is described as intermediate between the prokaryote and the eukaryote. It has a membrane-bound nucleus but no mitochondria or chloroplasts.

   Example:
   Prokaryotes—bacteria; Eukaryotes—human cells
   Intermediate—*Giardia*

Review Questions (pg. 8)

1. (a) He was the first to see and describe microorganisms.
   (b) He discovered dark-field illumination.

2. Some infusions from soil environments contained heat-resistant cells that boiling did not kill.

3. In the past, smallpox and influenza have been very destructive. Today, and in the foreseeable future, AIDS is especially deadly.

4. (a) Microorganisms can produce a wide variety of medically important products such as insulin, blood clotting factors and vaccines.
   (b) A bacterium is used to transfer a wide variety of genes into plants thereby giving them useful properties.
   (c) Bacteria can break down toxic pollutants, and may be useful in cleaning up oil spills and hazardous wastes.

Review Questions (pg. 14)

1. All organisms are composed of cells, which are the fundamental units of life.

2. (1) Reproduce exact copies of themselves.
   (2) Synthesize the components of living matter from the simple foodstuffs available to them.
   (3) Obtain and use energy.

3. Prokaryotes and eukaryotes. Prokaryotes lack a "true nucleus" and membrane-bound internal structures. Eukaryotes possess a "true nucleus." The nuclear material in eukaryotes is surrounded by a double-membrane that forms the nucleus. Eukaryotes also contain intracellular membrane-bound structures.

4. (1) The appearance of the cells.
   (2) The means by which cells move.
   (3) The nutritional characteristics of the cells.
   (4) The ability to multiply in the absence of living cells.

Review Questions (pg. 19)

1. The prokaryotic members of the microbial world are eubacteria and archaebacteria.

2. Two nonliving members of the microbial world are:
   (1) viruses—either RNA or DNA enclosed in a protein coat; obligate intracellular parasite.
   (2) viroids—obligate intracellular parasite consisting of RNA and no protein coat.

3. The three eukaryotic members are:
   (1) algae—gain energy from sunlight
   (2) fungi—not photosynthetic; mostly multicellular
   (3) protozoa—not photosynthetic; unicellular

4. Genus—*Escherichia*
   Species—*coli*
   Strain—K-12

Critical Thinking Questions (pg. 21)

1. The swan-necked flask experiment of Pasteur.
2. Louis Pasteur. He made so many very important contributions to microbiology and immunology.
3. Because they are obligately intracellular they share a fundamental similarity to viruses. They cannot be grown in the absence of living cells. However, they contain all of the structures and other properties of prokaryotic cells, and lack the structures of viruses as well as their mechanism of replication.

CHAPTER 2

Introduction

The basic units of matter are the atoms. Although there are about one hundred naturally occurring elements, only a fraction of them are of significance in the chemistry of life, which is also known as metabolism.

Both organic and inorganic molecules are involved in cellular metabolism. The inorganic molecules contribute to maintenance of the fluid and electrolyte balance, while the organic molecules are involved primarily in the structure and function of life.

Water is the most abundant product on earth and is intricately involved in every cellular metabolic process. This chapter surveys the basic molecules, their chemical activity and the results that set the pattern of the life of the cell.

Objectives

After completing this chapter, the student should be able to:

1. Discuss atoms as the basic units of matter.
2. Describe the types of chemical bonds found in living systems.
3. Describe the important molecules of life.
4. Describe the structure and function of proteins.
5. Describe the structure and function of carbohydrates.
6. Describe the structure and function of nucleic acids.
7. Describe the structure and function of lipids.

Important Terms

mirror image (isomer)
atom
element
electrons
protons
neutrons
molecules
atomic number
isotopes
autoradiography
molecular weight
mole

covalent bond
hydrogen bond
enzyme
polar
nonpolar
ions
ionic bond
complementary structure
anions
cations
organic
inorganic

macromolecules	monosaccharide
proteins	disaccharide
nucleic acids	lactose
polysaccharides	sucrose
lipids	cellulose
polymers	nucleotides
subunits (monomers)	deoxyribonucleic acid
carboxyl group	purine
amino group	pyrimidine
dehydration synthesis	ester linkage
hydrolytic reaction	double-stranded helix
unnatural form	ribose
polypeptide chain	fats
primary structure	fatty acids
hydrophobic bonds	glycerol
denaturation	polyunsaturated
quaternary structure	oils
multiprotein complex	steroids
substituted protein	phospholipids
glycoprotein	lipoprotein
lipoprotein	lipopolysaccharide
carbohydrate	

Essay

1. Discuss the differences among the types of chemical bonds.

   There are three types of chemical bonds that are important in life.

   (1) Covalent bonds are strong. They are formed when atoms share electrons, are very stable, and require a great deal of energy to break. Organic molecules maintain their organization with covalent bonds.

   (2) Ionic bonds are weak bonds typically found in small inorganic molecules. The severing of the bonds yields ions, which contribute to the fluid and electrolyte balance of the cell.

   (3) Hydrogen bonds are weak but biologically important. They hold the DNA strands together. They are important in maintaining the shape of proteins.

2. Describe the structure and function of macromolecules.

   (1) Proteins are complex molecular compounds of amino acids held together by peptide bonds. The shape of the protein is determined by hydrogen bonds that maintain the three-dimensional shape of the protein.

(2) Polysaccharides are polymers of monosaccharides. Their basic structure is 1:2:1, with the proportion carbon:hydrogen:oxygen. Different polysaccharides vary in size, degree of branching, and bonding of monosaccharides, and in the monosaccharides involved.

(3) Nucleic acids include DNA and RNA. These molecules are polymers of nucleotides, and function to determine the genetics of the cell and the expression of these genes.

3. How do lipids differ from the macromolecules?

Lipids are biologically important but are too small and heterogeneous in structure to be considered macromolecules. They are insoluble in water but soluble in organic solvents. Lipids may be simple or compound. They may have additional elements in addition to the glycerol and fatty acids found in neutral fats. Among the compound lipids are phospholipids, lipoproteins, and lipopolysaccharides.

Review Questions (pg. 30)

1. The 4 major elements that make up most all living matter are carbon, hydrogen, oxygen, and nitrogen.

2. The major components and their properties:
   a. protons—heavy and positively charged; in the nucleus.
   b. neutrons—heavy; neutral in charge; in the nucleus.
   c. electrons—light; negative charge; orbit the nucleus.

3. The covalent bonds are the strongest; hydrogen are the weakest. Ionic bonds are weak in liquid, but very strong when holding atoms together in solids. Covalent bonds require enzymes for their formation and breakage.

4. Hydrogen bonds hold the two strands of DNA together and determine the properties of water.

Review Questions (pg. 43)

1. Proteins, polysaccharides and nucleic acids. The three properties they share:
   (1) size
   (2) they are all polymers
   (3) biosynthetic similarities

2. (1) Proteins composed of amino acids
   (2) nucleic acids - nucleotides, and
   (3) polysaccharides - monosaccharides

Name of covalent bonds
   proteins - peptide bond
   nucleotides - ester linkage bond
   polysaccharide - glycosidic bond

3. Size of polymer, composition, degree of branching, orientation of bonds holding monosaccharides together.

4. RNA and DNA
RNA is usually shorter than DNA and single stranded. Uracil is found in place of thymine; ribose in place of deoxyribose.
DNA is composed of a long, double-stranded helix.

5. Lipids are divided into two groups: simple and compound. Lipids are only slightly soluble in water but very soluble in organic solvents.

Critical Thinking Questions (pg. 45)

1. Water is very polar and therefore is able to dissolve many polar compounds. Also, water molecules are attracted to one another because of their polar natures which explains the fact that water is liquid and a great deal of heat must be put into the system to convert it into a gas.
2. Carbon can form four covalent bonds with a variety of other atoms, in particular C, H, and O. It is also a small atom. The combination of these properties allows it to form a variety of very large molecules. Silicon can also form four bonds with other atoms but since it is large, it cannot in turn form large molecules.

CHAPTER 3

Introduction

The study of microbiology advanced significantly after the development of the light microscope. By using a variety of staining techniques, cellular detail was more discernible. The main limitation in the study of the cell is the resolving power of the microscope.

The structure and function of bacteria have been studied in detail by using the various types of microscopes available and by extensive laboratory scrutiny.

The prokaryotic and eukaryotic organisms share a great many similarities and also exhibit significant differences. The study of the nature of these similarities and differences has led to a greater understanding of life in the unseen world.

Objectives

After completing this chapter, the student should be able to:

1. Understand the historical events leading to the development of microscopy.
2. Describe the types of microscopes important in microbiology.
3. List the staining techniques used to differentiate bacteria.
4. Describe bacterial morphology and physiology.
5. Identify the bacterial structures and their functions.
6. Compare cell structures of prokaryotes and eukaryotes.
7. Understand how energy is generated in eukaryotic cells.
8. Describe how genetic information is stored in microbes.
9. Relate other bacterial structures to their function.

Important Terms

light microscope
simple microscope
compound microscope
objective
ocular
oil immersion lens
phase contrast microscope
darkfield microscope
interference microscope
fluorescence microscope

transmission electron microscope (TEM)
thin section
freeze etching
scanning electron microscope (SEM)
scanning tunneling microscope (STM)
positive stains
negative stains
differential staining techniques
Gram stain
acid-fast stain

cocci
bacilli
spirals
binary fission
cell wall
cytoplasmic membrane
capsule
cell envelope
glycocalyx
plasma membrane
endoplasmic membrane
rough endoplasmic reticulum
lysosomes
flagella
Golgi apparatus
mitochondria
microtubules
pili
chromosome
supercoiled
histonelike
lipoprotein
lipopolysaccharide (LPS)
phospholipid
outer membrane
endotoxin
porins
halophiles

lysozyme
protoplast
spheroblast
sterols
cytoplasmic membrane
periplasmic space
bilayer membrane
membrane-bound enzymes
semipermeable
permease
active transport
facilitated diffusion
osmosis
diffusion
oligotrophs
exoenzymes
nucleus
nuclear membrane
histones
chromatin
ribosomes
glycogen
endospore
vegetative cell
sporogenesis
cellular differentiation
germination

Essay

1. List the processes necessary for cells to function.

   (1) enclose the cell contents and separate it from the external environment
   (2) replicate genetic information
   (3) synthesize cell components
   (4) generate, store, and use energy-rich compounds
   (5) cytoplasmic and/or external movement

2. Discuss the properties of the cell wall.

   (1) composed of unique subunits
   (2) can produce symptoms of disease

(3) site of action of some of the most effective antibiotics

(4) determine the Gram staining properties of the cell

3. Distinguish between the cell walls of Gram-positive and Gram-negative bacteria.

Cell walls of Gram-positive bacteria consist of multiple layers of peptidoglycan, whereas cell walls of Gram-negative bacteria consist of a single layer of peptidoglycan with several other layers. These include lipoprotein, lipopolysaccharide, and phospholipid. The differences in cell wall chemistry account for the Gram staining differences and the effectiveness of antibiotics.

4. How is energy generated in eukaryotic cells?

Energy is generated by the degradation of glucose in the mitochondria of heterotrophs. Chloroplasts in photosynthetic organisms are responsible for the conversion of carbon dioxide and water into stored energy under the influence of sunlight. It is theorized that these unique structures originated from bacteria that were engulfed by primitive eukaryotic cells.

5. Describe how genetic information is stored in prokaryotic cells.

The major structure that stores genetic information is a long, double-stranded, circular DNA molecule.

The single chromosome contains the basic genetic information of the cell. It is not surrounded by a nuclear membrane. Nonessential information is coded into plasmids, which are capable of replicating independent of the chromosome.

6. Describe how genetic information is stored in eukaryotic cells.

Genetic information is contained in several chromosomes, each of which carries different information. Each DNA molecule is doublestranded and makes up structures called chromatin. Chromatin is found in the membrane-bound nucleus.

Review Questions (pg. 54)

1. The names of the lenses are ocular and objective. The magnification is the product of the magnification of each of the lenses.

2. Magnification and resolving power.

3. Gram stain; cell wall structure.

4. *Mycobacterium,* members of which cause leprosy and tuberculosis.

Review Questions (pg. 66)

1. (1) Enclose the internal contents of the cell and separate it from the external environment.
   (2) Replicate genetic information.
   (3) Synthesize components of the cell.
   (4) Generate, store, and utilize energy rich compounds.

2. Cell wall, cytoplasmic membrane, and the capsule.

3. Peptidoglycan.

4. The chemical composition of the cell wall determines:
   (1) different staining characteristics—Gram-negative and Gram-positive
   (2) the effectiveness of penicillin

Review Questions (pg. 78)

1. Regulates flow of molecules into and out of cell, and many enzymes of energy metabolism are located in the cytoplasmic membrane.

2. In prokaryotic cells, the flagella rotate, either clockwise or counter-clockwise; in eukaryotic cells the flagella have a whip-like motion.

3. Mitochondria and chloroplasts.

4. Flagella and pili.

Review Questions (pg. 85)

1. Unlike chromosomes, many plasmids may be found in the cell and they code for dispensable functions.

2. The ribosomes of eukaryotic cells contain different proteins and RNA than those of prokaryotic cells, and antibiotics cannot attach as readily to the eukaryotic proteins.

3. Bacteria contain the storage products: the polymer of ß-hydroxybutyric acid which provides carbon and energy;
   volutin, which is the storage form of long chains of phosphate; and glycogen, which is a storage form of carbon and energy.

4. The three ways endospores differ from vegetative cells are:
   (1) they are dormant and cannot multiply;
   (2) they do not degrade compounds to generate energy; and
   (3) they do not synthesize cell components.

Critical Thinking Questions (pg. 87)

1. The success of antibiotics depends on the selective toxicity of the antibiotic, and differences in structure between the host and pathogen. Ribosomes and cell walls differ between bacteria and their hosts; the cytoplasmic membranes are similar in structure and therefore it is difficult to find an antibiotic that affects the membrane of the bacterium and not the host.

2. Proteins in the periplasm would most likely be involved in the breakdown on the entry of molecules in the environment. For example, nucleases, proteases, and binding proteins involved in the uptake of sugars and other molecules into the inside of the cell are found in the periplasm. This makes sense since only small molecules that require specific mechanisms for their uptake can pass through the cytoplasmic membrane.

CHAPTER 4

Introduction

Until Robert Koch devised a means of growing organisms in pure culture, it was impossible to identify specific organisms. The opportunity to associate a particular disease with a single organism progressed from this discovery.

With time and experience, methods of determining the growth and nutritional factors for bacterial cultivation were established. Meeting these needs required patience and experience. Through these efforts, the problems associated with laboratory maintenance of bacteria were solved.

Both synthetic and complex media were developed with specific nutrients for each of the diverse groups. The growth phases of a bacterial population have been established for both the closed and the open systems.

Objectives

After completing this chapter, the student should be able to:

1. Explain the significance of pure cultures.
2. Determine the way cell growth can be measured.
3. Describe the factors that influence microbial growth.
4. Explain the nutritional aspects of bacterial growth.
5. List the problems associated with growth of microorganisms in the laboratory.
6. Compare the growth of bacteria in nature and in the laboratory.
7. Understand the stages of growth in a bacterial population.

Important Terms

mixed population
pure culture
colony
streak-plate
agar
Petri dish
pour-plate method
binary fission
generation time
plate count
direct microscopic count

colony-forming units
counting chambers
extremophiles
psychrophiles
mesophiles
thermophiles
extreme thermophiles
obligate aerobes
obligate anaerobes
facultative anaerobes
microaerophiles

aerotolerant

catalase

acidophiles

buffers

osmotic pressure

phototrophs

autotrophs

heterotrophs

photoheterotrophs

chemoautotrophs

chemoheterotrophs

lithoheterotrophs

growth factors

synthetic medium

undefined medium

enriched medium

pathogen

anaerobe jars

enrichment culture

differential media

selective media

growth curve

lag phase

log phase

closed system

open system

chemostat

Essay

1. Explain the observation that microorganisms can cause disease in some of the body parts and not others.

    Wide variations exist in the temperature of various parts of the human body. Organisms that cause disease can only do so when their growth requirements are met. Some parts of the body are too warm for them and others too cool. Leprosy usually involves the coolest regions of the body (ears, fingers, toes). *Treponema pallidum* also selectively grows and produces lesions more frequently on the scrotum and penis than on the lips, tongue, or throat due to the cooler temperature of the male reproductive organs.

2. Discuss the enzymatic differences between aerobes and anaerobes.

    Aerobes require oxygen because they have metabolic pathways that require oxygen to convert foodstuffs into useful energy. Some anaerobes are either killed or inhibited by the presence of oxygen. In anaerobes, oxygen may be converted into toxic products. The relationship of bacteria to the presence of oxygen is related to their enzymes of metabolism.

3. Explain how some bacteria can live in a highly acidic medium.

    Some organisms (*Thiobacillus species*) can pump out protons rapidly. Amino acids and proteins in the cell can serve as natural buffers in the cytoplasm. The ability of organisms to grow over a broad pH range depends on their ability to maintain neutrality on the inside of the cell and not because their cytoplasmic proteins can withstand a very high or very low pH.

4. Discuss the importance of Koch's discovery of pure culture techniques to the development of microbiology as a science.

Before Koch's discovery, there was no way to isolate organisms from each other and therefore to connect any single organisms with a specific disease.

Koch paved the way for the discovery of the causes of specific disease by the formulation of his postulates and therefore reproducible results in the field of scientific research.

5. What is the difference between an open and closed system when describing bacterial growth?

In the laboratory, bacteria are grown in flasks of liquid or on agar plates, which have limited amounts of nutrients that cannot be renewed. This is termed a closed system. In nature, the nutrients can be continuously replaced and waste materials removed. This is termed an open system since the bacteria may remain in the exponential phase of growth for a long time period.

Review Questions (pg. 93)

1. (1) Solidifies at temperatures below $50°C$, but must be heated above $95°C$ before it is liquid. Therefore, it is a solid over the range most bacteria grow. (2) Resistant to degradation by most bacteria. (3) Stable to high temperatures and can be sterilized by heating to high temperatures.

2. Binary fission

3. (1) Plating single cells on agar plates and counting colonies that arise.
   (2) Counting the number of bacteria in a counting chamber.
   (3) Measuring turbidity of cell suspension in a liquid culture.
   Living cells are measured only by counting colonies.

4. 64 cells

Review Questions (pg. 97)

1. Temperature, osmotic pressure, pH of environment, oxygen.

2. Psychrophilic, mesophilic, thermophilic, extremely thermophilic.

3. Store at low temperature (refrigeration); store in high salt or sugar.

4. Obligate aerobes, obligate anaerobes, facultative anaerobes, microaerophiles, aerotolerant.

5. In acid, the proton H^+ combines with a compound, such as HPO_4^{2-}, to form $H_2PO_4^-$, which is a weak acid. Thus, a weak acid is formed in place of the strong acid. A strong base, OH^-, combines with other compounds, such as $H_2PO_4^-$, to form HOH (water) and HPO_4^{2-}. Therefore, a strong base, OH^-, is converted into HOH.

Review Questions (pg. 107)

1. Photoautotrophs, photoheterotrophs, chemoautotrophs, chemoheterotrophs.

2. A growth factor is a small molecule that is required for growth because it is converted into cell constituents or takes part in cell metabolism. It is not degraded in the course of its metabolism. It must be supplied to the cell because the cell cannot synthesize it. A carbon source is broken down by the cell into other molecules, which are then converted into cell components. The carbon source is often degraded and thus provides energy to the cell.

3. The nutrients in the environment may not be duplicated in the laboratory and other organisms in the environment may be providing the environmental conditions required for growth.

4. A selective medium allows only a limited number of organisms to grow; a differential medium allows all organisms to grow but certain bacteria modify the medium in a specific way that can be recognized and serves to distinguish one species of bacterium from others that can also grow on the medium. Hektoen enteric agar or bismuth sulfite agar.

5. Lag, log, stationary, death. In lag phase the cell population is synthesizing molecules and enzymes needed for cell division. An exponential increase in the cell population occurs during log phase. The number of viable cells stops increasing during the stationary phase. Death phase is characterized by an exponential decrease in the number of viable cells.

6. In an open system, there is a continual recycling of the medium, addition of nutrients, and removal of waste products. In a closed system, nothing is added or taken away from the growth environment.

Critical Thinking Questions (pg. 109)

1. Take a soil sample from the environment that is likely to contain spore-forming organisms. Heat the soil to 100°C to kill organisms other than spores. Inoculate the soil onto a medium that contains $CaCO_3$ (CO_2) as a carbon source, and 3% NaCl along with other salts, except for ammonium salts. Shake the medium in the air, which contains N_2, which can be converted into NH_4^+ by the organism.
2. Probably not; cells growing in a bacterial colony are much more crowded than they ever are in a liquid medium and yet cells continue to multiply.

CHAPTER 5

Introduction

To reproduce, cells must degrade foodstuffs to manufacture energy. The nutrients also serve as a source of precursor metabolites for the synthesis of basic units of macromolecules. The synthesis of all cell components as well as reproductive compounds requires energy.

Energy can be extracted from cellular products either in the presence of oxygen or in its absence. Enzymes and vitamins are essential molecules in metabolism. Their activity can regulate the speed and direction of reactions by lowering the energy of activation.

An enzyme functions by temporarily combining with its substrate and then being released unchanged when the end products are formed.

The two major pathways for generating energy are fermentation, in which an organic compound is the final hydrogen acceptor, and respiration, in which an inorganic compound is the final hydrogen acceptor. About twenty times more energy is gained in respiration than in fermentation.

Some bacteria have pathways other than the glycolytic pathways. They can oxidize a variety of inorganic compounds to gain energy.

Objectives

After completing this chapter, the student should be able to:

1. Consider the historical significance of the scientific contributions of early researchers.
2. Understand the chemical kinetics and mechanisms of enzyme action.
3. Relate enzyme action to metabolic reactions.
4. List the environmental and chemical factors that influence the activity of enzymes.
5. Diagram the pathways of energy metabolism.
6. Differentiate between glycolysis and respiration.
7. Differentiate between anaerobic and aerobic respiration.
8. Understand the principles of bacterial photosynthesis.

Important Terms

precursor metabolites

anabolism

metabolism

catabolism

enzyme

product

substrate

kinetic energy

energy of activation

active (catalytic) site

competitive inhibitor

adenosine diphosphate

18

adenosine triphosphate

electron transport chain

end products

fermentation

respiration

glycolysis

oxidation

reduction

donor compound

acceptor compound

nicotinamide adenine dinucleotide (NAD)

homofermenters

heterofermenters

mixed fermentation

flavin adenine dinucleotide (FAD)

electron transport chain

quinones

cytochromes

proton motive force

chemiosmotic hypothesis

anaerobic respiration

Entner-Doudoroff pathway

pentose phosphate pathway

xenobiotics

transaminase

phototroph

oxygenic photosynthesis

anoxygenic photosynthesis

photophosphorylation

Essay

1. Discuss the expression "lock-and-key" in relation to substrate enzyme action.

    The arrangement of the enzyme and its substrate is commonly referred to as a "lock-and-key" arrangement. The substrate is the key and the enzyme, the lock. Since the key must fit into the lock precisely, a particular enzyme will act on only one substrate or a very limited number of substrates.

2. Describe the environmental and chemical factors that influence enzyme activity.

    The important features of the environment include temperature, pH, and salt concentration. A 10° rise in temperature doubles the speed of enzymatic reactions. Most enzymes function optimally near a pH of 7.

    Enzymes are inhibited by a variety of compounds, which may result in reversible or irreversible inhibition. An example of reversible inhibition is the action of sulfa drugs. An irreversible inhibitor may alter amino acids, which give the enzyme its shape.

3. Discuss the role of coenzymes in cell metabolism.

    Coenzymes are small, nonprotein molecules that can be in contact with, but can be readily separated from, enzymes. All coenzymes perform the same general function—the transfer of atoms from one molecule to another.

    As the enzyme attaches to the substrate, the coenzyme attaches to the enzyme, picks up a small molecule or atom from the substrate, separates from the enzyme and transfers the small molecule to the acceptor molecule.

4. Explain the importance of coenzymes in cell metabolism.

   Coenzymes are derivatives of vitamins and are essential in the function of all the different enzymes required in energy synthesis. If an organism lacks the vitamin, serious impairment of metabolism will result.

5. Describe the production of alcohol from pyruvic acid.

   Two steps are involved:

   (1) Pyruvic acid is converted to acetaldehyde and CO_2.
   (2) Acetaldehyde is reduced by NADH to ethyl alcohol.

   The two reactions are carried out by yeast and form the basis for wine and beer production.

   Overall equation:

   $$1 \text{ glucose} \rightarrow 2 \text{ ethanol} + 2CO_2$$

6. Describe the most advantageous manner in which facultative anaerobic cells can generate energy.

   Facultative anaerobes can generate the greatest amount of energy under aerobic conditions. A cell growing anaerobically must degrade about 20 times more glucose than a cell growing aerobically to provide the same amount of energy. The oxygen is the final electron acceptor and it can only be made available if the aerobic system is used.

Review Questions (pg. 120)

1. Lowering the energy of activation.

2. Temperature, pH, salt concentration.

3. In fermentation, the final electron receptor is an organic compound; in respiration, the final electron acceptor is an inorganic compound.

4. Vitamins are converted to coenzymes.

Review Questions (pg. 129)

1. CO_2 and water.

2. Nicotinamide adenine dinucleotide (NAD) and flavin adenine dinucleotide (FAD).

3. Glycolysis produces the least amount of energy, aerobic respiration the most, and anaerobic respiration produces less than aerobic respiration but more than glycolysis.
   Approximately twenty times more energy is produced in aerobic respiration than in glycolysis.

4. α-ketoglutaric acid + NH_4+ $\rightleftharpoons$ glutamic acid. This reaction is reversible.

Critical Thinking Questions (pg. 131)

1. Cyanobacteria always produce oxygen as an end product of photosynthesis, whereas photosynthetic bacteria use other reduced compounds, either inorganic or organic. The green and purple sulfur bacteria utilize a reduced inorganic sulfur compound such as H_2S and therefore the end product is sulfur. α-ketoglutarate is an especially important compound because it can be converted to glutamic acid by the addition of ammonia. Glutamic acid is an important amino acid.
2. α-ketoglutarate acid is a key organic molecule in the cell's ability to utilize ammonia or ammonium salts. Ammonia incorporation is a very important reaction in the bacteria cell's metabolism.

CHAPTER 6

Introduction

Deoxyribonucleic acid is the double-stranded molecule that is the core of the characteristics in every organism and accounts for their expression in every organism. DNA functions in the transcription of RNA (ribonucleic acid) and ultimately in its translation into proteins. Although most proteins function as enzymes, some function to control the expression of those enzymes. At this time, the understanding of these mechanisms is not fully understood but technology is progressing so rapidly that a greater knowledge will be possible when the mechanics are perfected.

Objectives

After completing this chapter, the student should be able to:

1. Realize the contributions of the early researchers to the science of genetics.
2. Understand the chemistry of DNA and RNA.
3. Understand how information is stored and transferred in nucleic acids.
4. Explain how genes are expressed.
5. Relate the differences between prokaryotic and eukaryotic transcription and translation.
6. Recognize the environment and control systems involved in transcription.
7. Explain the mechanisms of control in biosynthetic pathways.
8. Understand the mechanisms of control of degradative pathways.

Important Terms

nucleotides

DNA replication

messenger RNA (mRNA)

central dogma of molecular biology

template

origin of replication

replication fork

semiconservative

polypeptide chain (enzyme)

coding region

RNA polymerase

regulatory region

transcription

genetic code

nonsense, or stop, codons

start codons

promoter region

holoenzyme

core enzyme

template strand

polygenic mRNA

translation

transfer RNA (tRNA)

ribosomal RNA (rRNA)

"charged"

"uncharged"

ribosomal binding site
start signal
peptide (P, site)
acceptor (A, site)
polyribosomes
amino terminal end
carboxyl end
intron
mRNA splicing
ribozymes
end product repression
feedback (allosteric) inhibition
regulatory gene
regulatory (repressor) protein
operator region

operon
allosteric site
feedback inhibitor
inducible enzymes
enzyme induction
inducers
penicillinase
constitutive enzymes
diauxic growth
catabolite repression
global repressor
cyclic AMP (cAMP)
positive regulator
negative regulator

Essay

1. Describe the sequence of events in the printing of a molecule of mRNA.

   The DNA holds the genetic code for all the proteins of the organism. It transcribes this code by unzipping and printing on the opposite strands a molecule of RNA, which carries the complementary sequence of nucleotides. The mRNA elongates and then when complete is released from the DNA and migrates to the cytoplasm, where it functions to act as a model for the development of a protein.

2. Briefly explain the translation of mRNA into protein.

   A complete protein is synthesized on each ribosome. The ribosome attaches to the start codon at the beginning of the mRNA and continues as the mRNA moves along the ribosome three nucleotides at a time. The protein ends when the ribosome reaches a nonsense codon that serves as a "stop" signal. At this "stop" signal, the protein dissociates from the tRNA molecule.

3. Cite some major differences in transcription and translation in prokaryotes and eukaryotes.

Prokaryotes	Eukaryotes
(1) transcription closely tied to translation in time and space	(1) mRNA is synthesized completely before it is translated
(2) mRNA not processed	(2) mRNA is processed by removal of introns
(3) mRNA often transcribed as polygenic mRNA	(3) polygenic mRNA rare

Review Questions (pg. 142)

1. The sequence of nucleotides in one strand of DNA determines the sequence in the complementary strand since adenine can only base pair with thymine, and guanine with cytosine. Therefore, if one knows the sequence of one strand, one can deduce the sequence of the other.

2. DNA transfers information to a complementary sequence in messenger RNA by transcription and provides the information for its own replication.

3. Messenger RNA, transfer RNA, and ribosomal RNA. Transfer RNA is the decoder.

4. RNA polymerase binds to the promoter region of the DNA.

Review Questions (pg. 149)

1. End product inhibition. End product repression.

2. The synthesis of all of the enzymes of the pathway is repressed. In allosteric inhibition, only the first enzyme of a pathway is inhibited.

3. The two regions are the promoter region, which interacts with RNA polymerase, and the repressor, which interacts with the operator region.

4. Cyclic AMP.

5. The degradation of lactose is positively regulated in that cyclic AMP and a protein to which cAMP binds is required for degradation. The biosynthesis of tryptophan is negatively regulated.

Critical Thinking Questions (pg. 151)

1. DNA polymerase binds to the origin of replication to initiate DNA replication. RNA polymerase binds to the promoter region of genes to initiate transcription. The repressor protein binds to the operator region of DNA to regulate transcription.

2. Prokaryotes have much smaller genomes than eukaryotes and also must be extremely efficient in regulation of pathways in order to survive in nature in which organisms multiply rapidly and the environment changes frequently. The operon allows a single molecule of a regulatory protein to control an entire pathway. This is a very efficient use of a few regulatory genes.

3. It is unlikely that small molecules will bind to DNA directly to control gene expression. The binding of small molecules is nonspecific and would not recognize specific nucleotide sequences in DNA. Small molecules are important in the regulation of gene expression, but they do so by binding to proteins that modify the protein's shape. The proteins, in turn, recognize and then interact with specific base sequences of DNA to control the gene expression.

CHAPTER 7

Introduction

The haploid nature of the bacterial chromosome lends itself nicely to genetic study. The organisms multiply rapidly, are easy to care for, and show their genetic changes readily. The study of bacterial genetics began early in this century and continues today with the use of increasingly sophisticated techniques. Many scientists have contributed to the current knowledge of genetics, primarily due to bacterial investigation. From this experimentation has come extensive knowledge of human DNA, and the development of techniques to manufacture human chromosomes and enzymes for the treatment of human diseases. Knowledge of bacterial genetic information transfer has also led to information concerning the replacement of faulty human genes with functional ones in the hope of eliminating human genetic disorders.

The study of bacterial genetics has also helped scientists to understand more completely the patterns of disease and antibiotic resistance.

Objectives

After completing this chapter, the student should be able to:

1. Trace the significant events that led to the understanding of microbial genetics.
2. Identify the sources of diversity in microorganisms.
3. Understand the significance of nucleotide sequence alteration in DNA.
4. Understand the nature of mutagens.
5. Follow the process of mutation through several causes.
6. Identify transposons and their significance.
7. Identify radiation as a cause of mutation.
8. Recognize how damaged DNA can be repaired.
9. Define mutation rate, selection, and lethal genes.
10. Recognize the difference between direct and indirect selection.
11. Understand the commercial applications of mutants.
12. Understand the mechanisms of gene transfer.

Important Terms

genotype
genome
phenotype
mutagenesis

mutagens
spontaneous
induced
chemical agents

transposable elements

radiation

nitrous acid

alkylating agents

base analogs

intercalating agents

carcinogen

transposable elements (transposons)

frame shift mutation

insertion sequence (IS) element

inverted repeat (palindromic sequence)

X rays

ultraviolet (UV) light

SOS repair

SOS system

light repair

excision repair

dark repair

reversion

haploid

diploid

direct selection

indirect selection

auxotroph

prototroph

replica plating

penicillin enrichment

conditional lethal mutants

temperature-sensitive mutants

Ames test

genetic recombination

recombinants

DNA-mediated transformation

transduction

conjugation

homologous recombination

breakage & reunion

competent

electroporation

bacteriophage (phage)

male (donor) cell

female (recipient) cell

F plasmid

sex pilus

high frequency of recombination (H fr)

origin of transfer

episome

low copy number plasmid

high copy number plasmid

narrow host range plasmids

wide host range plasmids

conjugative plasmids

resistance (R) factor plasmids

resistance transfer factor (RTF)

resistance genes (R genes)

transposable elements

foreign DNA

restriction enzymes (restriction endonucleases)

modification enzyme

substrate

Essay

1. Discuss the early contributions that led to the discovery that genetic information could be transferred among organisms.

   In 1944, Avery, MacLeod, and McCarty identified DNA as the most important molecule in the cell. Prior to this, in the 1920s, Griffith, while studying pneumonia, discovered that he could transfer capsule production to nonencapsulated organisms by mixing the products of the dead organisms with those of the living. He then showed that the presence of the capsule was indicative of virulence and the genetic trait could "transform" nonvirulent organisms.

2. Discuss the sources of diversity in microorganisms.

    The two major sources of diversity in microbes are:
    (1) variations in sequences of nucleotides in DNA
    (2) the environment, which regulates gene function and enzyme activity

3. Differentiate between genotype and phenotype.

    Genotype represents the potential an organism has if all its genes are expressed. The phenotype describes the actual characteristics displayed by the organism at any one time.

4. Explain the development of antibiotic-resistant strains of a bacterium.

    When cells divide, there is statistical evidence that a certain number of cells will mutate and therefore not all the cells are identical. The mutations provide a mechanism by which organisms can respond to a changing environment. A spontaneous mutation to provide antibiotic resistance may be rare but will result in the mutants becoming the dominant organism in the environment where the antibiotic is present.

5. Describe the method used for the indirect selection of isolating mutants in the laboratory.

    Indirect selection must be used to isolate mutants that cannot grow on a medium on which the parent can grow.

    (1) Replica plating involves the transfer of all colonies on one plate to another plate simultaneously.
    (2) Antibiotic enrichment could increase the proportion of mutants in a population. It would allow only those organisms resistant to the antibiotic to grow.
    (3) They can be identified by careful placement of the original over the replica plate and identifying which colonies have grown.

6. Describe the principles applicable to the Ames test.

    (1) Reversion of a mutant gene in a biosynthetic pathway.
    (2) The frequency of reversions is increased by mutagens.
    (3) Most carcinogens, substances promoting the formation of cancers, are mutagens.

    The Ames test measures the rate of reversion of a histidine auxotroph of *Salmonella* to prototrophy in both the presence and absence of the chemical being tested. If the chemical is mutagenic, it will increase the reversion rate.

7. Describe the three different methods by which genes can be transferred between bacteria.

    (1) Transformation: Naked DNA is transferred from lipid organisms to living cells.
    (2) Transduction: DNA is transferred by a bacterial virus (bacteriophage) from one bacterium to another.

(3) Conjugation: DNA is transferred between bacteria that are in contact with one another.

8. Describe the significance of plasmids in bacteria.

    Plasmids are units of replication that are very common in the microbial world. Their functions are not understood, but the following facts are known:

(1) most bacteria contain one or more different plasmids
(2) plasmids vary in size from a few genes to several hundred
(3) traits coded by plasmids provide useful but not indispensable capabilities

9. How do bacteria prevent destruction of their own DNA?

    Bacteria contain a modification enzyme that adds methyl groups to certain bases in the fragments of DNA to which the restriction enzyme cleaves. Therefore, the restriction enzymes are unable to recognize the specific base sequences as a substrate and so do not cleave to the DNA.

10. How does gene transfer provide new genetic information to microorganisms?

    The major source of variation within a bacterial species is mutation. By using gene transfer, the recipient cell gains much more genetic information because many genes are transferred simultaneously. Transposable elements greatly expand the opportunity for new gene combinations to arise since their movement can inactivate and separate genes. Due to the insertion sequences that they frequently flank, many genes may be transferred to a broad range of both Gram-positive and Gram-negative organisms.

Review Questions (pg. 165)

1. Mutations alter permanently the properties of organisms in; changes in the environment of organisms; changes in the environment, change is that phenotypic but not genotypic properties of organisms.

2. The genotype is the potential for all of the properties of an organism; the phenotype is the actual properties of the organism growing under specific conditions.

3. Mutations can occur by:
   (1) mistakes in the copying of a nucleotide sequence by DNA polymerase;
   (2) the inactivation of genes by transposable elements;
   (3) the removal or addition of nucleotides;
   (4) the modification of preexisting bases into DNA leading to the misincorporation of nucleotides in the course of DNA replication.

4. In light repair, the covalent bond connecting adjacent thymine molecules is broken; in dark repair, a single-stranded fragment of DNA containing the thymine dimer is excised and a new single strand complementary to the undamaged DNA is synthesized.

Review Questions (pg. 172)

1. Conjugation

2. Bacteria grow very rapidly; can be grown to very large numbers in small volumes; have a very large number of genes whose properties can be readily determined; many bacteria undergo genetic recombination.

3. A donor cell of *E-coli* contains genes which code for mobilization functions and transfer functions for DNA. The transfer functions include the synthesis of a sex pilus for DNA-mediated transformation.

4. DNA-mediated transformation

Review Questions (pg. 177)

1. They must be able to replicate.

2. A broad host range plasmid.

3. Restriction endonuclease.

Critical Thinking Questions (pg. 179)

1. Conjugation. In conjugation, large fragments of DNA can be transferred at any one time and therefore it is possible to map the genes that are not closely linked together. In transduction and DNA-mediated transformation, small fragments are only transferred and therefore genes that lie on different fragments cannot be mapped relative to one another.
2. The gene on a chromosome must be first transferred onto a plasmid. This commonly occurs if the gene becomes a part of a transposable element containing inverted repeats at its ends. The transposable element containing the gene can then be transferred to a wide host range plasmid within the cell, which then can be transferred into a *Pseudomonas,* where it can replicate.
3. A number of techniques can be used, depending upon the properties of the plasmid. First, it might be possible to cure the strain of the plasmid and determine whether or not the antibiotic resistance characteristic was lost simultaneously. If it was, this would suggest that the plasmid carries the antibiotic resistance gene. More convincing evidence would be obtained if a plasmid could be introduced into the cured strain with the regain of antibiotic resistance. It might also be possible to look for the transfer of antibiotic resistance from antibiotic-resistant cells to antibiotic-sensitive cells. If the transfer occurs under conditions in which only a plasmid is transferred, this would suggest that the antibiotic resistance gene is carried on the plasmid.

CHAPTER 8

Introduction

Recombinant DNA technology involves a series of techniques that result in the manipulation of DNA to stimulate the formation of new DNA molecules. The techniques are fundamentally the same as those processes naturally occurring in cells but allow the construction of hybrid DNA from unrelated organisms. Shotgun cloning allows all the genes of an organism to be cloned and then the gene of interest to be identified. The expression of this gene is brought about by a promoter gene. Many human proteins are now manufactured by using recombinant DNA techniques.

Bioremediation has been applied to environmental stresses (oil spills) in an attempt to reclaim the territory with a minimum of damage. Natural insecticides of bacteria have been cloned into plants to protect them from harmful insects. The polymerase chain reaction allows researchers to duplicate exactly specific pieces of DNA.

Continuous research into DNA technology will result in many new discoveries that will help humans deal more effectively with their environment.

Objectives

After completing this chapter, the student should be able to:

1. Trace the development of biotechnology in the field of microbial genetics.
2. Define biotechnology, recombinant DNA, gene cloning, and genetic engineering.
3. Describe the steps and uses of shotgun cloning.
4. Describe the various methods of expressing cloned genes.
5. Understand the applications of biotechnology.

Important Terms

restriction endonuclease

vector

electroporation

agar

Southern hybridization

replica plating

ribosome binding site

interferons

spores

primers

DNA typing

restriction enzyme

gene library

agarose gel electrophoresis

carcinogen

probe

promoter region

intron

bioremediation

genetic engineering
polymerase chain reaction

Essay

1. Describe some of the advantages of genetic engineering in the medical field.

    The advantages are:
    1. the production of mammalian proteins in large quantities with quality control
    2. decreased risk of infection
    3. the production of vaccines for hepatitis B and foot-and-mouth disease. Cloned products do not cause allergic responses and prevent possible transmission of cadaver infections

2. Describe shotgun cloning and the steps involved.

    Shotgun cloning involves the separation of a chromosome into its genes. This allows their insertion into a bacterium with subsequent identification of the gene in the bacterium. The steps involved are:

    1) release of DNA from human cells
    2) construct recombinant DNA molecules
    3) introduce recombinant DNA into host
    4) select host cells that contain recombinant DNA molecules
    5) identify cells carrying the recombinant DNA clone

3. Describe the agarose gel electrophoresis process of DNA fragment separation.

    DNA molecules are negatively charged and will migrate toward a cathode if placed in an electric current. The variable pore sizes of the gel will sieve the DNA molecules, allowing the smaller ones to migrate faster than the larger ones. The DNA molecules can then be separated on this gel by using an electric field. Once separated, the molecules can be detected by soaking the gel in a fluorescent dye called ethidium bromide, which binds to nucleic acid molecules that can be observed by exposing the gel to ultraviolet light.

4. Explain the process by which genetic engineering is used to confer resistance to insect pests to plants.

    Agrobacterium is a plant pathogen capable of producing crown gall via its tumor-inducing (Ti) plasmid. The bacterium *B. thuringiensis* carries a gene that can make the plant toxic to the insect pest.
    By using genetic engineering, the toxin-producing gene from *B. thuringiensis* can be substituted for the Ti plasmid of *Agrobacterium*. This gene transfer protects the plant, which will express the genes under the influence of the promoter genes.

Review Questions (pg. 188)

1. The DNA that is introduced into cells to modify them can come from any source. Second, the isolation of the genes and their introduction into the cells that are modified is done in the laboratory in test tubes and not in the natural environment of the organism.

2. Restriction enzymes and DNA ligase.

3. A gene library is the entire genome of the organism cloned into a cloning vector, such as a plasmid or a bacteriophage.

4. A single-stranded DNA that is identical to DNA sequences whose presence is being searched.

Review Questions (pg. 193)

1. The gene of interest can be cloned onto a cloning vector, which is then introduced into *E. coli.* Growing these organisms under conditions in which the gene of interest will be expressed will increase the level of gene product over what it would be in its natural environment.

2. The *Bacillus thuringiensis.*

3. The B.T toxin is highly specific for insects and has no effect on higher animals and plants. The toxin is readily degraded in the soil within a few days of application.

4. *Agrobacterium tumefaciens.*

5. The nucleotide sequences at the two ends of the target DNA are needed to generate primers.

Critical Thinking Questions (pg. 196)

1. Sporulating bacteria in an insect gut can multiply and the toxin synthesized dissolves the gut, thereby releasing the bacteria. Therefore, the ability to synthesize the toxin might allow the organism to be readily disseminated.
2. Foreign DNA can be inserted into the *A. tumefaciens* tumor inducing (Ti) plasmid and subsequently integrated into a plant chromosome. The result is a plant with new properties.
3. Their small size so large fragments can be cloned; their ability to be transferred into cells readily; and their ability to replicate in a variety of different bacteria.
4. The bacteria may develop resistance through the B.T toxin and therefore it may be necessary to genetically engineer plants through a variety of different genes.

CHAPTER 9

Introduction

One of the greatest problems facing health concerned individuals is the maintenance of a medically safe and wholesome environment for the population. A wide variety of physical and chemical methods of controlling microbial growth have been developed since Lister first introduced the concept of asepsis.

The difficulty of supplying the population with a viable ecology while ensuring its safety has been met with increasingly sophisticated methods of sterilization, disinfection, and sanitation. A variety of physical and chemical means along with radiation are presented here as well as their appropriate uses.

Objectives

After completing this chapter, the student should be able to:

1. Understand the relationship between microbial growth and infection.
2. Describe the development of aseptic techniques.
3. Discuss the differences between disinfection and sterilization.
4. Rank the different types of microorganisms and viruses according to their resistance to destruction.
5. Discuss the conditions that could alter the effectiveness of microcidal agents.
6. Discuss the uses of heat in the destruction of microorganisms.
7. Understand the uses and limitations of pasteurization.
8. Determine alternate ways of sterilizing delicate materials.
9. Compare the various chemical substances used in sterilization, disinfection, and preservation.
10. Compare the uses of metal compounds in disinfecting contaminated objects.
11. Differentiate among the variety of germicidal chemicals used in the medical field.
12. Describe the uses of radiation in controlling microbial growth.

Important Terms

antiseptic	iodophors
decontamination	mercury
germicide	sterilants
fungistatic	gamma rays
acid-fast stain	disinfection
denaturation	sanitize
flash autoclaving	bacteriostatic
caramelization	logarithmic death curve
membrane filters	endospores

33

autoclave
biological indicator
sterilization
available chlorine

quaternary ammonium compounds
ethylene oxide
electromagnetic radiation
dimer formation

Essay

1. Discuss the work of Lister as it relates to the development of asepsis.

   Joseph Lister used principles suggested by Pasteur that minute organisms were responsible for fermentation. He extrapolated from this that pus in wounds could be caused by the same type of organisms. Lister used carbolic acid in wound dressings to prevent gangrene and improve the recovery rate of his patients.

2. Discuss or describe the differences among sanitation, disinfection, and sterilization.

   Sanitation refers to the reduction of bacterial population to a safe number.
   Disinfection refers to the use of chemical substances that kill microorganisms on inanimate objects, significantly reducing their number although some survive.
   Sterilization refers to the complete destruction of any microorganism or infectious agent on a fomite.

3. Discuss the uses and limitations of heat in controlling infectious agents.

   Heating is fast, reliable, and inexpensive compared to most other methods, and it does not introduce toxic substances into the material being treated.
   Dry heat is slow and requires high temperature to penetrate the larger materials.
   Boiling may kill vegetative cells but is unreliable for eliminating spores.
   Steam under pressure is a frequently used sterilizing method but is unreliable with materials that are impenetrable to steam.
   Pasteurization may rid food of vegetative pathogens but many microorganisms survive.
   Incineration will accomplish destruction of the microbes but the product is also destroyed.

4. The autoclave is the classic sterilization tool. Discuss some practical aspects of its use.

   An autoclave is used in medical facilities to safely sterilize contaminated objects. Some of the common uses are:
   (1) rapid sterilization of surgical instruments
   (2) complete destruction of microbes and their spores in their growth medium or contaminated objects
   (3) sterilization of nutrient media for the growth of microbes
   (4) sterilization of liquids and laboratory tools that will not be harmed by using steam under pressure.

5. Rate the chemical agents useful in sterilization, disinfection, or preservation.

Ethylene oxide gas is a highly effective means of sterilization. Of the chemicals listed, it is probably the best.

The aldehydes are good disinfectants. The alcohols, halogens, phenols, and ammonia and mercury compounds may be used as disinfectants and in some cases antiseptics but only within certain limitations.

6. Why has silver nitrate been discontinued as the method of preventing infection in the eyes of newborns?

Silver nitrate is an irritant and is not effective against all microorganisms. It has been replaced by antibiotic treatment.

7. What is the significance of the three categories of germicides labeled by the EPA?

The three categories are:
(1) "sterilants"—germicides capable of destroying large numbers of endospores and vegetative cells
(2) "hospital disinfectants having tuberculocidal activity"—not sporicidal but effective against the most difficult vegetative cells
(3) "hospital disinfectants"—even less antimicrobial activity and no tuberculocidal activity

8. Describe a reasonable approach for handling equipment that has been contaminated with blood and/or other body fluids.

(a) Blood and/or body fluids are cleaned up using: 1) gloves and 2) towels soaked with a tuberculocidal disinfectant.
(b) Disinfection is completed by following the specifics of the product used or by using a properly diluted bleach solution.

Review Questions (pg. 202)

1. Sterilizing means to kill or remove all the microorganisms and viruses on it.

2. Disinfecting means to reduce the number of pathogens to a safe level.

3. An antiseptic is a disinfectant that is safe to use on human tissues.

4. A bactericidal substance kills bacteria while a bacteriostatic substance inhibits the growth of bacteria.

5. One-half.

6. Clean objects have fewer microorganisms.

7. Heat coagulates their essential proteins.

8. Boiling may not kill bacterial endospores.

9. The temperature in a pressure cooker is higher than boiling.

10. Heat of vaporization is the additional amount of heat energy that is taken up by water without a change in temperature, whereupon boiling begins to occur. Release of the heat of vaporization by steam condensing on organisms may play a role in the coagulation of proteins and in rapid killing.

Review Questions (pg. 210)

1. Steam is lighter than air. Air in objects must be able to escape as steam displaces it downward.

2. Steam might not penetrate the material.

3. Pasteurization is the process of controlled heating at temps. below boiling to kill pathogens or spoilage organisms. It does not necessarily kill all microorganisms.

4. Porcelain: large irregular passageways, microorganisms trapped by electrostatic charges as well as sieve action.
   Membrane: paper thin, mainly sieve action.
   Advantages:

	Porcelain	*Membrane*
	Inexpensive	Chemically inert
	Fast filtration	Won't absorb enzymes
		Can be made transparent
		Some can remove viruses

5. Organic material.

6. High: kills bacterial endospores, vegetative organisms, viruses
   Intermediate: kills vegetative organisms, many viruses
   Low: kills many vegetative microorganisms and viruses, but not *Mycobacterium tuberculosis.*

7. Alcohol diluted with water.

8. Sodium hypochlorite, iodine solutions.

9. Iodophors are compounds of iodine linked with surface-active agents. Limitations—(1) Dilution with alkaline water results in inadequate microbial activity. (2) Not reliable for disinfecting instruments contaminated by *M. tuberculosis.*

36

10. Glutaraldehyde: No special equipment needed and easily removed, but high concentration and long sterilizing time required.

Ethylene oxide: Penetrates well, but explosive, carcinogenic and requires lengthy aeration after sterilization is complete.

Review Questions (pg. 214)

1. Environmental Protection Agency (EPA).

2. The germicide of choice depends on whether the use is for critical, semicritical, or noncritical materials.

3. Preservatives are used in soaps, medicines, deodorants, and contact lens solutions to prevent or retard spoilage by microbes.

4. Ultraviolet light and γ-rays.

5. Bacterial endospores and some *Deinococcus* bacteria species.

6. Ultraviolet light is more optimally absorbed by DNA than are shorter and longer wavelengths.

Critical Thinking Questions (pg. 315)

1. The temperature of the solution might be above the boiling point at the lower pressure.
2. They only reflect conditions at the point where the indicator was placed and not necessarily at all those points where microorganisms might exist. Also, microorganisms can be reintroduced into sterilized materials by air, water, insects, or other means, and so indicators cannot always guarantee the sterility of a material.
3. Conditions under which disinfectants are used may differ from the conditions of the laboratory tests in important ways, such as the temperatures, pH, duration of application, and presence of organic material.

CHAPTER 10

Introduction

Classification schemes have changed significantly over the centuries from the first differentiations made by early humans—Do I eat it or does it eat me? Is it food or poison? The classification systems progressed in a similar fashion until Carl von Linné used binomial nomenclature. His methodology began scientific cataloguing, which continues today in the form of sophisticated technology that is able to differentiate organisms by using their DNA and/or their metabolic products.

Taxonomy allows for a universal language to exist among scientists so in any language or any context, communication is possible. *Bergey's Manual of Systematic Bacteriology* lists the known bacteria and the characteristics that are used for their identification. Future editions of the manual will contain the greatest advancements of technology. Classification technology will continue and fine differences among organisms along with their evolutionary roots will be available to researchers, bacteriologists, and students.

Objectives

After completing this chapter, the student should be able to:

1. Trace the development of bacterial classification.
2. Understand the problems of classifying bacteria.
3. Describe the present day classification schemes.
4. List the molecular approaches to taxonomy.
5. Describe methods for identifying bacteria.

Important Terms

classification
taxonomy
species
subspecies
genus
family
order
class

kingdom
numerical taxonomy
similarity coefficient
GC content
nucleic acid hybridization
signature sequence
DNA probes

Essay

1. Describe the problems in classifying microorganisms.

    The purpose of classification schemes is to provide easy and universal methods of identifying organisms. The best classification schemes try to group organisms with similar properties and that are relation through evolution, and separate those that are different. Since bacteria have only a few different shapes and gene transfer is rare, the most effective method of classification is not by morphology but by physiology, cell metabolism, and staining properties.

2. Discuss the modern approaches to the evolutionary taxonomy.

    Modern approaches to the evolutionary taxonomic scheme involve:
    (1) comparison of DNA or RNA base sequences
    (2) comparing the amino acid composition of a specific protein or a group of proteins in different organisms

3. Discuss the discoveries of Dr. Carl Woese in relation to solving the evolution puzzle.

    Dr. Carl Woese and his colleagues have divided all organisms into one of three groups based on their sequences of 16S ribosomal RNA (signature sequences) and their related properties. The three different groups (Archaebacteria, Eubacteria, and Eukaryote) are the results of development from a common ancestor.

4. Discuss the uses of DNA probes in identification of pure cultures of microorganisms.

    Probe technology is extremely useful to:
    (1) detect and identify an organism that is difficult to culture by conventional methods;
    (2) detect and quickly identify an organism if a reliable laboratory procedure is unavailable; and
    (3) detect and identify an organism when its toxin or its product of interest is difficult to detect by standard methods.

5. List some common methods for identifying bacteria:

    A. Colonial appearance on media
    B. Microscopic analysis
       1. unstained preparations—shape, arrangement, motility
       2. stained preparations—Gram, acid fast, spore
    C. Biochemical tests
    D. Growth requirements
       1. nutritional requirements
       2. O_2 requirements
    E. Reactions to other biological reagents

F.   Detection of microbial by-products

G.   Nucleic acid hybridization

Review Questions (pg. 225)

1.   The basic taxonomic unit is the species.

2.   *Bergey's Manual of Systematic Bacteriology* is the basic reference text.

3.   Nucleic acid hybridization is used to compare the sequence of DNA bases.

4.   eubacteria, archaebacteria, and eukaryotes

Review Questions (pg. 235)

1.   Direct examination of specimens is inexpensive, gives information very quickly, and requires minimal laboratory space and equipment.

2.   This method may fail because it is relatively insensitive, and 10^4–10^5 organisms per gram must be present to be detected.

3.   The most frequently used stain is the Gram stain.

4.   DNA probes can be used successfully on an organism that is difficult to culture and to identify a single species in mixed species cultures.

Critical Thinking Questions (pg. 236)

There are several possible approaches.

1.   The sample containing the suspected organism can be inoculated onto enriched broth continuing radiolabeled glucose (^{14}C). The release of $^{14}CO_2$ indicates an organism is present and metabolizing glucose. If the identity of the organism is suspected, the sample may be inoculated onto broth containing a radiolabeled compound that it and very few other organisms are able to degrade to form detectable end products. A positive result would indicate both growth as well as presumptive evidence of the identity of the organism. Also, if the identity is suspected, a radiolabeled probe could be used to look for complementary DNA sequence in the sample.
2.   Blood and urine could be used. Intestinal contents could not be used since there are too many different bacteria already present as normal inhabitants. Some of the bacteria would undoubtedly be able to degrade the radiolabeled compound.

CHAPTER 11

Introduction

In this chapter, the various types of prokaryotic organisms are considered, and are classified into taxonomic divisions for identification. The bacteria can be classified by shape, Gram reaction, sugar fermentation, oxygen and temperature requirements, and pH.

Included in this chapter is an abbreviated treatment of microbial evolution with a discussion of the development of symbionts.

The classifications used in the latest edition of *Bergey's Manual of Systematic Bacteriology* have relied on the use of sophisticated biochemistry, which allows for the precise identification and separation of closely related organisms.

Objectives

After completing this chapter, the student should be able to:

1. Understand the significance of the Gram stain in bacterial identification.
2. Differentiate among the four divisions of the prokaryotes.
3. Describe the spirochetes.
4. Describe organisms that have a Gram-negative cell wall.
5. Describe organisms that have a Gram-positive cell wall.
6. Describe organisms that lack cell walls.
7. Discuss organisms whose cell walls lack peptidoglycan.
8. Distinguish between the gliding bacteria and the myxobacteria.
9. Distinguish between the phototrophic and chemolithotropic bacteria.
10. Describe the various types of Gram-positive cocci and rods.
11. Distinguish among the non-spore-forming Gram-positive rods.
12. Describe the prokaryotes that lack cell walls.

Important Terms

anaerobic respiration

axial filaments

cyst

nitrogenase

endosymbiotic theory

microcysts

phototrophic bacteria

chemotrophic bacteria

gas vesicles

cyanobacteria

phycobiliprotein

thylakoids

heterocysts

prosthecae

carboxysomes
∝-(alpha) hemolysis
β-(beta) hemolysis
γ-(gamma) hemolysis
viridans streptococci
snapping division
metachromatic staining
hyphae

mycelium
conidia
sporangia
pleuropneumonia
extremophiles
methanogens
halobacteria
thermoacidophiles

Essay

1. How do the spirochetes and spirals differ?

    The spirochetes have a flexible cell wall and axial filaments, which contribute to their peculiar manner of movement. They are capable of pathogenicity in areas of the body where the temperature is lowest.
    The spirals have rigid cell walls, move by flagella, and require oxygen for growth. Very few of them are capable of human pathogenicity.

2. What is unique about the pseudomonads?

    The *Pseudomonas* genus contains a large group of organisms that produce non-photosynthetic pigments. These pigments are frequently water soluble and can color the medium in which they are growing. Most pseudomonads are aerobic; those that are anaerobic use nitrate as a substitute for molecular oxygen. The human pathogenic pseudomonads are highly virulent and resistant to antibiotics.

3. How are the rickettsias and the chlamydias unique among the bacteria?

    The rickettsias and the chlamydias are, with the exception of one genus, unable to grow outside the cells of a host animal. They are transmitted from one host to another by a variety of arthropod vectors. Genetically, they are unrelated though they may share some characteristics.

Review Questions (pg. 243)

1. Spirochetes move by means of axial filaments, unique motility structures.

2. Small spirochetes are best seen using dark-field microscopy.

3. Spirochetes are found in aquatic habitats, ticks, and in human saliva.

4. No, they require small amounts of oxygen.

5. Bdellovibrios prey on other bacteria.

6. These bacteria differ from spirochetes in possessing a rigid cell wall and flagella for motility.

7. *Rhizobium* species fix atmospheric nitrogen.

8. *Pseudomonas* species degrade complex organic compounds that cannot be degraded by other microorganisms.

9. *Azotobacter* species can fix nitrogen under aerobic conditions in the absence of plants.

Review Questions (pg. 247)

1. Enterobacteria cause harmful diseases in plants and animals.

2. Enterobacteria are distinguished by differences in their capsules, cell walls, and flagella.

3. *Vibrio* sp. are fermentative, facultative anaerobes some of which are luminescent.

4. Luciferase mediates a light-producing reaction, resulting in luminescence.

5. These bacteria are anaerobic, and oxygen in the air can kill them.

6. These bacteria use sulfur compounds as electron receptors, reducing rather than assimilating them.

7. These bacteria are found in mud containing abundant organic material.

8. An obligate intracellular bacterium can only reproduce inside a eukaryotic cell, which acts as the host cell.

9. Rickettsias survive poorly in the environment and depend on arthropods for transmission.

Review Questions (pg. 254)

1. Endosymbionts are microorganisms that have a stable intracellular relationship with eukaryotic host cells. Eukaryotic organelles such as mitochondria and chloroplasts may have evolved from prokaryotic endosymbionts.

2. Non-photosynthetic gliding bacteria are found in bodies of water and in the human mouth.

3. Fruiting bodies are structure produced by some myxobacteria in response to nutrient depletion. They contain microcysts, spherical cells that are more resistant to heat, drying, and radiation than are the vegetative cells.

4. Sheathed bacteria often produce scum on the surface of polluted streams.

5. A phototrophic bacterium derives energy from sunlight.

6. Anoxygenic phototrophs grow under anaerobic conditions and do not release oxygen as a by-product of photosynthesis, whereas oxygenic phototrophs release oxygen during photosynthesis.

7. Cyanobacteria are oxygenic phototrophs that contain phycobiliproteins and many species fix nitrogen.

8. A prostheca is a cytoplasm-filled extension from a bacterial cell. Unlike other projections, it contains a cell wall.

9. Chemolithotrophs use reduced sulfur compounds, reduced nitrogen compounds, hydrogen gas, and reduced metals as sources of energy.

Review Questions (pg. 266)

1. α-hemolytic streptococci cause partial red cell destruction; β-hemolytic streptococci cause complete red cell destruction; and γ-streptococci produce no hemolysis.
   Viridans streptococci are α-hemolytic and cause a green discoloration.

2. Streptococci are anaerobic or aerotolerant. Staphylococci grow best under aerobic conditions.

3. *Bacillus* and *Clostridium* form endospores. *Bacillus* species are aerobic or facultatively anaerobic, while *Clostridium* species are obligate anaerobes.

4. Their metabolic by-product is lactic acid.

5. The term refers to bacteria with branching filaments.

6. Mycobacteria differ in their susceptibility to antibacterial medicines and are more resistant to disinfectants.

7. An aerial hypha is a filament that projects into the air above the medium on which the bacteria are growing.

8. *Frankia* has a wide host range, in contrast to *Rhizobium*.

9. Several species of *Streptomyces* produce antibiotics used in treating infections.

10. Both genera of bacteria lack cell walls.

11. The three groups of archaebacteria are: methanogens, halobacteria, and thermophiles.

Critical Thinking Questions (pg. 268)

1. Sources of energy for biosynthesis are organic and inorganic chemicals, and sunlight.
2. These adverse conditions may resemble those of an earlier time in the earth's history.
3. Bacteria discussed in this chapter include those that can obtain nitrogen from N_2, carbon from CO_2, energy from sunlight, and can grow at temperatures as high as 105°C and at a pH as low as 1. Some grow in the absence of oxygen and use H_2 or H_2S as an energy source. Of course, all earth's life forms require water and a source of elements such as Na, K, Ca, and P.

CHAPTER 12

Introduction

Although the major scope of a microbiology course focuses on bacteria, other microscopic life is also studied. The algae, fungi, and protozoa deserve close scrutiny because they are important to the environment. The fungi are a large group of organisms important to ecological recycling. The algae are the primary producers of the planet. The protozoa are important as producers and decomposers.

Members of each of the three eukaryotic groups also cause a variety of human diseases that are important economically.

Objectives

After completing this chapter, the student should be able to:

1. Describe the contributions of early scientists to our knowledge of vectors and research techniques.
2. Compare and contrast the characteristics of eukaryotic organisms with prokaryotic organisms.
3. Use Table 12.1 to describe the major characteristics of the algae groups mentioned here.
4. Discuss the various algae mentioned and include their economic and/or medical importance.
5. List several differences between microscopic and macroscopic algae.
6. Understand the differences between sexual and asexual reproduction in eukaryotic cells.
7. Describe the toxicity of some algae to humans.
8. Describe the basic characteristics of fungi.
9. Discuss the variable habitats of fungi.
10. Discuss the various types of fungi.
11. Recall the human pathogenic fungal diseases, along with their causes and symptoms.
12. Describe the basic characteristics of protozoa.
13. Describe the variable habitats of protozoans.
14. List and describe the protozoal diseases of humans.

Important Terms

algae	zygote
phytoplankton	gametophyte
diatoms	cysts
blade	heterotrophs
pectin	dimorphic fungi
carotenoids	zygospores
haploid	mycologist

aflatoxins
mycorrhizas
microtubules
flagella
phagocytosis
binary fission
malaria
giardiasis
symbiotic
zooplankton
stipe
cell wall
chloroplasts
fragmentation
diploid

gamete
sporophyte
fungi
mycelium
fruiting bodies
ascospores
asthma
mycoses
protozoa
cilia
polymorphism
pinocytosis
toxoplasmosis
sleeping sickness

Essay

1. Discuss the contributions of Smith and Kilbourne to the development of epidemiology.

   Smith and Kilbourne investigated the possibility that ticks were the vectors spreading Texas cattle fever. They designed experiments that showed that ticks brought south by the northern cattle infected the southern cattle and transferred the infective agent to them.

2. Compare the characteristics of eukaryotic cells with those of prokaryotic cells.

eukaryotic cells	*prokaryotic cells*
membrane-bound nucleus	no membrane
mitochondria	no mitochondria
	mesosomes
mitosis apparatus	divide by fission

3. Compare mitosis and meiosis.

   Mitosis produces cells that are genetically identical, but contain one half the mass.
   Meiosis produces cells that are genetically dissimilar and contain one half the chromosome number as the parent cell.

4. Describe the basic characteristics of fungi.

   Fungi are non-photosynthetic, heterotrophic, eukaryotic organisms that are usually saprophytic. Many are microscopic, while others are macroscopic.

5. Discuss the variable habitats of fungi.

   Fungi are a highly diverse group found in almost every habitat on earth where organic materials are found. They can grow on/in both natural and synthetic products. They can grow in environments with a wide range of pH, temperature, and humidity but most prefer moist, humid conditions. Their spores are highly resistant to destruction.

6. Discuss the reasons for the classification change of *Pneumocystis carinii* from protozoa to fungus.

   P. carinii is the most frequent cause of pneumonia in AIDS patients. Its direct inhalation into the lungs without need of an intermediate host, its ribosomal RNA sequences, and its direct transmission from person to person are more characteristic of fungi than protozoa, and have directed its reclassification.

7. How can fungi cause disease in humans?

   Fungal diseases can be caused by the following four ways:
   (1) allergic reaction to spores or vegetative cells
   (2) reaction to fungal toxins
   (3) growth on or in tissues
   (4) destruction of human food supply, causing starvation and death

Review Questions (pg. 278)

1. Algae are eukaryotic organisms and bacteria are prokaryotic organisms.

2. Classification is primarily based on the principal photosynthetic pigment (in addition to chlorophyll *a*) displayed by each group, Phaeophyta (brown algae) have xanthophyll, rhodophyta (red algae) have phycobillin and chrysophyta (golden brown algae) have carotene.

3. Algae do not directly cause disease in humans but some produce toxins that cause shellfish poisonings (red tides) in humans who eat contaminated shellfish.

4. Algae are important economically because some contain carrageenan and agar, which are used as stabilizing compounds in food. Agar is an important medium for growing microorganisms in the laboratory. Diatoms are used in filters and insulation, and as abrasives in polish. Algae also serve as a food source for the zooplankton.

Review Questions (pg. 286)

1. There are three divisions of fungi:

   (a) Amastigomycota, the terrestrial fungi, which have both sexual and asexual spores. This group includes most of the fungi that cause diseases in humans and food.

(b) Mastigomycota, the water fungi, which have flagellated reproductive cells and cause serious food crop diseases.

(c) Gymnomycota, the slime molds, which live on decayed vegetation and serve as food for larger predators.

2. Yeast are single-celled fungi. Molds are filamentous fungi. A single filament is called a hypha and a group of hyphae are called a mycelium. Mushrooms are the fruiting bodies of filamentous fungi.

3. Fungi affect humans in the following ways:

(a) A person can be allergic to the fungal spores or vegetative cells.
(b) A person may react to the toxin produced by the fungus.
(c) A fungus can grow directly on or in the human body.
(d) Fungi can destroy the food supply, causing starvation.

4. Fungal diseases are called mycoses.

5. Fungi form symbiotic relationships with algae or cyanobacteria to form an organism known as a lichen. The alga or cyanobacterium produces food for the relationship through photosynthesis. The fungus, in turn, protects the alga or cyanobacterium. Mycorrhizas are fungi that are associated with certain trees, and increase the ability of the roots to absorb nutrients.

Review Questions (pg. 290)

1. The four groups of protozoa are:

(a) Mastigophora have one or more flagella and divide by longitudinal fission.
(b) Sarcodina have pseudopodia and divide by binary fission.
(c) Sporozoa are often nonmotile and divide by multiple fission.
(d) Ciliata have cilia and divide by transverse fission.

2. Pinocytosis is the mechanism by which large, soluble substances that will not diffuse through the cell membrane are moved into the cell via the formation of a vacuole. Phagocytosis is the mechanism by which large particulate materials are moved into the cytoplasm of cells by a phagocytic vacuole.

3. Protozoa that exist in different forms during different stages of their life cycle are said to be polymorphic. The trophozoite form is the vegetative growing form. The cyst is the resting form. The cyst can withstand adverse conditions waiting for a change in the environment and the opportunity to grow. The cyst is usually the form which exists outside the body and is the form which is usually passed from one person to another.

Critical Thinking Questions (pg. 292)

1. Algae, fungi, and protozoa are all eukaryotic organisms often found in single cell form. They lack some of the tissue development found in higher animals and plants. Most can reproduce asexually as well as sexually. Fungi and protozoa lack photosynthetic capabilities, while algae are photosynthetic. Fungi and protozoa directly cause disease in humans, and algae indirectly cause diseases in humans.

2. Since algae are photosynthetic, it would be difficult for them to cause diseases inside the human body. Also, the usual habitat for many groups of algae is salt water.

3. Fungi are able to grow under many conditions of heat, cold, alkalinity, and pH.

4. Since sewage disposal depends heavily on the action of protozoa, we would have great difficulty disposing of sewage without the protozoa. A single paramecium eats five million bacteria a day, for example. Since protozoa serve as food for many small animals that are the first links in the food chain, the entire food chain would be severely affected.

CHAPTER 13

Introduction

Viruses have been described in various ways since Beijerinck first suggested that they were filterable disease agents. It is accepted that they are submicroscopic agents capable of attaching to and invading host cells. Receptor sites on the cell's surface act as attachment locations at which spot the virus adsorbs.

Viral study was indirect and biochemical until development of the electron microscope and tissue culture. With these two technological advances, viruses could be studied directly. Much progress has been made in the last thirty years, and the future of virology will depend on technological advances that will allow closer analysis of viral and subviral components.

Objectives

After completing this chapter, the student should be able to:

1. Understand the historical significance of viruses.
2. Describe the general characteristics of viruses.
3. Understand the differences between viruses and cells.
4. Describe the structure of viruses.
5. Discuss the host-virus relationship.
6. Describe viral replication.
7. Describe transduction and its significance.
8. List the methods used to study viruses.

Important Terms

bacteriophage

capsid

virion

segmented virus

capsomeres

nucleocapsid

naked capsid virus

matrix protein

enveloped virus

spikes

prion protein

scrapie

Kuru

productive response

latent state

phage-induced protein

vegetative reproduction

self-assembly process

burst size

one-step growth curve

plaque

eclipse period

latent period

bacterioplankton

virulent

temperate phage

prophage

lysogenic cell

phage lambda

repressor protein

induction

immune

lysogenic conversion

filamentous phage

extrusion

carrier cells

budding

transduction

generalized transduction

specialized transduction

defective phage

episomes

embryonated chicken egg

cell (tissue) culture

primary cultures

monolayer

lysate

cytopathic effect

quantal assays

titer

hemagglutination

Essay

1. Summarize the differences between cells and viruses.

    Cells have both DNA and RNA. Most cells are free-living. They are capable of independent metabolism and reproduction, and contain many enzymes.

    Viruses have either DNA or RNA, never both. They only grow within cells, and are 100 to 1,000 times smaller than cells. Viruses can neither metabolize nor reproduce without the metabolic enzymes of the host cell.

2. Discuss the three major types of relationships that can exist between a virus and its host cell.

    (1) The productive response is described as one in which virus particles are made inside the cell and are released by lysis.
    (2) The nonlytic productive response occurs when the completed viruses escape the cell by extrusion rather than lysis.
    (3) The latent state occurs when the viral nucleic acid becomes integrated into the chromosome of the host cell and then replicates as part of the host chromosome.

3. List and describe the stages in the life cycle of the virulent phage.

    The stages in the life cycle of the virulent phage include all the steps from invasion to lysis.
    (1) viral attachment to host cell receptors
    (2) viral nucleic acid enters host cell
    (3) phage DNA is transcribed, leading to production of specific proteins
    (4) phage DNA replicates and phage structural proteins are synthesized separately

(5) phage DNA and protein assemble to form mature virions

(6) virions are released from the host cell

4. Describe four features of viruses that distinguish them from cells.

(1) Viruses are large molecules, not complete cells and cannot live on their own.

(2) Viruses are not able to metabolize and/or reproduce.

(3) Viruses have either DNA or RNA, not both.

(4) Viruses are obligate intracellular parasites.

5. How is tissue culture important in animal virology?

It is important for propagating viruses in the laboratory. The virus is mixed with the susceptible cells and the mixture is incubated until the cells are lysed.

While the viruses are in their culture medium, they can be studied both quantitatively and qualitatively. The commonly used method for detecting and quantitating the amount of virus present is the plaque assay.

6. Describe the host range of viruses.

The organisms that a virus can infect is its host range. Viruses can infect bacteria, protozoa, algae, fungi, plants, and animals. The host range can involve almost any other life form but a particular virus can generally infect cells of only one or very few species. Limitation of host range is usually due to the fact that the virus must adsorb to a specific receptor site on the host cell surface. Without the matching receptor site, the virus cannot invade the host cell.

7. Describe a few of the methods used to study viruses.

Since viruses cannot be studied directly, one method is to cultivate them in host animals or in specially prepared laboratory media by use of tissue culture.

In tissue culture, the host cells grow as a monolayer in which they can be infected by the viruses, which will multiply in the cells and finally be released by lysis. Lysis of the cells results in a readily observable plaque.

Quantal assays measure the number of infected cells.

Hemagglutination, the ability of some viruses to agglutinate (clump) red blood cells, can also be used to measure virus concentration.

Review Questions (pg. 302)

1. Every virus has a protein coat (capsid) and nucleic acid.

2. Viruses grow only within cells, whereas most cells are free-living. Viruses contain either DNA or RNA, never both, whereas cells always contain both DNA and RNA. Viruses contain very few if any enzymes, whereas cells contain many enzymes. Viruses are much smaller than the cells they invade—between one hundred to one thousand times smaller.

3. Viruses occur in polyhedral (spherical) and helical shapes, and in a combination of the two.

4. Host cells may die, multiply with continuous release of virions, or multiply in a modified form.

Review Questions (pg. 311)

1. A temperate phage has DNA that can be integrated into host chromosome.
    A lysogen, or lysogenic cell, is the name of the host cell.

2. Lysogenic conversion is the process by which prophages confer new properties on host cells.

3. In generalized transduction, any piece of the bacterial genome can be transferred. In specialized transduction, only DNA near the site at which the virus is integrated is transferred.

4. Virions can be counted using the electron microscope or by using quantal assays.

Critical Thinking Questions (pg. 312)

1. Viruses have properties that are associated with both living cells and nonliving agents. However, they are best considered nonliving because they lack some very important features that are common to all known living cells. These features are: (1) they contain only one type of nucleic acid and (2) they contain few if any enzymes. Therefore, they can best be considered chemicals that can multiply inside living cells.
2. Plaques indicate sites at which a single phage has infected a cell. The size of plaques depends on the virus, host, and culture conditions.

CHAPTER 14

Introduction

Much has been learned about viral replication and transmission by the study of animal and plant viruses. The first virus discovered and named, the tobacco mosaic virus, belongs to the latter group. The pathogenic properties of viruses have been studied by extensive use of experimental animals. The sequence of events from absorption through release has been determined and evaluated. Much knowledge concerning the onset of human disease and the interaction of the oncogene can be attributed to continuous investigation into the nature of viruses. The economic impact of viral diseases continues to be an area of active research. Will this lead to greater knowledge of viroid diseases and possibly those caused by the slow viruses?

Objectives

After completing this chapter, the student should be able to:

1. Describe the classification schemes of animal viruses.
2. Understand how viruses interact with animal cells.
3. Describe the manner of viral multiplication.
4. Discuss the interactions of animal viruses with their hosts.
5. Relate animal tumors to viral causes and the manner in which they transform cells.
6. Relate human diseases and tumors to viral causes.
7. Understand the relationship between plant and insect diseases with viruses.
8. Understand the functions of viroids.

Important Terms

productive

persistent

latent

permissive

attachment proteins

endocytosis

maturation

self-assembly

budding

matrix protein

balanced pathogenicity

carriers

acute infections

persistent infections

retroviruses

chronic infections

chronic carrier

latent infections

tumor

neoplasm

benign tumor

malignant tumor (cancer)

leukemia

transformed cells

transforming viruses
reverse transcriptase
provirus
oncogenes
protooncogenes

temporary transmission
circulative transmission
baculovirus
viroid

Essay

1. Describe the common taxonomic criteria employed for animal viruses.

    The four common taxonomic criteria are:
    (1) nature of the nucleic acid enclosed in the capsid—DNA or RNA, single stranded or double stranded.
    (2) virus particle structure—spherical, cylindrical, or complex
    (3) presence or absence of a viral envelope
    (4) size of the viral particle

2. Describe the dependency of viruses on the host enzymes for their replication.

    The dependency of the virus on the host enzymes varies greatly. The larger the viral genome, the fewer host enzymes are involved in replication. The large poxviruses are totally independent of the host enzymes for their replication. In contrast, the very small parvoviruses are so dependent on the biosynthetic machinery of the host that they can only enter the cell and have their viral DNA synthesized if the host cell is synthesizing its own DNA. Most animal viruses are between these two extremes.

3. Distinguish between normal cells and cancer cells in cell culture.

    Normal cells
    (1) grow in single layers
    (2) stay attached to each other
    (3) have limited reproduction
    (4) do not form tumors

    Cancer cells
    (1) grow in unorganized, multilayer pattern
    (2) attach loosely or not at all
    (3) grow indefinitely
    (4) most form tumors in susceptible animals

Review Questions (pg. 320)

1. The three types of viral-host cell interactions are: productive, persistent, and latent.

2. In enveloped viruses two mechanisms exist: the envelope may fuse with the cytoplasmic membrane of the host cell and the nucleocapsid is released into the host cytoplasm; or the host cell cytoplasm surrounds the whole virion in a vesicle, after which the two membranes fuse and the nucleocapsid is released into the cytoplasm. This is termed endocytosis. In the case of naked virions, the virions enter by endocytosis and cause the vesicle to dissolve, thereby releasing the virion into the cytoplasm.

3. Phage DNA codes for the enzyme lysozyme, which dissolves the bacterial cell wall. Animal viruses, on the other hand, do not code for this enzyme but lyse cells because virus-infected cells do not synthesize their own structures and therefore die and then lyse.

4. An animal virus that has the same relationship to its host cells as lambda phage to *E. coli* is the poliovirus or the herpesvirus.

5. Animal viruses are classified by the following four criteria:
    (a) Type of nucleic acid (RNA or DNA), and whether the nucleic acid is single stranded or double stranded.
    (b) Shape
    (c) Size of virus particle
    (d) Whether or not the virion is surrounded by an envelope

Review Questions (pg. 328)

1. (a) Normal cells grow as a monolayer; transformed cells grow in multiple layers in a disorganized pattern
    (b) Normal cells attach firmly to glass surfaces; tumor cells do not attach firmly to glass
    (c) Normal cells grow for a limited number of generations and then die; transformed cells are immortal and grow indefinitely.

2. The substrate of reverse transcriptase is single-stranded RNA, and the product is double-stranded DNA.

3. The oncogene is a transforming gene that retroviruses integrate into the host chromosome. Protooncogenes are similar to oncogens, but are found in normal host genes. The protein coded by the oncogene probably disrupts the function of its similar protooncogene.

4. Insects are responsible for spreading plant viruses.

Critical Thinking Questions (pg. 329)

1. No, not all retroviruses of animals cause cancers. Some retroviruses such as HIV, which causes AIDS, kill the cells they infect rather than cause their abnormal growth.
2. Viruses that infect bacteria and animal cells first bind to specific receptors on the host cell, whereas plant viruses do not bind to specific receptors but enter through wounds in the cell wall. In plants, the cell walls are very tough and rigid, and therefore they cannot fuse with the viral envelope.

CHAPTER 15

Introduction

The establishment of disease in the human is a complex interaction between the host's defense mechanisms and the organism's ability to overcome these defenses. The resident flora protect humans from invasion by pathogens but under certain circumstances the host's defenses are decreased, and the pathogens can become opportunistic. The microorganism must be able to attach to the host tissue, invade them, and produce toxins. The host may resist invasion and infection by both specific and nonspecific defense mechanisms.

Phagocytosis and the inflammatory response are nonspecific mechanisms by which the host limits access of the body to invading organisms. If these efforts are successful, the host is protected and the specific immune system is not mobilized.

Objectives

After completing this chapter, the student should be able to:

1. Describe the means by which the host defends itself against invaders.
2. Describe and discuss the normal flora of the human body.
3. Understand the relationship between infectious agents and human disease by using Koch's Postulates.
4. Discuss the mechanisms of pathogenesis used by microorganisms.
5. Compare exotoxins and endotoxins.
6. List the host factors that influence resistance to disease.
7. Describe the steps of the inflammatory response.

Important Terms

nonspecific defense	infection
specific defense	infectious disease
ecosystem	pathogen
normal flora	pathogenicity
transient flora	opportunist
symbiosis	virulent
commensalism	avirulent
mutualism	infectious base
parasitism	epithelial cells
parasite	basement membrane
host	ligands
colonization	receptors

glycocalyx

endocytosis

exocytosis

extracellular

facultatively intracellular

obligately intracellular

leukocytes

monocytes

macrophages

neutrophils

phagosome

lysosomes

phagolysosome

leukocidins

exotoxin

endotoxin

toxins

erythrocytes

platelets

hematopoietic cells

colony-stimulating factors

basophil

eosinophil

neutrophil

monocytes

inflammation

interleukin-1

Essay

1. Describe the ability of a microorganism to establish an infectious disease.

   The establishment of an infectious disease depends upon a variety of factors that include the ability to:
   - (1) attach to host tissue and overcome normal flora
   - (2) invasion host tissues and proliferate there
   - (3) manufacture and release products toxic to the host
   - (4) overcoming the defenses of the host

2. List the steps in the production of an infectious disease:

   - (1) transmission to a susceptible host
   - (2) adherence to a target tissue
   - (3) colonization
   - (4) damage to the host
   - (5) exit from the host
   - (6) survival outside the host long enough to be transmitted to another host

3. Describe some ways that pathogens interfere with phagocytosis and establish infection.

   Some organisms and the ways in which they evade phagocytosis are:
   S. pneumoniae has a large capsule that inhibits attachment and killing.
   S. aureus produces the enzyme coagulase, which causes plasma to clot, impeding phagocyte movement.
   M. tuberculosis changes the phagosome surface, preventing fusion with the lysosome.

Yersenia pestis can reproduce inside the phagolysosome, protected from the host's immune system and antimicrobial medicines.

4. Differentiate betweens exotoxins and endotoxins.

Exotoxins are produced by both Gram-positive and some Gram-negative organisms. They are protein in nature and capable of being used to form toxoids. They are heat labile. In addition, they cause diverse effects in the body.

Endotoxins are produced by Gram-negative species only. They are lipopolysaccharides released upon cell death from part of the cell wall. They are heat stable and cannot be used in toxoid manufacture. They produce the same symptoms in the body.

5. Describe the effects of endotoxins on the host.

The endotoxins are released when the cell dies. They bind to macrophages and stimulate them to release pyrogens. Pyrogens are protein mediators that are responsible for other symptoms of infection—weakness, aching, and possible shock.

6. Describe the effects of fever in the body.

Fever enhances the inflammatory response. It stimulates phagocytes, the proliferation of lymphocytes, and the production of interferon and antibodies. Fever increases the bacterial requirements for iron and decreases the host's absorption of iron from food.

7. Describe the principal events of inflammation.

Tissue injury is followed by the release of chemical mediators that stimulate blood vessel dilation and the increased flow of plasma. The signs of redness, pain, and fever in the area signal leukocyte migration to the site of injury, which is followed by debris removal and the repair of the injured tissue.

Review Questions (pg. 337)

1. The two primary lines of defense are:
   (a) The nonspecific line of defense includes the physical barriers such as the skin, mucous membranes, and mucus, and the flushing action of the urinary tract and the physiological barriers such as fever, inflammation, and the action of phagocytes.
   (b) The specific line of defense includes specialized cells and chemicals that respond very specifically to a foreign invader.

2. (a) Commensalism is a relationship between two organisms in which one partner benefits and the other is unaffected.
   (b) Mutualism is a relationship between two organisms in which both partners benefit.
   (c) Parasitism is a relationship between two organisms in which one organism benefits at the expense of the other.

3. Normal flora is established at the time of birth. As the fetus passes through the birth canal, it picks up microorganisms that are present in the mother's vagina. In addition, the baby acquires microorganisms from his environment, such as from the food he eats, and from contact with other humans.

4. (a) Normal flora often excrete chemicals that provide a hostile environment for other organisms to live in.
    (b) Normal flora help stimulate the immune system and through cross reactions help to ward off invasion of harmful microorganisms.
    (c) Normal flora often provide a surface incompatible for the attachment of the invader.

5. Koch's Postulates
    (a) First a particular microorganism must be present in all cases of the disease.
    (b) The microorganism must be grown in pure culture.
    (c) The microorganism must then be injected into healthy subjects and they must come down with the disease.
    (d) The microorganism must then be recovered from the experimentally infected host.
    Applications of all of Koch's Postulates to every situation of an infectious disease establishes the causal relationship between the microorganisms and the infectious disease.

Review Questions (pg. 345)

1. The six steps necessary for producing an infection are:
    (1) transmission to a susceptible host
    (2) adherence to susceptible tissues
    (3) colonization
    (4) damage to the host
    (5) exit from the host
    (6) survival outside the host long enough for transmission to another host.

2. These areas are lined with tightly packed epithelial cells. In addition, skin cells are shed regularly, removing microorganisms in the process. The ciliated cells in the respiratory tract along with the mucus are able to move the microorganisms out of the lungs and into the throat and stomach. The urinary tract is cleansed frequently by the flushing action of urine. The digestive tract moves microorganisms out of the body through peristaltic action. The stomach has a very acid environment, which is very hostile to most microorganisms.

3. Colonization implies that the organism has attached, multiplied, and established itself in or on the host. It is a prerequisite to the production of infectious diseases.

4. Steps that phagocytes use to destroy microorganisms are:
    (a) The phagocyte is attracted to the microorganism by certain chemicals it secretes, by products of the complement system, or by phospholipids that are released.
    (b) Once the phagocyte makes contact, it attaches to the microorganisms and engulfs it.
    (c) Once the phagocyte has engulfed the microorganism, it resides in a vacuole known as a phagosome.

(d) Other vacuoles called lysosomes, which are filled with digestive enzymes, fuse with the phagosome and release their enzymes. The vacuole resulting from the fusion is called a phagolysosome.

(e) Various enzymes of the phagolysosome kill and destroy the microorganisms, and the contents of the phagolysosome are then excreted.

5. Exotoxins are found in both Gram-positive and Gram-negative species.
They are synthesized in the cytoplasm and released from the cell, often carried to distant sites in the body, where they cause damage. They are protein and are destroyed by heating. A toxoid useful for vaccination can be made from most exotoxins.
Endotoxins are found only in Gram-negative species. They are a part of the cell wall and are lipopolysaccharides. They usually are a contributing factor in disease rather than playing a primary role. They are heat stable, not destroyed by heating. Toxoids cannot be made from endotoxins.

Review Questions (pg. 354)

1. The different kinds of leukocytes are:
   (a) Polymorphonuclear neutrophils (PMNs or polys) have lobed nuclei. They phagocytize and destroy foreign matter.
   (b) Eosinophils have large granules that stain red. Their nuclei are nonsegmented or bilobed. They play a role in allergic reactions and in rejecting parasitic worms.
   (c) Basophils have lobed nuclei and stain dark blue. They secrete histamine and serotonin, and contribute to the inflammatory response.
   (d) Mononuclear phagocytes have a single nucleus. They phagocytize foreign matter.
   (e) Lymphocytes are small and have a single nucleus. Most have a small amount of cytoplasm in the cell. Lymphocytes are the principal cells of the lymphoid system, which are primarily responsible for the specific immune response.

2. The principal events of the inflammatory response are:
   (a) Following tissue injury, there is a release of chemical mediators that act on blood vessels in the area.
   (b) The tissues swell and there is an increase of temperature in the area, often accompanied by redness and pain.
   (c) White blood cells (leukocytes) migrate through the vessel walls and phagocytize the foreign material and tissue debris, and initiate antibody production.

3. The body is able to mobilize the metabolism in several ways. First, interleukin-1, a protein released by phagocytic cells, causes the body to raise the temperature, inducing fever and the body's defenses. Phagocytic cells remove much of the readily available iron, keeping it away from bacterial cells. Rapid synthesis of proteins is necessary during infection for such things as phagocytosis and antibody production.

Critical Thinking Questions (pg. 355)

1. *E. coli* is a pathogen when it is in the wrong place at the wrong time. For example, it sometimes invades the urinary tract, causing an inflammation of the ureter and bladder. Some *E. coli* strains have a strong attachment for cells in the intestinal tract and causes serious disease with bloody diarrhea and release of potent enterotoxins.

2. An infection occurs when a microorganism enters the body and multiplies, whereas disease implies that there is an impairment to body function. Thus, there can be infection in the body but not disease.

3. Normal microbial flora often is a threat to human health if the host is immunologically compromised, in which case the normal flora can then promote disease. In some cases an organism is in a place where it is not normally found or the host defense mechanism does not work well enough to keep the organism under control, which occurs in people with AIDS.

CHAPTER 16

Introduction

The specific immune system is an elaborate interplay of cells and chemical mediators that defend the body against invasion by foreign antigens. The nonspecific defense mechanisms function to limit the invasiveness of infectious agents, but if they are unsuccessful then further measures must be taken. Humoral and cell-mediated immune structures respond to the antigenic determinants and neutralize their toxic effects before eliminating them from the body. The anamnestic response ensures long-term protection from damage by these same invaders.

Objectives

After completing this chapter, the student should be able to:

1. Describe the characteristics of innate immunity.
2. Describe the various types of acquired immunity.
3. Differentiate between nonspecific and specific immunity.
4. Describe antigens.
5. Describe antibody production and response.
6. Describe the characteristics of the antigen-antibody response.
7. List the classes of immunoglobulins.
8. Understand clonal expansion.
9. Describe cell-mediated immunity.
10. Understand the interactions of the two divisions of the immune response.
11. Describe the relationship between the anamnestic response and vaccination.
12. Describe immunological tolerance.

Important Terms

variolation
vaccinia
acquired immunity
antibodies
naturally acquired active immunity
naturally acquired passive immunity
artificially acquired active immunity
artificially acquired passive immunity

gamma globulin
nonspecific immune response
specific immune response
immunogen (antigen)
autoimmune disease
antigenic determinants (epitopes)
haptens
antigen-processing cells

major histocompatibility complex

human leukocyte antigen

humoral immune system

cell-mediated immune system

B cells (B lymphocytes)

T cells (T lymphocytes)

cluster designation markers

helper T cells

precursor B cells

plasma cells

clonal expansion

globulins

immunoglobulins

heavy chains

light chains

variable region

constant region

hypervariable regions

secretory Ig A

secretory component

serum Ig A

hybridoma

monoclonal antibodies

memory cells

opsonization

opsonins

T dependent antigens

lymphokines

cytokines

natural killer cells

killer cells

colony stimulating factor

monocytes

superantigens

autoimmunity

anamnestic response

attenuated

immunological tolerance

Essay

1. Describe the term "attenuated" as it is used in this chapter.

   Attenuate means to weaken or alter. Infectious agents that are used to induce immunity are altered so they are not pathogenic. This process removes the virulence of the agent but allows it to maintain its ability to produce an immune response.

2. Describe the nonspecific immune responses.

   The nonspecific defense system is designed to protect the individual from invasion by physical (splinter) or biological (virus or bacteria) invaders. The nonspecific defenses include physical barriers such as the skin, cilia, and flushing action of the urine, and the physiological barriers such as inflammation, fever, and phagocytosis.

3. What is meant by innate immunity?

   Innate immunity is inborn due to the genetic constitution of the species or individual. It is independent of prior exposure. It confers immunity to one species for diseases that affect other species. Innate immunity is affected by the individuals health, age, nutrition, and sex.

4. Distinguish between active and passive immunity.

Active immunity develops after exposure and response to the antigen. The host recognizes the antigen and stimulates the cells and lymphokines necessary to provide long-term protection.

Passive immunity is the result of acquisition of immunoglobulins, which offer short-term protection against specific diseases. The immunoglobulins are not self-made, but are introduced via the placenta, breast milk, or gamma globulin.

Review Questions (pg. 361)

1. Most of the biological world has some type of nonspecific immune response that includes barriers and chemicals that can prevent invasion by microorganisms and viruses. Vertebrates have developed a very specific immune response that recognizes previous invaders via memory cells.

2. Innate immunity is natural immunity that is independent of previous exposure to foreign substances and depends on the activities of phagocytic cells, nonspecific antimicrobial substances, and metabolic changes such as fever. Acquired immunity occurs as the result of exposure to foreign materials and is dependent on a highly specific response to that foreign material.

3. Immunization with polio vaccine is an example of artificially acquired passive immunity.

4. Macromolecules such as proteins and polysaccharides make good antigens. Nucleic acids and lipids are not good antigens.

5. Antigenic determinants are sequences of amino acids on a protein molecule to which an antibody is made.

6. Antigens must be processed by antigen-processing cells such as macrophages before they can be presented to the immune cells that produce an immune response.

Review Questions (pg. 372)

1. See the diagram in Figure 16.7.

2. The constant region of the IgG molecule attaches to phagocytes and helps the IgG molecule cross the placenta. IgM constant region is active in the complement cascade. IgA constant region aids in protecting the mucous membranes. IgE constant region is involved in the specific attachment to mast cells and basophils, and is therefore involved in the hypersensitivity response to antigens. IgD constant region functions during the development of the antibody response.

3. When a B cell comes into contact with an antigen that has a close enough fit chemically, it triggers that B cell to divide and mature in a clone of plasma cells all producing antibodies of that particular antibody configuration.

4.	The variable region of the antibody molecule has the possibility to have many different configurations. The close fit of the configuration to the antigenic determinant (lock-and-key relationships) determines whether or not a B cell will become a clone of turned on plasma cells all producing antibody molecules of the same configuration.

5.	During the clonal expansion, some cells do not become plasma cells but become memory cells. Thus, the pool of B cells that have the same or similar antibody configuration is greatly increased, and are primed and ready to respond when they reencounter the same antigen. This accounts for the secondary immune response. The theory behind vaccination is that if the immune system is primed with memory cells, they will be ready to respond quickly to subsequent exposure to the same antigen.

Review Questions (pg. 379)

1.	Similarities: Both B cells and T cells have receptors on their membranes that are complementary for antigenic determinants on antigens. B cells and T cells both produce memory cells that trigger an enhanced response when challenged a subsequent time by the same antigen. Clonal expansion is present in both B cells and T cells.
	Differences: B cells produce antibodies that have two antigen recognition sites, whereas T cells have only one antigen recognition site. B cells mature into plasma cells that produce antibodies, which are secreted. T cells display their antigen recognition site on their cell membrane.

2.	Cytokines are hormones secreted by cells, such as macrophages and B cells, that signal various cells to act in the immune system. Examples include interleukins and interferon.

3.	The secondary immune response occurs following the clonal expansion of B cells or T cells into memory cells, which give the immune system a larger number of cells primed to react to the second or subsequent contact with an antigen. Due to the larger number of cells, the response is more rapid and intense compared to the first exposure to antigen.

4.	Some disease-causing agents such as influenza and malaria are not easily controlled by vaccination. In the case of influenza, the problem is caused by the fact that the virus antigenic determinants mutate frequently, causing new stains to develop. In the case of malaria, the parasite has a number of stages of development, all of which have different antigens. Thus, determining which antigens to use for immunization purposes has been difficult.

5.	First, the presence of the antigen is needed to continue the immune response. Once the antigen has been eliminated, the response is turned off. Second, plasma cells and antibodies have finite life spans, and disappear if no longer stimulated. Also, since antibodies bind to antigenic determinants, they become unavailable to interact and stimulate lymphocytes. Additionally, suppressor T cells prevent B cells from maturing and T cells from dividing, thus shutting down the immune response.

Critical Thinking Questions (pg. 381)

1. Deficiencies in either the humoral or the cell-mediated immune response would affect the body's ability to defend itself against foreign invaders. If the **B** cells are defective, antibodies cannot form and thus cannot attach to various bacteria. Since people are prone to infections, if the T cells are defective, there cell-mediated immunity is suppressed or inhibited, and virus infections and cancers are more common. If both **B** and T cells are defective or absent, no immune response is possible and people usually succumb to massive infections.

2. Vaccines are often difficult to make because it is not clear which antigens present are the ones that provoke a good immune response. Also, the surface antigens of some viruses such as influenza can change genetically, making the production of a specific vaccine difficult. Since these changes often occur quickly, vaccine production often cannot keep up with demand.

CHAPTER 17

Introduction

Specific immunity is designed to protect the body from the ravaging effects of infectious disease. When the immunoglobulins and the cells organized for this purpose respond to harmless foreign antigens, hypersensitivity results. Some forms seem to be familial, while others are stimulated by the transfer of cells or immunoglobulins from sensitized individuals into the susceptible hosts.

In many cases, the reactions may be mild and even protective; in others, the violent response may be fatal.

Autoimmune disease results when the body becomes sensitized to its own proteins, resulting in severe damage. The more information that is learned about the immune system, the greater the degree of speculation about its uses in therapy. Since so much of the body's homeostasis is dependent upon the interaction of all aspects of immunity, interfering with one facet could affect other facets. Researchers need to be aware of the far-reaching effects of their discoveries on the body as a whole.

Objectives

After completing this chapter, the student should be able to:

1. List and describe the primary immunodeficiency disease.
2. Describe the secondary immunodeficiency diseases.
3. Distinguish among the Type I hypersensitivities.
4. Describe the cytotoxic reactions and their results.
5. Describe the inflammatory mechanisms associated with the Type III hypersensitivities.
6. Understand the types of Type IV hypersensitivities.
7. Describe graft rejection and antigenic incompatibility.
8. List the autoimmune diseases.
9. Describe the various types of diagnostic testing used in the laboratories.

Important Terms

immunodeficiency
hypersensitivity
autoimmunity
deficiency
primary, or congenital, immunodeficiency disease
secondary, or acquired, immunodeficiency disease
DiGeorge syndrome

infantile X-linked agammaglobulinemia
severe combined immunodeficiency disease
   (SCID)
Chediak-Higashi disease
anaphylaxis
hemolytic disease of the newborn
immune complex

disseminated intravascular coagulation
Arthus reaction
graft versus host disease
major histocompatibility complex (MHC)
human lymphocyte antigen (HLA)
T dependent antigen
T independent antigen
immunosuppressant medicines
myasthenia gravis
sympathetic ophthalmia
rheumatoid arthritis
lupus erythematosus

precipitins
primary complexes
Ouchterlony technique
immunoelectrophoresis
countercurrent immunoelectrophoresis (CIE)
microtiter plates
complement fixation test
indicator system
contact hypersensitivity
patch tests
zone of optimal proportion

Essay

1. Differentiate between primary and secondary immunodeficiency diseases.

    The primary immunodeficiency diseases develop as a result of a genetic defect. A failure of the thymus to develop results in DiGeorge syndrome, whereas a lack of ß-lymphocytes results in X-linked agammaglobulinemia. If the bone marrow stem cells are defective, then severe combined immunodeficiency disease (SCID) can occur.
    Secondary immunodeficiency disease is acquired as a result of infection, aging, or other factors. Some examples are AIDS and myelomas.

2. Briefly describe the types of hypersensitivity, with a specific example of each.

    Type I hypersensitivity involves B cells and IgE, with histamine and other mediators released from sensitized mast cells. Skin reactions include flare and wheal. Examples are local anaphylaxis, hay fever, and generalized anaphylaxis.
    Type II hypersensitivity involves B cell, IgG, IgM, and complement. There is no distinctive skin reaction and the response may be slightly delayed. Examples are transfusion reaction and hemolytic disease of the newborn.
    Type III hypersensitivity involves B cells, IgG, IgM, and complement. It may be transferred in serum. Examples are serum sickness, farmer's lung, and rheumatic fever.
    Type IV hypersensitivity involves sensitized T cells, not antibodies. The mediators are lymphokines. It can only be transferred in whole blood and may take several days to develop. Examples are contact dermatitis and transplantation rejection.

3. Compare the two classes of MHC.

    MHC identification is important in the successful matching of tissue donors with recipients.
    Class I MHC molecules are found on all nucleated cells. It is responsible for the recognition of virus-infected cells and is involved in antigen presentation to cytotoxic T lymphocytes.

Class II MHC molecules are found principally on macrophages and are important in recognition of antigens on the cell surface. They are involved in presenting the antigens to helper T lymphocytes.

4. Discuss the problems associated with autoimmune diseases.

Autoimmune diseases result from the body's inability to differentiate between self and non-self. Some may be mediated by humoral antibodies and others by delayed hypersensitivity. Immunosuppressant drugs and thymectomy are helpful in many cases of myasthenia gravis. However, these treatments can lead to lowered resistance to disease and therefore increased susceptibility to infectious agents. The cause of the breakdown of the recognition system is not understood completely and may be triggered by drugs or an earlier infection in which remaining antibodies attach to host tissues, resulting in damage. Other examples of autoimmune diseases are rheumatoid arthritis, Graves' disease, multiple sclerosis, and sympathetic ophthalmia.

Review Questions (pg. 393)

1. A primary immune deficiency disease occurs during fetal development and results in a defect in the lymphoid system. For example, the thymus does not develop correctly, resulting in a lack of T cells. A secondary immunodeficiency disease is acquired later in life as the result of an infectious disease, immunosuppressive drugs, malnutrition, or advanced age. An example is AIDS, which is characterized by the destruction of helper T cells.

2. In IgE-mediated allergies, the antigen to which the person is sensitive makes contact with T cells and then B cells, which produce IgE antibodies instead of IgG antibodies. These IgE molecules then attach themselves to mast cells and basophils, which are loaded with histamine and leukotrienes. When the antigen is presented to these cells on a second or subsequent time, the mast cells and basophils release their histamine and leukotrienes, causing hives, hay fever, and other allergic reactions.

3. Type I hypersensitivities are rapid, and involve B cells and IgE. Histamine and other similar chemicals are released. Hay fever and hives are common manifestations of Type I.
   Type II hypersensitivities take a longer time to develop, and involve B cells, IgG, and IgM. Complement is involved, causing lysis of red cells. Transfusion reactions and hemolytic disease of the newborn are examples.
   Type III hypersensitivities take a longer time to develop, and involve B cells, IgG, and IgM. Antibodies react with specific antigen and fix complement, resulting in immune complexes. Immune complexes are not engulfed by phagocytes and often accumulate in certain tissues such as joints. Examples are serum sickness and farmer's lung.
   Type IV hypersensitivities take several days to develop and involve T cells and no antibody. Contact dermatitis such as poison ivy is an example.

4. Immune complexes cause inflammatory response in joints, kidneys, and the skin. Disease states in which they play a role are bacterial endocarditis, German measles, malaria, and farmer's lung.

5. MHC compatibility tests are done before transplantation to avoid rejection of the tissue graft. Graft versus host disease occurs when a tissue such as a heart or kidney is transplanted into a person who does not have the exact same major histocompatibility antigens. As a result the body views this organ as foreign, and a delayed hypersensitivity reaction is mounted against this organ. The T cells try to destroy the cells of the graft.

Review Questions (pg. 400)

1. Autoimmunity occurs when the immune system reacts to self antigens as if they were foreign. The inability to differentiate self from non-self may be a result of the development by some bacteria and viruses of amino acid sequences that closely resemble self antigens. Thus, an immune response including inflammation and activity by phagocytes is mounted, resulting in destruction of normal tissue.

2. Lupus erythematosus affects the joints, skin, heart, and kidneys.
   Diabetes mellitus affects the pancreas and its production of insulin.
   Multiple sclerosis affects the brain and spinal cord.
   Sympathetic ophthalmia affects the eyes.

3. Precipitin reactions: When antibodies and soluble antigens are mixed, they form large antibody-antigen complexes that precipitate and can be seen.
   Agglutination reactions: These reactions are similar to precipitin reactions except that they involve particulate rather than soluble antigen, so much larger aggregates of antigen and antibody.
   Complement fixation: This test involves antibody-sensitized red blood cells to detect the presence of complement and cause red cell lysis.
   Virus neutralization test: This test involves mixing a serum with the known virus. If the antibody is present, it will bind to the virus and neutralize it. When the suspension is plated on a cell culture or injected into a chick embryo, the virus will be unable to grow.
   ELISA test: Antigen is first placed into a test tube. It absorbs to the tube and the excess is washed away. Serum is then added to the tube and is incubated. If the antibodies are present, they will absorb to the antigen. The excess is washed away and an enzyme-linked molecule that can react with a ligand on the antibody as well as a color reagent is added. A color change can then be observed and measured.

Critical Thinking Questions (pg. 401)

1. Everyone does not have allergies because not everyone has mast cells and neutrophils that are ready to produce IgE when they come in contact with antigen.
2. Usually a person has detectable antibody levels when they have had a particular infectious disease. These antibodies show that the body has mounted an immune attack against that particular virus or microorganism. The antibodies are produced by the plasma cells that have been primed by a particular virus or microorganism to expand as a clone and produce a particular configuration of antibody.

CHAPTER 18

Introduction

The study of the incidence and spread of infectious disease in a population is termed epidemiology. Specialists in the field attempt to describe and possibly predict the direction of communicable infections. Susceptible populations can be identified and suggestions made to ensure their protection.

Epidemics may be transmitted personally or through a common source. The rate of spread depends on the characteristics of the population in question. A variety of approaches can be taken to control transfer of infectious disease agents among the group.

Hospital patients are among the most susceptible victims of contagious diseases due to the environment in which they are confined and the nature of their treatment. Hospital epidemiologists aspire to reduce the incidence and spread of nosocomial infections. The National Centers for Disease Control and Prevention continuously offers aid directed toward the protection of the medical personnel and the population in general.

Objectives

After completing this chapter, the student should be able to:

1. Determine the historical significance of epidemiology as an investigative science.
2. Identify the types of epidemics.
3. Distinguish among the methods of identifying organisms involved in epidemiology.
4. Understand the factors that are involved in the spread of disease through a population.
5. Determine the characteristics of a population that could limit or enhance an epidemic.
6. Describe nosocomial infections.
7. Elaborate on the infectious disease control network.

Important Terms

epidemic

pandemic

endemic disease

epidemiology

bacteriocin typing

antibiograms

biotyping

gel electrophoresis

reservoir

vectors

vehicles

sentinel

index case

droplet nuclei

attack rates

infectious dose

fomite

transient carrier

chronic carrier

universal blood and body fluid precautions
body substance isolation

Essay

1. Discuss the types of epidemics and give examples of diseases identified with each type.

    An epidemic is an infectious disease that affects a large number of people in a geographic area. Chicken pox and measles occur as epidemics.

    A pandemic is described as an epidemic that spreads to include most of the world population, as with influenza and AIDS.

    An endemic disease occurs sporadically in a population over a long period, as with the common cold.

2. List the techniques involved in identifying agents thought to be responsible for epidemic diseases.

    A variety of techniques may be employed, depending on the nature of the agent.
    (1) Bacteriophage typing is used to identify strains of *Staphylococcus.*
    (2) Bacteriocin typing is used for species of *Pseudomonas.*
    (3) Biotyping and serotyping are used to identify specific metabolic products, such as those found in *E. coli.*
    (4) Herpes simplex viruses may be identified by gel electrophoresis.

3. What influences are responsible for the characteristics of an epidemic?

    Characteristics that influence the nature of epidemics are:
    (1) The age of the population. The young and the elderly are among the most susceptible, due to lack of exposure in the former and depressed immunity in the latter.
    (2) Genetics. Some populations appear to have a greater susceptibility or resistance based on innate immunity.
    (3) Nutrition. Malnourished people tend to have poor defense mechanisms and reduced resistance to disease.
    (4) Crowding and hygiene. Overcrowded conditions foster the spread of disease agents, especially those that are airborne. Poor hygiene patterns can aid in the transfer of microbes by the fecal-oral route.
    (5) Microbial virulence. The greater the virulence, the lower the dose necessary for the disease to become established.

4. Discuss the nature of nosocomial infections.

    Nosocomial infections are those acquired in the hospital. These are extremely dangerous due to several aspects of both the agent and the population.

    Since the hospital environment is saturated with antimicrobial agents, the surviving microbes are especially virulent and frequently antibiotic resistant.

The population of the hospital is composed of compromised individuals with increased susceptibility and reduced resistance to infection. The illness and the treatment of it contribute to increased availability of entry site for the microbes.

Hospital personnel may also contribute to the transmission of disease agents by poor aseptic techniques and acting as carriers of the disease.

5. Describe the nature of an infectious disease control network.

The duty of an infectious disease network is to identify, control, and stop the spread of epidemics. Immunization and nutrition programs increase the general health of the population. Identification of responsible agents, vectors, and carriers of a disease is an important step in controlling the spread of disease. The National Centers for Disease Control and Prevention in Atlanta continuously monitors infectious disease and suggests means of preventing their recurrence.

Review Questions (pg. 411)

1. Epidemiology is the study of the factors that influence the frequency and distribution of diseases.
   An epidemic results when an unusually large number of cases of an illness occurs in a population.
   A pandemic results from an epidemic that spreads worldwide.
   An endemic disease results when scattered cases of an illness appear in a population over a long period.

2. Four features of our modern world underscore the importance of epidemiology:
   (1) Increasing population density fosters the spread of infection.
   (2) Social unrest undermines the prevention, detection, and control of epidemics.
   (3) Commercial food production and distribution can spread pathogens widely.
   (4) Rapid transportation can disseminate pathogens worldwide.

3. Identification of specific strains is necessary to distinguish an epidemic strain from other strains. Methods of identification include bacteriophage typing, bacteriocin typing, antibiograms, biotyping, using antisera, and using restriction nucleases.

4. The reservoir of an infectious agent is the sum of all the places where an infectious agent is found.

5. A vector is an insect or other animal that spreads an infectious agent, whereas a vehicle is a nonliving substance, such as air or water, that spreads an infectious agent.

6. An index case is the first case of an epidemic.

7. The attack rate is the number of cases developing per hundred people exposed to the epidemic agent.

8. A fomite is an inanimate object, such as a door knob or towel, that can transmit an infectious agent.

Review Questions (pg. 417)

1. Older people may be immune or partially immune because of prior exposure to the same or related infectious agents. The relative lack of susceptible individuals in the population may prevent an epidemic from occurring, or shorten its course.

2. The virulence of an infectious agent varies with different strains.

3. Resistance to infection is enhanced by maintaining general good health. Employing cleanliness and other infection control measures, for example in day care centers help curtail infections. Identifying and treating carriers, and other measures to decrease the reservoir of the disease agent can help prevent the spread of infections.

4. Nosocomial infections are a big problem since virulent pathogens are continually introduced in hospitalized patients who have low resistance to infection.

5. Prevention of nosocomial infections includes using "universal blood and body fluid precautions" and "body substance isolation." Other measures include using good ventilation and aseptic techniques. Employing a hospital epidemiologist can also limit the spread of nosocomial infections.

6. (1) Public schools report absentee rates.
   (2) The news media inform the public of the existence of epidemic disease and preventive measures.
   (3) State public health laboratories help detect the presence of an epidemic, and precisely identify its cause.
   (4) The Centers for Disease Control and Prevention supports state laboratories, keeps track of national and international disease, and helps local public health officials in the control of epidemics.

Critical Thinking Questions (pg. 419)

1. Hand washing can decrease the quantity of pathogens below an infectious dose, or to a level easily defeated by the body's defenses.
2. A person with mild illness or in the incubation phase of a disease in a Laotian village could carry the causative agent to any part of the world by air travel.
3. There should be an effective treatment or immunizing agent against the disease. The reservoir should be easily defined.

CHAPTER 19

Introduction

The skin is the body's first line of defense against microorganisms and any other invading agent. When it is breached, the inner tissues may be invaded by both the normal flora of the skin and opportunists present in the area of the break. The major function of the normal flora is to maintain a constant population, which is itself protective, both by its presence and its by-products.

All types of pathogens can invade the skin and mucous membranes, causing diseases that range in severity from mild to fatal, and from local to systemic. Immunizations to several of the skin pathogens are available and should be encouraged to increase herd immunity and prevent epidemics.

Objectives

After completing this chapter, the student should be able to:

1. Understand the concept of the skin as an organ and the body's first line of defense.
2. Describe the normal flora of the skin.
3. Describe the skin diseases attributable to *S. aureus*.
4. Describe streptococcal involvement in skin infections.
5. Describe Rocky Mountain spotted fever and other tick-borne diseases.
6. Describe skin diseases caused by viruses.
7. Describe the skin diseases caused by fungi.

Important Terms

epidermis
keratin
dermis
sebum
acne
eczema
tinea versicolor
folliculitis
furuncle
carbuncle
coagulases
protein A
penicillinase
pyoderma

impetigo
S. aureus
M-protein
zoonosis
erythema chronicum migrans
intranuclear inclusion bodies
subacute sclerosing panencephalitis
Koplik's spots
congenital rubella syndrome
erythema infectiosum
exanthem subitum
papillomas
dermatophytes

Essay

1. Describe the major characteristics of the skin that help it protect the body against infection.

   The skin resists infection by disease causing organisms by being tough, dry, acidic, and toxic. The normal flora of the skin generally prevent colonization of pathogens by the production of fatty acids. The staphylococci maintain the balance among flora of the skin.

2. Compare the two most common pathogens of the skin.

   The two most common pathogens of the skin are *Staphylococcus aureus* and *Streptococcus pyogenes. S. aureus* is a Gram-positive cocci arranged in clusters. It produces several extracellular products that enhance its virulence. These extracellular products include hemolysins, leukocidins, and hyaluronidase. They are prominent causes of boils, abscesses, impetigo, food poisoning, and toxic shock syndrome.

   Streptococcus pyogenes is a Gram-positive cocci in chains. Several extracellular products that enhance its virulence are hemolysins, streptokinase, hyaluronidase, and erythrogenic toxins. Among the diseases caused by *S. pyogenes* are impetigo, pharyngitis, scarlet fever, and puerperal fever.

3. Describe the two most prominent diseases transmitted by ticks.

   The two major tick-borne diseases in the United States are Rocky Mountain spotted fever and Lyme disease. Rocky Mountain spotted fever is caused by *Rickettsia rickettsii* and is transmitted by a bite of an infected member of the *Dermacentor sp.* The symptoms include headache, muscle and joint pain, and fever, followed by a hemorrhagic rash. The organisms multiply at the site of the bite and invade the bloodstream. Vascular lesions and endotoxin account for the pathologic changes. Prevention of the disease is through such measures as avoidance of tick-infested areas, tick repellent, and removal of ticks within four hours of exposure.

   Lyme disease is caused by *Borellia burgdorferi* and is spread by the bite of ticks of the *Ixodes sp.* infected with the spirochete. The symptoms include an enlarging red, ringlike lesion at the site of bite, fever, malaise, headache, and enlargement of lymph nodes near the bite. The spirochetes disseminate to body tissues by the bloodstream, and tissue damage is caused by the immune reaction to the organisms. Prevention includes wearing protective clothing and using tick repellents when in areas inhabited by the white-footed mouse and white-tailed deer.

4. Compare measles with German measles.

   Measles is caused by the Rubeola virus, and RNA virus. The symptoms, which develop after 10–12 days, include rash, malaise, fever, conjunctivitis, cough, and nasal discharge. The disease is highly contagious and is spread by the respiratory route. The human is the only source. The virus multiplies in the respiratory tract and is carried by blood to various parts of the body where serious damage may develop. The disease can be prevented by immunization with the attenuated virus at 15 months of age, with a second vaccine given at kindergarten age or adolescence.

   German measles is caused by the rubella virus, an RNA virus. Symptoms develop after an incubation period of 14–21 days and include malaise, headache, fever, and a rash that begins on the forehead and face. The virus replicates in the respiratory tract and may be disseminated to all parts of the body. It can cross the placental barrier and cause serious CNS damage to developing fetuses. Live attenuated virus vaccine is administered to children at 15 months of age to prevent occurrence of the disease.

Review Questions (pg. 424)

1. Skin injuries such as cuts and burns are two routes by which microorganisms and viruses invade the skin.

2. The normal skin surface is too dry, unstable, acidic and toxic for most pathogens.

3. (1) Diphtheroids are Gram-positive rods of the *Corynebacterium* and *Propionibacterium* genera which are varied in shape and nonmotile. *P. acnes* grows primarily within hair follicles where conditions are anaerobic.
    (2) Staphylococci are Gram-positive cocci arranged in packets or clusters. These common skin bacteria are facultative anaerobes which help to prevent pathogens from colonizing the skin.
    (3) *Malassezia* species are small, generally harmless yeasts that require oily substances for growth. These species can, however, cause face rashes, dandruff and tinea versicolor in some people.

Review Questions (pg. 436)

1. *Staphylococcus aureus* is found in furuncles and carbuncles. It is a Gram-positive coccus that is facultatively anaerobic. Also, it grows on a high-salt medium, ferments mannitol, and is coagulase positive.

2. Bacteria attach to cells of hair follicle, multiply in the pilosebaceous gland, and reach subcutaneous tissue, where an abscess forms. Increasing pressure in the abscess forces the bacteria into the bloodstream. The blood then carries the organism to the bone, where it establishes infection.

3. Protein A binds to the Fc region of antibody molecules, thereby preventing attachment of the phagocyte to the antibody.

4. Leukocidin, hyaluronidase, protease, and lipase all contribute to the virulence of *S. aureus.*

5. The blister is caused by the circulating toxin exfoliation, not by skin infection by the bacterium.

6. Similarities: Gram-positive, cocci, produce hemolysins.

Differences:	*S. aureus* arranged in clusters, produce catalase, protein A in cell wall	*S. pyogenes* arranged in chains, catalase-negative, M-protein and group A polysaccharides in cell wall

7.

	Rocky Mountain spotted fever	Lyme disease
Cause:	*Rickettsia rickettsii,* an obligate intracellular, Gram-negative coccobacillus	*Borrelia burgdorferi,* a spirochete
epidemiology:	spread by wood ticks or dog ticks of the genus *Dermacentor*	Spread mainly by deer ticks of the genus *Ixodes.*

Review Questions (pg. 446)

1. Herpes zoster, or shingles, is a typically localized disease resulting from activation of a latent varicella-zoster virus. It represents a reactivation of the varicella virus infection in the distribution of a sensory nerve.

2. Infectious varicella-zoster virus can persist indefinitely in certain populations, causing recurrent epidemics of chicken pox in people who lack immunity to the disease.

3. The rubeola virus can cause irreversible damage to the lungs and nervous system. It can also be responsible for serious secondary bacterial infections.

4. A telltale sign of rubeola is Koplik's spots in the mouth.

5. During the period of viremia, the rubella virus crosses the placenta and infects the fetus, often resulting in infants with the congenital rubella syndrome and prolonged viral excretion.

6. Warts are skin tumors caused by papillomaviruses, which are spread by such fomites as wrestling mats, towels, and shower floors.

7. Ringworm is caused by various dermatophytes, including members of the *Trichophyton, Microsporum,* and *Epidermophyton* genera.

Critical Thinking Questions (pg. 448)

1. One possible complication of rubella for boys is sterility.
2. Since no vaccines are available, the only preventative measures one can take involve avoiding tick infested areas, wearing protective clothing, using tick repellents, checking for body ticks, and careful removal of any attached ticks. These measures are more difficult to follow than a vaccination regimen.

CHAPTER 20

Introduction

The upper respiratory system is the most susceptible area of the body to infectious disease. In a population dense environment, infectious agents can enter this system with every inhalation. The non-specific defense mechanisms are extremely efficient in protecting the deep respiratory system, but occasionally they fail, and severe diseases result.

Most diseases of the upper respiratory system are mild with few complications. However, diphtheria can be fatal in one out of ten cases.

Objectives

After completing this chapter, the student should be able to:

1. Understand the historic contributions to the development of diphtheria toxoids.
2. Describe the anatomy of the upper respiratory system.
3. Describe the normal flora of the upper respiratory system.
4. List and describe the diseases caused by *Streptococcus pyogenes.*
5. Describe diphtheria as a toxin-mediated disease and its prevention.
6. Describe the differences between sinusitis and otitis media.
7. Describe conjunctivitis, its causes and epidemiology.
8. Discuss the various viral infections of the upper respiratory system.

Important Terms

toxoid	acute rheumatic fever
cilia	antitoxin
normal flora	otitis media
Lancefield grouping	sinusitis
scarlet fever	pink eye
acute glomerulonephritis	pharyngitis

Essay

1. Describe the normal flora of the upper respiratory system.

   The normal flora of the upper respiratory system includes a variety of Gram-positive and Gram-negative bacteria. The staphylococci are usually present, including the potential pathogen *S. aureus*. Many nonpathogenic Gram-positive rods, commonly referred to as diphtheroids may be cultured from this area. Gram-negative cocci that resemble the *Neisseria* sp. are found, along with the small, pleomorphic, Gram-negative rods of the genus *Haemophilus*. Occasionally, Gram-negative, strict anaerobes of the genus *Bacteroides* are isolated. Many streptococci are found in the area, including the potential pathogen *S. pneumoniae*.

2. Discuss the characteristics of strep throat.

   After an incubation period of two to five days, the throat appears red, with tiny hemorrhages and exudate. Swollen lymph nodes in the neck are also characteristic. Occasionally, such complications as tonsillitis, rheumatic fever, and glomerulonephritis occur as sequels to the disease.
   The disease is spread through direct contact with infected hosts and occasionally, by ingesting contaminated food. The disease can be prevented by avoiding crowded places and by daily treatment with penicillin for people with a history of rheumatic heart disease.

3. Discuss the progress of diphtheria as a potentially fatal disease.

   Diphtheria is an upper respiratory disease characterized by sore throat, fever, fatigue, and malaise after an incubation period of less than one week. The disease is spread by inhalation of infected droplets or direct contact with patients or carriers. The severe complications of diphtheria are due to the release of a powerful exotoxin manufactured by lysogenic conversion, which is absorbed by the bloodstream. The toxin is the cause of severe damage to the cells of the heart, kidney, and nervous tissue. The pseudomembrane formed in the throat is the result of an inflammatory response. It frequently is responsible for asphyxiation of the patient. Treatment with antibiotics is not fully successful since they have no effect on the toxin, which causes the severe damage to the body.

4. Describe the results of an infection with the rhinoviruses.

   Epidemics are spread by contact with infected individuals or with objects they have contaminated. The symptoms of scratchy throat, nasal discharge, headache, and malaise last for about seven days. Treatment with antibiotics is counterproductive in disease caused by the rhinoviruses due to elimination of normal flora, which could be protective.

5. Why are vaccines against the common cold unavailable?

   More than eighty-nine types of rhinovirus are responsible for the common cold. Because of the variety of infectious agents, a vaccine is unavailable. Interferon may be used as a treatment but it is too expensive and difficult to obtain.

Review Questions (pg. 453)

1. *Staphylococcus aureus* commonly colonizes the nostrils.

2. The cilia of the mucous membranes propel a film of mucus containing any entrapped microorganisms and viruses into the nose and throat. The mucus is swallowed and the entrapped microorganisms and viruses are exposed to the killing action of the stomach's juices.

3. Viruses and microorganisms adhering to the moist membranes covering the eyes are almost immediately swept into the nasolacrimal duct and nasopharynx by tears and the eyelid's blinking reflex.

4. *Streptococcus pneumoniae, Haemophilus influenzae,* and *Neisseria meningitidis.*

Review Questions (pg. 463)

1. Group A ß-hemolytic streptococci cause scarlet fever. Carbohydrates on the cell wall determine the groupings.

2. Streptolysin O is only active under anaerobic conditions, which suppress the growth of competing flora.

3. M protein in the cell walls accounts for the disease-causing ability.

4. Scarlet fever is a disease characterized by a red rash and whitish coating on the tongue. It is caused by infection with *S. pyogenes* strains that produce erythrogenic toxin.

5. Acute glomerulonephritis and acute rheumatic fever can develop after recovery from strep throat.

6. Diphtheria can be fatal because a potent toxin diffuses into the bloodstream from the site of infection and damages the heart, kidneys, and nervous system. The disease is prevented by immunization with diphtheria toxoid.

Review Questions (pg. 468)

1. Viruses that infect the upper respiratory system impair host defenses, thereby allowing secondary bacterial infection.

2. The common cold is caused by any of more than 200 viruses and a few bacteria, but mainly rhinoviruses.

3. Colds are transmitted by inhaling infectious aerosols, or by touching the nose or eyes with fingers contaminated by cold viruses.

4. Colds can be prevented by hand washing, keeping the hands away from the nose and eyes, and avoiding public places during cold and flu season.

5. Adenoviruses cause illnesses resembling the common cold and strep throat.

Critical Thinking Questions (pg. 469)

1. Receptors for the B component of diphtheria toxin exist only on certain cell types such as heart muscle or nervous tissue.
2. One advantage is that tissue damaged by toxin could release iron that is required by the bacterium.

CHAPTER 21

Introduction

Typically, the lower respiratory system is protected from infectious invaders by the normal flora and the special anatomy of the upper respiratory system. Occasionally, however, infectious agents colonize the area and serious disease can result. Tuberculosis and pneumonia are two frequent bacterial infections that can cause serious, long-term effects. Tuberculosis is an ancient disease that has continued through the centuries to plague urban populations. Pneumonia is a serious disease of the lungs manifested by cough, fever, and respiratory distress.

When the infectious agents and/or their toxins are disseminated through the body, the most serious complications result.

Objectives

After completing this chapter, the student should be able to:

1. Explain how the historical perspective contributes to an understanding of lower respiratory diseases.
2. Describe the anatomy and physiology of the lower respiratory system.
3. Differentiate among the bacterial pneumonias.
4. Describe some other bacterial infections of the lungs.
5. Discuss tuberculosis as a chronic infection.
6. Describe influenza, its progress, and complications.
7. Differentiate between the fungal infections of the lungs.

Important Terms

BCG vaccine
mucociliary escalator
normal flora
pleurisy
cold agglutinins
granuloma
tubercles
caseous necrosis
cavity
reactivation tuberculosis
tuberculin

tuberculin (Mantoux) test
hemagglutinin (H)
purified protein derivative (PPD)
neuraminidase (N)
pandemics
antigenic drift
antigenic shift
Reye's syndrome
Guillain-Barré syndrome
spherule
arthrospores

coccidioidin

histoplasmin

macroconidia

spelunkers disease

Essay

1. Describe the main features of pneumococcal pneumonia.

    Pneumococcal pneumonia is caused by Gram-positive, encapsulated diplococci. After a short incubation period, the symptoms of cough, fever, chest pain, and shortness of breath develop. Full recovery usually follows the acute disease unless complications of endocarditis or meningitis develop. The most severe cases develop in persons whose health is impaired by alcoholism or chronic disease of the heart or lung.

    The disease is spread by inhalation of infected droplets. Prompt treatment with penicillin, erythromycin, or tetracycline is usually successful. A vaccine is now available, which is helpful in preventing the disease, especially in high-risk patients.

2. Describe the pneumonia caused by *K. pneumoniae*.

    The pneumonia of *K. pneumoniae* is quite virulent and can result in a mortality rate of 50%. After a short incubation period, chills, fever, cough, and bloody mucoid sputum may appear. Occasionally complications of shock and delirium develop. The infection is spread by inhalation of infected droplets. The alveoli of the lungs become congested with blood, plasma, and inflammatory cells. Destruction of lung tissue and abscess formation are common. Infection can spread to other body tissues via the blood. There is no vaccine available to prevent spread, but treatment with cephalosporins and gentamicin is usually successful.

3. Describe tuberculosis as a chronic disease.

    Tuberculosis is a chronic bacterial infection of the lungs, characterized by slight fever, weight loss, night sweating, and a chronic, productive cough of blood-streaked sputum. Infection usually occurs by inhaling aerosolized organisms from a person with tuberculosis. The organisms resist destruction and multiply within the pulmonary macrophages, which carry them to the lymph nodes. In this fashion, dissemination to other parts of the body can be widespread. Delayed hypersensitivity to the tubercle bacilli develops after about two weeks. Intense reactions occur around the bacilli, and fusion with the macrophages result in the formation of "giant cells." The infected areas become walled off from the surrounding tissue, forming granulomas, or tubercles, a characteristic response of the body to foreign substances that resist digestion and removal. When these tubercles rupture, the organisms spread through the system and set up colonies elsewhere. Tuberculosis can spread to the bone and blood with symptoms of delayed hypersensitivity being expressed.

4. Describe influenza as an infectious disease.

Infection with influenza virus is acquired by inhaling aerosolized respiratory secretions from a person with influenza. The virus spreads rapidly to nearby cells, including the mucus-secreting cells and cells of the alveoli. The mucociliary escalator is destroyed when infected cells die and slough off, thus severely impairing one of the body's major defenses. Outbreaks of influenza occur every year, with variations in the surface antigens diminishing the effectiveness of immunity.

Direct transmission of the orthmyxovirus frequently takes place in crowded areas via inhalation of infected droplets. The symptoms of fever, muscle aches, malaise, cough, and headache may continue for up to two weeks, while the lingering cough and malaise may continue for several weeks longer. Vaccines are moderately effective when produced from the epidemic strain, but lose their effectiveness due to the antigenic shift of the viruses.

Review Questions (pg. 481)

1. The capsule of *S. pneumoniae* is responsible for its virulence.

2. Antibiotic resistance may occur as a result of the acquisition of resistant genes by DNA transformation.

3. Alcohol depresses ciliary action, the cough reflex, and phagocytosis, while tobacco smoke damages the mucociliary escalator. The cough reflex, mucociliary escalator, and phagocytosis are all mechanisms by which the lung defends itself against pneumonia.

4. *Klebsiella* species become antibiotic resistant by selection of rare antibiotic-resistant mutants present in each population of klebsiellae, and by acquiring R factor plasmids.

5. *Mycoplasma* species lack cell walls and are therefore not affected by penicillin.

6. *M. pneumoniae* is too slow growing for culturing to be a useful diagnostic aid, whereas tests for cold agglutinins can give a fast answer and help guide early treatment.

7. Immunization against *B. pertussis* is started early since most deaths from the disease occur in infants less than one year of age.

8. Since adolescents and adults with pertussis usually have mild symptoms, they constitute a major reservoir for *B. pertussis,* and foster transmission of the virus to nonimmune infants.

Review Questions (pg. 495)

1. Routine milk pasteurization kills the organism.

2. Strong acids and alkalis kill the rapidly growing microorganisms in the mixture, leaving the more resistant *M. tuberculosis* intact.

3. Each strain of *M. tuberculosis* contains mutants resistant to the different antitubercular medicines, which would continue to multiply if only one medication were used in treatment. Mutants simultaneously resistant to two antimicrobials occur at a very low frequency, making treatment with two medications much more likely to be successful.

4. *Legionella pneumophila* was difficult to detect since it does not stain well with the Gram stain and it fails to grow on the usual laboratory culture media.

5. The percentage of fatal influenza cases is small, but the total number of infected people is very high, accounting for the large number of deaths.

6. Antigenic drift consists of a series of minor mutations in the H antigen, which result in sufficient antigenic change that antibody to the original antigen affords only partial protection.
Antigenic shift consists of abrupt genetic changes, which result in totally different antigens. Antibody to the original antigen offers no protection.

7. The causative agent of coccidioidomycosis lives in the soil of certain arid regions of the Western Hemisphere. The causative agent of histoplasmosis lives mainly in soils contaminated by bird or bat droppings, and is distributed in temperate and tropical areas around the world.

Critical Thinking Questions (pg. 496)

1. No, because immunity depends on antibody to the *S. pneumoniae* capsule and there are many different capsular types.

2. Tuberculosis outbreaks are rising because immunodeficiency diseases such as AIDS can cause activation of latent tuberculosis. Treatment of these cases is difficult because the body's defenses are weak and unable to assist the antitubercular medicines. Also, a marked increase in resistance to the medicines has reduced the possibility of effective therapy.

3. It would take many years to completely eradicate tuberculosis, because millions of people with latent infections can suffer reactivation of their disease at any time during their life.

CHAPTER 22

Introduction

The alimentary tract along with the upper respiratory system are major routes of entry into the body for pathogenic microorganisms.

The breakdown of the tooth enamel is related to diet and the acid-forming bacteria, which colonize the hard, smooth surfaces of the mouth. If dental caries develops, it can interfere with the mechanical breakdown of food and allow pathogens in the food products to pass into the stomach and small intestine, where gastroenteritis may result.

Esophageal infections are rare, and if they occur suggest immunodeficiency.

The acid of the stomach is inhospitable to microorganisms and unless they are in a protected state, are destroyed. Peptic ulcers have been related to colonization and infection with *Helicobacter H. pylori,* a Gram-negative bacterium.

Objectives

After studying this chapter, the student should be able to:

1. Describe the structures of the alimentary canal and their contributions to protection of the body from infectious disease.
2. Understand the relationship of the normal flora to oral health.
3. List infections caused by bacteria of the mouth.
4. List viral diseases of the mouth and salivary glands.
5. Describe the infections that can be found in the esophagus and stomach.

Important Terms

dental caries
enamel
gingivitis
periodontal disease
amylase
dental plaque
glucan
pellicle
calculus

hand, foot, and mouth disease
herpangina
coxsackie parotitis
herpes simplex
mumps
intranuclear inclusion body
herpetic whitlow
esophagitis
gastritis

Essay

1. Discuss the role of saliva in preventing tooth decay.

   Saliva helps to maintain healthy teeth due to several of its properties:
   1. its high buffering capacity can neutralize the acid by-products of mouth bacteria
   2. it is supersaturated with calcium and phosphate ions, which can slow the advance of dental caries by preventing the destruction of tooth enamel
   3. contains antibacterial substances, including lysozymes, lactoferrin, and specific IgA antibodies

2. Describe the formation of dental plaque and its contribution to tooth decay.

   The first step is the adherence of the streptococci or actinomycetes to the tooth pellicle. If dietary sucrose is present, *S. mutans* produces glucan, which binds the organisms together and to the tooth, providing attachment sites for additional species, such as aerobic organisms. Their metabolic activities consume oxygen, thus making the plaque a suitable location for anaerobic species. When an acidic environment is produced, the lactobacilli colonize the plaque and can be cariogenic. As the process continues, *Actinomyces* colonize the plaque and make up the bulk of it. If it remains in place, caries will result.

3. Describe the sequence of events that produce the herpes simplex virus infections.

   The organism is usually acquired from contact with lesions or saliva from infected patients. The virus can infect almost any body tissue, but most frequently infects and multiplies in the epithelium of the mouth and throat, and destroys cells. Vesicles form that contain large numbers of infectious virions. Some of the virus particles may be carried by lymph vessels to nearby lymph nodes. The production of interferon limits the progress of the infection. During the acute infection, the virions enter the sensory nerves in the area. Viral DNA persists in these nerve cells in a noninfectious form. Stresses such as sunburn, fever and menstruation can stimulate the virus to revert to its infectious form and, if immunity is weak, a fresh outbreak of vesicles will appear. The virus continues to produce infections for long periods.

4. Describe some of the complications that can arise with mumps.

   Mumps is a viral disease that characteristically involves the parotid glands. The acute viral infection can spread throughout the body and attack certain types of glands. In addition to parotid swelling, symptoms can arise in other parts of the body. Headache and stiff neck are indicative of meningitis. Mumps in postpubescent males is complicated by rapid swelling of one or both testicles, accompanied by intense pain. Although atrophy of the infected testicle is common after recovery, sterility is rare. In postpubescent females, involvement of the ovaries along with pelvic pain is common. Pregnant women commonly miscarry, but birth defects do not result from mumps. Serious consequences of mumps are rare; however, if they occur, they include deafness and death from encephalitis.

Review Questions (pg. 505)

1. Saliva neutralizes acids, is high in calcium and phosphate, which slow the process of dental caries, and contains antibacterial substances.

2. Different species of bacteria attach to the differing receptors on the various mucosal areas of the mouth.

3. The teeth constitute a nonshedding surface to which bacteria attach and multiply to high numbers that consume the available oxygen. The pockets created by dental fissures and gingival crevices also provide an anaerobic environment.

4. It produces glucan, an extracellular polysaccharide that binds the organisms to the teeth. Also it stores intracellular polysaccharide that is metabolized, producing acids that create a suitable acidic medium for the growth of lactobacilli.

5. Methods for preventing dental caries include limiting the intake of dietary sucrose, drinking fluoridated water, sealing dental fissures with methacrylate, and regular brushing and flossing of teeth.

6. Periodontal disease is a chronic inflammatory process that involves the tissues around the roots of the teeth.

7. ANUG is an acute form of gingivitis associated with a marked increase in the numbers of fusiforms and spirochetes. It is associated with poor dental hygiene, malnutrition, and stress.

Review Questions (pg. 511)

1. The saliva of a seemingly normal person might contain herpes simplex virus.

2. The incubation period of the mumps virus is relatively long (15–21 days) because it first infects the upper respiratory tract, spreads throughout the body via the blood stream, and then produces symptoms once it infects the parotid glands, meninges, or testicles.

3. No, people with allergies to eggs should not receive the mumps vaccine since it contains traces of egg protein.

4. The development of *Candida* esophagitis suggests immune deficiency, such as AIDS.

5. *Helicobacter pylori* is implicated in gastritis, peptic ulcers, and stomach cancer.

6. *H. pylori* is spiral-shaped with sheathed polar flagella that produce the enzyme urease.

Answers to EOC Critical Thinking Questions (pg. 512)

1. The transmission of the mumps virus would cease completely as a result of widespread vaccination, and mumps would quickly disappear because latent infections do not occur. With herpes simplex, however, no vaccine exists, and many people infected with the virus continue to have reactivations of outbreaks and to secrete the virus throughout their life.

2. If the antibiotic is given too soon, a population of resistant bacteria will have replaced the sensitive ones at the time of the dental treatment.

CHAPTER 23

Introduction

The lower alimentary system is victimized by a variety of pathogens frequently introduced by food or water. Many of the infectious agents are destroyed by the enzymatic action of the digestive secretions. However, many of the pathogens have devised means of avoiding their destruction by these products. Viruses, bacteria, fungi, and protozoans are all capable of causing diseases of the alimentary tract, which range from mild to severe, and acute to chronic. In some cases, they can even be life threatening.

Food poisonings can be caused either by actively metabolizing infectious agents or by ingested toxic products of the agent. The most frequent results are vomiting and diarrhea—two means by which the system can attempt to rid itself of the damaging poison.

The hepatitis viruses in their various forms can do serious damage to the liver and therefore to its functions in maintaining homeostasis.

Objectives

After studying this chapter, the student should be able to:

1. Describe the historical significance of the contributions of early scientists.
2. Describe the anatomy and its significant physiology to the development of diseases of the lower alimentary tract.
3. Discuss the normal flora of the lower alimentary tract and their contributions.
4. Discuss cholera as an infectious disease.
5. Discuss bacillary dysentery.
6. Describe the characteristics of salmonellosis.
7. Describe gastroenteritis and diarrhea as caused by infectious agents.
8. Describe typhoid fever as an example of "enteric fever."
9. List the causes of food poisonings and the different types.
10. Describe the viral infections of the lower alimentary tract.
11. Describe the protozoan diseases of the digestive system.

Important Terms

jaundice	Shiga toxin
cholecystitis	enterotoxigenic
bile salts	enteroinvasive
villi	enteropathogenic
microvilli	enterohemorrhagic

adhesins

antibiotic-associated colitis

blind loop syndrome

dysentery

oral rehydration solution

choleragen

hemolytic-uremic syndrome

aflatoxins

amebiasis

amebic dysentery

chromatoidal body

chitin

Essay

1. List the major functions of the lower alimentary tract.

   1. degrading of macromolecules into simpler subunits
   2. absorption of these simple subunits, minerals, and vitamins into the blood stream
   3. recycling of water in digestive secretions into the blood stream

2. List the ways in which microbial growth in food can cause illness.

   Illness can be caused chiefly by two different mechanisms:
   1. Contaminating microorganisms can infect the person who ingests the food. Gastroenteritis caused by *Salmonella* species is an example.
   2. Products of microbial growth in the food can be poisonous. A variety of illnesses can be caused by species of *Clostridium, Bacillus,* and *Staphylococcus.*

3. Describe the main features of bacterial dysentery.

   The cause of bacterial dysentery is due to species of *Shigella,* Gram-negative, nonmotile enterobacteria. After an incubation period of 3–4 days, the symptoms of fever, diarrhea, vomiting, and pus and blood in the feces occur.

   The disease is transmitted by infected persons via the fecal-oral route. It tends to occur in overpopulated areas with poor sanitation. Humans generally are the only source. The organism can be isolated from culture of feces. Treatment is with the antibiotics ampicillin and trimethropim-sulfa. The disease can be prevented by sanitary precautions, which include careful hand washing.

4. Describe the main features of typhoid fever.

   Typhoid fever is caused by *Salmonella typhi,* a Gram-negative, motile enterobacterium. Following an incubation period of 7–21 days, the symptoms of fever, severe headache, and abdominal pain are common. Less frequently intestinal rupture and shock may occur. Organisms can be recovered from feces, blood, or bone marrow. Treatment with several antibiotics may lessen the effects of the infection.

   Prevention is through administration of an oral attenuated or killed injected vaccine. However, these methods are only partially protective.

5. Describe botulism as a severe bacterial intoxication.

   Botulism is caused by *Clostridium botulinum*, an anaerobic, Gram-positive, spore-forming bacillus. After a short incubation period of less than 36 hours, the symptoms of blurred or double vision, weakness, nausea, vomiting, and diarrhea occur. Nerve involvement can lead to generalized paralysis and respiratory insufficiency. The disease is due to ingestion of food contaminated with botulism toxin, which can be identified in blood and food. The toxin acts by blocking the transmission of nerve signals to the muscles, producing paralysis. Intravenous antitoxin can arrest the paralysis but the nerve damage is irreversible. In some cases, death results from respiratory failure.

6. Compare the pathogenic protozoan diseases of humans.

   There are four noteworthy protozoans responsible for human disease:
   Entamoeba histolytica is a medium size trophozoite infectious in its mature cyst form. It produces a mild gastroenteritis with occasional violent symptoms.
   Giardia lamblia is a pear-shaped, binucleate trophozoite infectious in its cyst form. It is the most frequent cause of protozoan diarrhea in the world and frequently leads to fatalities.
   Balantidium coli and *Cryptosporidium* species are occasional causes of protozoan diarrhea. The latter has frequently been associated with AIDS victims.

Review Questions (pg. 527)

1. Cholecystitis is an infection of the gall bladder caused by intestinal bacteria.

2. Bile inhibits the growth of many bacteria, and can be used to select certain bile-resistant pathogens.

3. They possess adhesins that allow them to attach to certain intestinal cells.

4. Clostridium difficile is an anaerobic bacterium that produces a toxin lethal to intestinal epithelium. It causes antibiotic-associated colitis.

5. Overgrowth of bacteria in the small intestine can cause blind loop syndrome.

6. The source of flatus (gas) is the metabolism of undigested food material by intestinal bacteria, which produce carbon dioxide, hydrogen, and methane gas.

7. The toxin of *V. cholerae* interferes with the normal secretory mechanism of the cells, causing diarrhea but not harming the colon. The normal shedding of cells rids the body of the toxin.

8.

	Epidemiology	Pathogenesis
Shigellosis	human source: infected persons via fecal-oral route	causative bacterium invades and multiplies within intestinal epithelial cells, resulting in cell death and ulceration of the intestinal lining.
Salmonellosis	animal source: food contaminated by animal feces	causative bacterium penetrates the intestinal lining and causes inflammation of the underlying tissues.

9. Isolation of *C. jejuni* requires cultivation on a special medium incubated under microaerophilic conditions.

10. The reservoir of *S. typhi* is human, while that of *S. enteritidis* is other animals.

Review Questions (pg. 541)

1. The bacteria itself in the food can cause food poisoning as can the toxins it produces.

2. *S. aureus* produces toxin while growing in the food, thus cooking can kill the bacteria without destroying its toxin.

3. *B. cereus* is aerobic, while *C. perfringens* is anaerobic.

4. Paralysis is the main distinguishing feature of botulism.

5. Aflatoxin in foods is produced by certain molds of the genus *Aspergillus.* They cause cancer in some experimental animals.

6. Hepatitis A is caused by an RNA virus of the enterovirus group of the picornavirus family. It is transmitted by the fecal-oral route.
   Hepatitis B is caused by a DNA virus of the hepadnavirus family. It is transmitted mainly by blood and semen.

7. Giardiasis is caused by *Giardia lamblia,* a flagellated protozoan with an adhesive disk. Transmission is usually by the fecal-oral route, especially via waters contaminated with the feces of various animals. Amebiasis is caused by *Entamoeba histolytica,* an ameba, and its source is generally food or water contaminated by human feces. Both diseases can be transmitted by person-to-person contact.

Answers to EOC Critical Thinking Questions (pg. 542)

1. Some of the things to consider include: 1. estimates of the risk of introducing *V. cholerae* and other chlorine-sensitive pathogens into the water supply of a population 2. the effect of preventive measures such as vaccines and laws controlling pollution of water by ocean-going ships 3. the time it would take to control an epidemic using available treatment and control facilities 4. the expected rates of death and disability 5. the risk of cancer in the population over a long period of time 6. the availability of early detection and treatment of cancer 7. the possible contribution of chlorination-related carcinogens to the cancer risk.

2. One can compare the rates of primary liver cancer in a large group of people vaccinated in infancy against hepatitis B with another large group matched according to geography, age distribution, diet and lifestyle, but not vaccinated. Assessment of the two populations would be continued for many years. The population would best reside in an area with known high rates of hepatitis B and primary liver cancer.

CHAPTER 24

Introduction

Genitourinary infections are among the most debilitating of the infectious diseases treated today. Urinary tract infections can be caused by normal flora transferred to an abnormal body area. Bladder infections can progress to affect the kidneys and later other organs of the body. The sexually transmitted diseases are usually caused by bacteria or other infectious agents that survive poorly in the environment, and therefore transmission from one person to another requires direct physical contact. Many of the diseases can be prevented by educational programs and the careful selection of sexual partners. Development of resistant strains of gonorrhea and syphilis is making treatment of these epidemic diseases extremely difficult. The increasing incidence of AIDS in the population has aroused public awareness of the problems associated with sexual promiscuity but has not as yet curtailed the spread of the deadly disease.

Objectives

After studying this chapter, the student should be able to:

1. Relate historical contributions to an understanding of diseases of the genitourinary system.
2. Describe the major structures and functions of the genitourinary system.
3. Understand the contributions of the normal flora.
4. Describe the types of urinary tract infections and their causes.
5. Describe the sexually transmitted bacterial diseases.
6. List the sexually transmitted viral diseases.
7. Compare the other lesser known genital tract infections.

Important Terms

ureter
urethra
pelvic inflammatory disease
ectopic pregnancy
ophthalmia neonatorum
primary syphilis
hard chancre
secondary syphilis
immune complexes
tertiary syphilis
general paresis

non-treponemal
indirect immunofluorescence test
immunoblot techniques
IgM capture method
human papillomaviruses
bacterial vaginosis
candidiasis
trichomoniasis
puerperal sepsis
clue cells
germ tubes

pseudohyphae chlamydospores
axostyle

hydrogenosomes
toxic shock syndrome

Essay

1. List the symptoms suggestive of a sexually transmitted disease:

    1. abnormal discharge from the vagina or penis
    2. pain or burning sensation with urination
    3. painful or painless blister or sores on the genitals; swellings in the groin
    4. abnormal vaginal bleeding or unusually severe menstrual cramps
    5. itching in the vaginal or rectal area
    6. pain during sexual intercourse; pain in the lower abdomen in women
    7. skin rash or mouth lesions

2. Describe the main features of gonorrhea.

    Gonorrhea is a sexually transmitted disease caused by *Neisseria gonorrheae*. After an incubation period of two to five days, inflammation of the urethra with pain on urination and discharge occur in men. In women, the symptoms may be slower to develop but can include painful urination and vaginal discharge. Later, infection of the fallopian tubes or sterility can result. The diagnosis is revealed by the presence of Gram-negative diplococci in smears of purulent discharge. Immunity following infection fails to develop due to variation in surface proteins. Treatment can be complicated by the discovery that the gonococci are resistant to several antibiotics. Prevention is through education, early treatment of sexual contacts, and the use of condoms.

3. Describe syphilis as a venereal infection.

    The spirochete *Treponema pallidum* is the causative agent of syphilis. The organism can be identified from the chancre of primary syphilis or the rash of secondary syphilis. Definitive tests are dependent upon the identification of specific treponemal antigens. The first symptoms of the hard, painless chancre occur ten to ninety days after infection. In the secondary stage, the organism invades the vascular system causing fever, rash, and mucous membrane lesions. Years later tertiary syphilis may develop with serious damage to the central nervous system, peripheral nervous system, and arteries. Treatment is usually with an antibiotic such as penicillin.

4. Briefly describe the viral STDs.

    1. Genital herpes simplex, caused by herpes simplex virus type 2, is a recurring disease that is extremely painful. The disease can be transmitted to the newborn and can be fatal. The disease is incurable but can be suppressed by medication.
    2. Genital warts are caused by the human papillomaviruses, which are DNA viruses. Some strains have been associated with cervical cancer. There is no cure.

3. HIV disease, which culminates in AIDS and is caused by the human immunodeficiency virus, is currently pandemic and almost universally fatal. No vaccine or medical cure is available but spread of infection can be halted by applying current knowledge and careful choice of sexual partners.

Review Questions (pg. 559)

1. The urinary tract is protected from infection by muscle groups that keep the system closed, the flushing action of urination, organic acids and antibodies in urine, and phagocytes.

2. Estrogen causes glycogen deposition in vaginal epithelium. The glycogen is then converted to lactic acid by lactobacilli, resulting in an acidic pH that makes the vagina resistant to infection by inhibiting the growth of pathogens.

3. Most bladder infections in women are caused by *Escherichia coli,* originating from the patient's intestine.

4. Diagnosis of gonorrhea is made by enzyme stains and culturing the organism on special media, such as chocolate agar, under aerobic conditions. Complications include sterility, pelvic inflammatory disease, ectopic pregnancy, and disseminated gonococcal infection, which can cause infection of the joints and heart valves.

5. Immunity to one strain of *N. gonorrheae* does not necessarily give immunity to another strain.

6. Newborns are given antimicrobial eyedrops to prevent ophthalmia neonatorum, a gonococcal disease transmitted by infected mothers during childbirth.

7. Clinical specimens can be inoculated onto the tissue cell cultures, and *C. trachomatis* is then identified by fluorescent antibodies two to three days later. A faster method is to use a monoclonal antibody that detects the organism in pus or other clinical specimens directly.

8. Syphilis is called the "great imitator" because its manifestations can mimic many other diseases.

9. Stage 1: hard chancre
   Stage 2: rash, aches and pains, mucous membrane lesions
   Stage 3: blood vessel and central nervous system damage

10. Yes, since *Treponema pallidum* can cross the placenta in asymptomatic mothers with syphilis.

11. Ulcerative lesions such as occur with chancroid predispose the person to HIV infection.

Review Questions (pg. 567)

1. Genital herpes simplex is much more serious than other herpes infections. It can be very painful and occur at frequent intervals for a lifetime, since there is no cure. Persons with active infections can transmit the disease to a sexual partner and should take the proper precautions to avoid embarrassment and conflict. Primary maternal infections acquired near the time of delivery are a major threat to the newborn infant. Additionally, genital herpes has been associated with cancer of the uterine cervix in women.

2. Certain papillomaviruses have been associated with prostate disease and anal cancer in homosexual men, and with cervical cancer in women. Babies may acquire infection at birth, which is manifested by warts in their respiratory tract.

3. Laser treatment removes the wart but infection persists in adjacent tissue.

4. HIV is primarily transmitted by sexually promiscuous homosexual or bisexual men, intravenous drug abusers, sexually promiscuous heterosexuals, women who have had sexual intercourse with such men, and infected mothers transmit the disease to their babies near the time of delivery.

5. Initially the disease involved only homosexual men and intravenous drug abusers who shared needles. The epidemic has since spread to the general population, involving an increasing percentage of heterosexual men and women, and children.

6. Bacterial vaginosis is characterized by a grayish white, slightly bubbly vaginal discharge with an unpleasant odor. Laboratory findings include the presence of clue cells and an elevated pH. Cultures show large numbers of *Gardnerella vaginalis* and *Mobiluncus* species.

7. Bacterial vaginosis primarily occurs in sexually active women and sexually abused children. The risk of acquiring the disease increases with the number of sex partners.

8. Vaginal candidiasis is associated with antibiotic therapy, pregnancy, the use of birth control pills, and diabetes.

9. The cytoplasm of *T. vaginalis* contains hydrogenosomes, which are responsible for the organism's generation of hydrogen gas.

10. A *T. vaginalis* infection in a child suggests the possibility of sexual abuse.

11. Puerperal fever, or childbirth fever, is an infection of the uterus following childbirth or miscarriage, due to contaminated instruments or contact with the physician's hands.

12. Toxic shock syndrome is caused by *S. aureus,* which can multiply in the vagina in high numbers in the presence of an intravaginal tampon, which promotes the production of staphylococcal toxins.

Critical Thinking Questions (pg. 568)

1. Some factors responsible for the increased incidence of sexually transmitted disease are: 1. Relaxation of social taboos that restricted extramarital sexual intercourse led to an increase in the number of sexual partners of an individual. 2. The availability of oral contraceptive pills fostered the abandonment of barrier devices such as condoms. 3. Adolescents have had inadequate instruction in the responsibility they assume if they engage in sexual intercourse. 4. Male and female prostitution has increased, along with poverty and drug addiction.

2. Some sexually transmitted diseases such as gonorrhea and *Chlamydia* infections are responsible for ectopic pregnancy. Others such as AIDS and syphilis can be passed to the fetus even when the mother is asymptomatic. (Bacterial vaginosis is associated with an increased risk of premature birth, but the condition has not been proven to be sexually transmitted.)

3. Antibody in the blood of a newborn may actually be transplacentally acquired IgG from the mother and does not necessarily indicate infection of the infant. The absence of antibody does not rule out infection because the infected infant may not have produced detectable antibody at the time of birth.

CHAPTER 25

Introduction

Infectious diseases of the central nervous system are uncommon but can be quite serious. The infectious agents do not easily penetrate the protective mechanisms of this vital body area, but when they do, these barriers interfere with the ability of antimicrobial agents to reach the affected areas.

The accidental entry of viruses, bacteria, fungi, and protozoans into the CNS can cause a variety of diseases, some of which are treatable. Recent investigation into the chronic degenerative brain diseases have led to the discovery of infectious agents even smaller and simpler than viruses. This has led to the conviction that the central nervous system, although the most protected area, is still vulnerable to debilitating diseases.

Objectives

After studying this chapter, the student should be able to:

1. Discuss the historical information that led to a greater understanding of the human nervous system.
2. Describe the anatomy and physiology of the nervous system.
3. Describe the ways by which infectious agents are able to enter the nervous system.
4. Describe the various types of bacterial meningitis.
5. Describe leprosy as a disease of the nervous system.
6. Describe the viral causes of encephalitis.
7. Discuss polio as an infectious disease.
8. Describe rabies as a nervous system disease.
9. Describe the fungal diseases of the central nervous system.
10. Describe the protozoan diseases of the central nervous system.

Important Terms

axons
cerebrospinal fluid
meninges
meningitis
bacterial meningitis
blood-brain barrier
serogrouping
Schwartzman phenomenon
satellitism

lepromatous type
tuberculoid type
sporadic encephalitis
epidemic encephalitis
hydrophobia
inclusion bodies
Negri bodies
Seller's stain

allergic encephalitis
serum sickness

Essay

1. Describe the reasons that CNS diseases are uncommon.

    Diseases of the CNS are uncommon due to the well-protected environment of these organs. The microorganisms that infect this area of the body do so accidentally. The organisms most frequently enter through the blood stream, although the passage of harmful agents is usually successfully prevented by the blood-brain barrier. The special cells of these capillaries are so close together that they prevent the entrance of even small molecules from passing into the meninges unless their concentration is very high. Although infectious agents can enter the CNS through nerves or bone, this is rare. The blood-brain barrier is extremely effective in protecting the CNS.

2. Describe the main features of meningococcal meningitis.

    Meningococcal meningitis is caused by *Neisseria meningitidis.* Infection is acquired by inhaling airborne droplets. The symptoms appear only a few days after infection occurs. These begin as a mild cold and progress to severe headache with stiffness of the neck and back. The organisms colonize in the upper respiratory tract and then enter the blood stream, where they are carried to the meninges and the spinal fluid. An inflammatory response follows, which can obstruct the normal outward flow of cerebrospinal fluid and increases brain pressure. Damage to motor nerves produces paralysis, and shock can result due to the release of endotoxins.

3. Describe leprosy as a progressive disease.

    Leprosy is a progressive disease with a very long incubation period, usually more than two years and often 10 years or longer. It begins gradually, with increased or decreased sensation in certain areas of the skin, and associated changes in pigmentation. Later, these areas lose their hair, sweating ability, and all sensation. Subsequently, the nerves of the arms and legs enlarge and are painful, eventually become numb, the muscles atrophy, and there can be a loss of fingers and toes. The organism first invades the small nerves of the skin and multiplies in macrophages. Much of the resulting nerve damage is due to the attack of the immune cells against the infected nerve cells. Immunity is frequently overwhelmed by the accumulated bacterial antigen and results in unrestrained growth of *M. leprae.*

4. Describe poliomyelitis as a paralytic disease.

    The characteristic feature of poliomyelitis is selective destruction of motor nerve cells, usually of the spinal cord, resulting in permanent paralysis of one group of muscles, such as in an arm or leg. Polio begins with symptoms of meningitis: headache, fever, stiff neck, and nausea with accompanying pain and muscle spasms, later followed by paralysis. Severe cases result in paralysis of respiratory muscles.

5. Discuss poliomyelitis. Include its cause, progression.

Poliomyelitis is a viral infection of the motor nerve cells. It is caused by three types of polio viruses which are introduced into the body by the fecal-oral route. The symptoms include headache, fever, stiff neck, nausea, pain and spasm of muscles followed by paralysis. The virus infects the throat and intestine, circulates via the blood stream and then enters some motor nerve cells of the brain and spinal cord. The infected nerve cells lyse upon release of the mature virus and therefore cannot transmit the impulses necessary for response. Hence paralysis. Vaccine of oral active or injected inactive polio virus is used to prevent the disease.

6. Explain the progress of rabies as a fatal disease.

Rabies is a viral disease that can be fatal if not detected before symptoms appear. The disease is acquired by the bite of an infected animal. The symptoms of fever, headache, nausea, and cough follow a one to two month period of incubation. The early symptoms rapidly progress, and include encephalitis, confusion, hallucinations, seizures, and increased sensitivity to light, sound, and touch. Later, spasms of the muscles of the throat and respiratory tract develop and are followed by coma and death. The virus multiplies at the site of the bite and travels via the nerves to the CNS where it multiplies and spreads to vital organs. Although there is no effective treatment, there are effective post-exposure measures with the human diploid cell vaccine as soon as possible after exposure that can stop spread of the virus.

Review Questions (pg. 580)

1. Since spinal fluid is produced in the ventricles and then spreads to the spinal cord, any disease-producing microorganism is carried in it and can be extracted in cerebrospinal fluid in the spinal canal.

2. Infections generally enter the CNS via the blood stream, skull bones, or nerves.

3. Meningitis is an infection involving the meninges, the protective membranes of the brain.

4. Unlike other types of meningitis, meningococcal meningitis is more likely to occur in epidemics.

5. Endotoxin in the cell walls of *N. meningitidis* accounts for a drop in blood pressure, which can result in shock and death.

6. *Haemophilus influenzae* is the leading cause of bacterial meningitis. It is a tiny, Gram-negative anaerobic coccobacillus bacterium that requires X and V factors, two blood components, for growth.

7. Most mothers have protective antibodies that cross the placenta, giving protection against these forms of meningitis during early infancy.

8. *Mycobacterium leprae,* an acid-fast rod that cannot be cultivated *in vitro,* is the only human pathogen to attack the peripheral nerves.

Review Questions (pg. 591)

1. Poliomyelitis has almost been eradicated from the western hemisphere, but since it can be reintroduced from other areas of the world, it is important for people to stay up-to-date on immunization.

2. Post-polio syndrome is a deterioration in strength of certain muscles years after recovery from poliomyelitis.

3. Immunization of dogs and cats against rabies has practically eliminated the disease in humans, although rabies in wild animals is increasing and can be transmitted to humans.

4. A Negri body is a type of cytoplasmic inclusion body representing the site of rabies virus replication in brain cells.

5. *T. brucei* variants with different surface antigens repeatedly arise during infection.

6. Such illness is very unlikely, although cryptococcal meningitis can be contracted from dust containing pigeon droppings, especially if a person is immunodeficient. Also, *Naegleria fowleri* meningoencephalitis can be contracted by swimming in warm freshwater, although millions of people have been exposed to it without harm.

Critical Thinking Questions (pg. 593)

1. Live polio vaccine provides better local immunity in the throat and intestine, is inexpensive, and does not require an injection. However, live vaccine occasionally causes poliomyelitis. Inactivated (killed) vaccine provides good immunity against poliomyelitis without the risk of causing the disease. However, it does not immunize the intestine, is expensive, and requires an injection.
2. Since the rabies virus does not kill infected cells, if the patient can survive the acute infection, recovery is complete.

CHAPTER 26

Introduction

Any break in the skin can be an entry site for human pathogens. The skin is the body's first line of defense against infection and once this barrier is breached, the sterile environment of the tissues is susceptible to the damage that can result from the multiplication of the invader or the diffusion of its toxin.

A wide variety of pathogens are capable of infecting wounds, some of which are members of the body's own normal flora and are carried to the interior of the body by a puncture or laceration. Many of the diseases are mild and localized, while others can be severe, systemic, and even fatal. The anaerobic environment created by the damage to healthy cells can result in gangrene or tetanus if the conditions are appropriate.

Objectives

After studying this chapter, the student should be able to:

1. Determine the contributions of early researchers to the current knowledge of wound infections.
2. Distinguish among the bacterial causes of wound infections.
3. Describe the wound infections caused by *P. aeruginosa.*
4. Describe the clostridial wound infections.
5. Distinguish among the nonclostridial infections of wounds.
6. Describe the bacterial infections that result from bites.
7. Describe the fungal infections of wounds.

Important Terms

tetanus
tetanus immune globulin
sulfur granules

cat scratch disease
L-form variants
Haverhill fever

Essay

1. Describe an abscess, including the positive and negative aspects of it.

    An abscess is a localized collection of pus, which is composed of living and dead leukocytes, tissue breakdown products, and infectious agents. No blood vessels are found in abscesses because they have either been destroyed or pushed aside. The area surrounding the abscess is inflamed, and the capsule that forms localizes the infection, preventing its spread. However, the microbes within the abscess may not be killed by an antimicrobial medicine because frequently they are dormant and the antimicrobials work best when the infectious agents are actively multiplying. Also, microorganisms in the abscesses are a potential source of systemic infection if they escape the abscess and enter the blood or lymph vessels.

2. What are the chief factors responsible for the changes in the type of organism expected to invade wounds?

    Among these factors are:
    1. the use of antibiotics given at the time of surgery suppresses some of the common pathogens
    2. the increased number of patients on immunosuppressive drugs that foster infections of normally nonpathogenic bacteria
    3. the increased use of foreign materials in tissue
    4. prolonged surgical procedures allow microorganisms from the air or medical personnel to invade surgical wounds

3. Describe the principal characteristics of tetanus.

    After an average incubation period of 8 days, there are symptoms of restlessness, irritability, and muscle pain and jaw spasms. There is usually difficulty swallowing. The disease process results from the effects of the organism's exotoxin, which enters the blood stream and travels to the brain and spinal cord. The exotoxin interferes with relaxation of major muscle groups and the uncontrolled spread of nervous impulses, which result in spasms. Breathing is labored and death is frequently due to pneumonia or aspiration of stomach contents into the lung. Prevention is with tetanus toxoid administered to children at regular intervals and to adults every 10 years.

4. What are the main factors in the development of gas gangrene?

    Gas gangrene is most commonly caused by *Clostridium perfringens,* type A. The two main factors that contribute to the establishment of the disease are:
    1. the presence of large amounts of dirt in the wound
    2. long delays before the wound is carefully cleaned of dirt and dead tissue.

5. Describe the major characteristics of gas gangrene.

The most common cause of gas gangrene is the entrance of *Clostridium perfringens* into a dirty wound with impaired circulation of muscle tissue. Symptoms of severe pain, gas and fluid seepage from the wound, and blackening of the overlying skin can develop usually 1 to 5 days following the invasion. If the offending tissue is not surgically removed, shock and death can result. The organism grows in dead and poorly oxygenated tissues and releases an α toxin, which kills leukocytes and adjacent normal tissue cells. Prompt cleaning and debridement of wounds can prevent gas gangrene from developing. No vaccine is available.

Review Questions (pg. 602)

1. Bacteria in abscesses may not be multiplying, which is a requirement for microbial killing by antibiotics. Some antibacterial medicines are inactivated by pus, and some diffuse into abscesses poorly, since there are no blood vessels.

2. Anaerobic bacteria thrive in puncture wounds, dirty wounds, and wounds with crushed tissue.

3. Wounds with dirt and dead tissue, and wounds in immunodeficient subjects favor such infections.

4. *Staphylococcus aureus,* a bacterium normally carried in the nose, is the primary cause of surgical infections.

5. *Pseudomonas aeruginosa* produces the pigments pyocyanin and fluorescein that diffuse into the pus, making it greenish in color.

6. Very few antibacterial medications are effective against *P. aeruginosa,* which can grow in aerobic and anaerobic conditions.

7. Tetanus is characterized by severe, uncontrollable painful contractions of one or more muscle groups, often beginning with the jaw muscles (hence the name "lockjaw"). Tetanus is prevented by maintaining immunity with injections of tetanus toxoid every 10 years.

8. Cultures are not helpful in diagnosing tetanus because the number of organisms present were too few or because little or no wound drainage occurred.

9. The antitoxin does not reverse the effect of toxin already bound to nerve tissue.

Review Questions (pg. 608)

1. *Clostridium perfringens* is the most common cause of gas gangrene, originating from soil, dust, and the human intestine.

2. The two main factors fostering development of clostridial myonecrosis are:
   1. wounds containing dirt and dead tissue, and
   2. a long delay before the wound can be cleaned.

3. Actinomycosis is a chronic disease with recurrent draining abscesses, caused by *Actinomyces israelii,* an anaerobic, branching, Gram-positive bacterium. It is likely to result from an anaerobic wound associated with dental or intestinal surgery.

4. *Pasteurella multocida* infects wounds caused by animal bites.

5. *Rochalimea henselae,* which causes cat scratch disease, is the most common cause of chronic lymph node enlargement in children.

6. Symptoms of rat bite fever are relapsing fevers and rash.

7. L-form variants develop in cultures of *Streptobacillus moniliformis.*

8. Human bites are dangerous because they can transmit diseases or allow opportunistic pathogens a mode of entry into other areas of the body, such as the tendons, joints, and bones.

9. Sporotrichosis results from thorn or splinter wounds.

Critical Thinking Questions (pg. 610)

1. Antimicrobial medicines suppress some species and allow resistant ones to increase, thus changing the etiology of wound infections. Immunosuppressed patients are susceptible to species that are not normally pathogenic. Also, various new surgical procedures permit infections that otherwise would not occur.
2. A noninvasive bacterium could cause disease if the bacterium produced a toxin.
3. Actinomycosis is caused by an anaerobic, branching, filamentous bacterium, and is usually characterized by chronic draining abscesses that originate from oral or intestinal lesions.

   Sporotrichosis is caused by *Sporothrix schenkii,* a dimorphic fungus, and arises from puncture wounds of the extremities caused by thorns or other vegetation. Draining chronic abscesses follow the course of a lymphatic vessel.

CHAPTER 27

Introduction

Any infection of the cardiovascular system can lead to devastating effects or fatal complications. The types of bacterial endocarditis described in this chapter affect both damaged and healthy organs, causing devastating results if the endotoxins are distributed to vital organs.

Several other infectious conditions affect the cardiovascular system; plague and brucellosis for example. Since the cardiovascular system carries nutrients and oxygen to the cells, it can also distribute soluble toxic substances to all parts of the body. The bite of a vector can introduce infectious agents directly into the blood stream, thus setting the stage for development of a cardiovascular disease.

Objectives

After studying this chapter, the student should be able to:

1. Understand the historical data that has led to a greater understanding of blood and lymphatic infections.
2. Describe the anatomy and physiology of the cardiovascular and lymphatic systems.
3. Describe bacterial endocarditis.
4. Describe septicemia and septicemic shock.
5. Describe tularemia as a zoonosis.
6. Discuss brucellosis as an infectious disease.
7. Describe plague as an ancient and now modern disease.

Important Terms

pericardium
atrium
ventricle
aorta
interstitial fluid
lymph nodes

lymphangitis
glomerulonephritis
septicemia
tumor necrosis factor
disseminated intravascular coagulation
V antigen

Essay

1. Describe the characteristics of subacute bacterial endocarditis.

SBE is an infection of the inner lining of the heart, usually localized in one of the heart valves. It is usually found in hearts that have congenital defects or have been damaged by disease and are infected with α-hemolytic streptococci or *S. epidermidis.* The symptoms of fever and energy loss are slow to develop, typically over a period or weeks or months.

Organisms enter the blood stream through dental procedures, surgery, or intravenous drug abuse. The blood flow through a damaged heart valve can cause the organism to becaught in a thin clot around the valve. The clot then traps circulating organisms, making them inaccessible to phagocytosis. Pieces of the clot may break free, block important vessels, and cause tissue death. Prophylactic antibiotic therapy prior to invasive procedures is advised for people with suspected SBE.

2. Discuss septicemic shock as a life-threatening event.

Septicemia generally develops after the invasion of the blood stream by Gram-negative bacteria or their endotoxin. An uncontrolled localized infection precedes the septicemia. The endotoxin released by the infectious agents trigger responses from the body's defense mechanisms, which result in fever, decreased cardiovascular efficiency, increased leakage of plasma from blood vessels, and several other actions that cause blood pressure to fall and oxygen exchange in the lung to be impaired. Tumor necrosis factor is then released from macrophages, and is largely responsible for causing septic shock and tissue death. The effects of the combined actions of the body's response to the bacterial endotoxin can be so great that multiple organs become involved and the extensive damage frequently causes death of the patient.

3. Describe the characteristics and pathogenicity of tularemia.

Tularemia is caused by the pleomorphic, nonmotile, aerobic, Gram-negative rod *Francisella tularensis.* The characteristic skin ulceration and enlarged lymph nodes accompanied by fever, rash, and malaise develop usually 2–5 days after being bitten by an infected insect or tick, or after skinning an infected rabbit. The organisms then spread throughout the body via the blood and lymphatic vessels. Pneumonia can develop in some patients, with a 30% mortality rate. They are ingested by phagocytic cells and grow within them, making it very difficult for the organism to be destroyed. Cell-mediated immunity is probably responsible for ridding the host of this infection. Even without treatment, over 90% of the cases survive. Vaccination against tularemia is administered only in special cases.

4. Describe the changes that occur in *Yersinia pestis* when it changes hosts from the flea to the human.

These changes are:
1. the production by the organism of an antiphagocytic capsule when it grows at human body temperature (37° C)
2. production of a plasmid-dependent cytotoxic factor at 37° C that kills macrophages

3. the synthesis of a plasmid-dependent cytoplasmic protein, V antigen in response to the low intracellular calcium concentration of macrophages.

The V antigen is thought to be the important virulence factor in the development of the plague symptoms.

Review Questions (pg. 617)

1. A "systemic infection" is an infection of the circulatory system, and can be carried to all parts of the body.

2. Bacteria in infected lymph first reach the right side of the heart, then the lungs, then the left side of the heart, and finally the tissues.

3. Bacteriological diagnosis of SBE is made by culturing the bacteria from the blood obtained from an arm vein.

4. Immune complexes in the circulatory system can lodge in the skin, eyes, and kidneys. They incite an inflammatory response in the kidneys, producing glomerulonephritis.

5. Subacute bacterial endocarditis is usually caused by bacteria of the normal skin or mouth flora. It is generally a chronic illness. Acute bacterial endocarditis is caused by more virulent organisms than in SBE, such as *S. aureus* and *S. pneumoniae,* originating in an infection elsewhere in the body. The disease progresses rapidly, often causing heart valve destruction.

6. Bacteremia is a condition in which bacteria circulate in the blood stream, with or without resulting illness. Septicemia usually results from bacteremia. It almost always originates from a localized infection elsewhere in the body. Microorganisms or their toxins are released into the blood stream, causing such effects as endotoxemia and shock.

7. Circulating endotoxin reacts with the cell membrane of macrophages, causing them to release cytokines such as TNF. TNF causes fever, and leukocytes to adhere to capillary walls, especially in the lungs. Complement is activated by endotoxin, and its components cause leukocytes to release tissue-damaging lysosomal enzymes. In addition, complement causes capillaries to leak excessive amounts of plasma. Endotoxin also causes decreased muscular tone of the heart and the blood vessel walls. The sum of these effects is a profound drop in blood pressure, so that vital organs fail to receive an adequate blood supply.

Review Questions (pg. 622)

1. Tularemia is contracted when the organism penetrates mucous membranes or breaks in the skin, as can occur when skinning rabbits, or through the bites of blood-sucking arthropods such as flies and ticks.

2. Cell-mediated immunity is important in recovery from tularemia because the organisms live intracellularly and are safe from humoral immunity.

3. Tularemia is diagnosed by testing for a rise in antibody against *F. tularensis,* cultures using cysteine-containing media, or fluorescent antibody stain of pus.

4. Workers in the meat-packing industry are most susceptible to brucellosis. They can be protected by wearing goggles and rubber gloves.

5. Milk pasteurization helps control brucellosis since the bacterium *B. abortus* is found in cows milk and can contaminate it.

6. In bubonic plague, the main manifestation is large, pus-filled lymph nodes; in pneumonic plague, the lungs are mainly involved and characteristic symptoms are cough and bloody sputum.

7. In humans, *Yersinia pestis* produces an antiphagocytic capsule and a plasmid-dependent cytotoxin. It also produces the cytoplasmic protein V antigen, another important virulence factor.

8. Wild rodents in the western states are the main vector of plague in the U.S.

Critical Thinking Questions (pg. 622)

1. Such patients are given antibiotics just prior to dental work to reduce oral bacteria to a low level without causing a large antibiotic-resistant population to develop.
2. Endotoxic shock as a result of septicemia causes death despite killing of the bacteria.
3. Infected fleas from the prairie dog could be present on the dog or could have infected the dog, which can act as a host. Since the fleas can bite humans, there is a risk of contracting plague.

CHAPTER 28

Introduction

The blood and lymphatic infections of humans can be highly debilitating. Frequently, the long-term effects of the disease interfere with the homeostatic mechanisms associated with the healthy state.

Infectious mononucleosis and HIV disease can result in life-long problems and destruction of some essential elements of the immune system. When the balance of the cellular components is disturbed, the entire system is immobilized. Yellow fever can result in permanent liver damage and therefore systemic disturbances. Malaria is a well-known disease that occurs worldwide. Attempts to eliminate it have been far less successful than those for yellow fever. Millions of people are still infected each year and immunity is not yet available.

Objectives

After studying this chapter, the student should be able to:

1. Discuss the historical contributions to an understanding of the diseases of the blood and lymphatic systems.
2. Describe infectious mononucleosis as an infection of the lymphatic system.
3. Describe AIDS as a manifestation of infection with HIV.
4. Describe yellow fever as a zoonosis of the tropics.
5. Discuss malaria as a worldwide infectious disease.

Important Terms

heterophile antibody
hairy leukoplakia
Kaposi's sarcoma
pneumocystosis
tuberculosis
toxoplasmosis
merozoites

gametocytes
zygote
cyst
sporozoites
cerebral malaria
exoerythrocytic cycle

Essay

1. What discoveries concerning yellow fever are attributed to Dr. Reed and the members of the Yellow Fever Commission of the United States Army?

    As a result of experiments by the Commission to determine the possibility of yellow fever being spread by mosquitoes, the following discoveries were made:
    1. mosquitoes were the vectors of the disease
    2. an interval of about 12 days must elapse between the time the mosquito becomes infected and the time it can transmit the disease to an uninfected person
    3. the disease can be transmitted from an acutely ill person to an uninfected person by injecting the healthy person with a small amount of blood from the acutely ill person
    4. the disease is not transmitted by soiled linens or any other item that comes in contact with the infected person
    5. the disease is caused by an infectious agent that can pass through a filter that excludes bacteria.

2. Describe the main features of infectious mononucleosis.

    Infectious mononucleosis, or "mono," is caused by infection with the Epstein-Barr virus. After an incubation period of 1–2 months, the symptoms of fatigue, fever, sore throat covered with pus, and enlargement of lymph nodes in the neck and the spleen develop. The infectious condition can be spread by saliva, hence the common name "kissing disease." There is lifelong recurrent shedding of virus into saliva of asymptomatic latently infected individuals due to infection of B lymphocytes. Most patients recover within a few weeks without specific treatment. Prevention is through careful hygiene and avoidance of sharing articles that could be contaminated with virus from saliva.

3. List some ways to reduce the risk of HIV infection:

    1. Abstain from sexual intercourse
    2. Maintain a monogamous sexual relationship
    3. Use latex condoms with a spermicide from start to finish
    4. Do not engage in anal intercourse
    5. Avoid trauma to the genitalia and rectum
    6. Avoid needles contaminated with someone else's blood
    7. Avoid pregnancy if you are a woman infected with HIV
    8. Do not have sexual intercourse with a person you do not know well
    9. Do not have sexual contact with anyone suspected to be at risk for HIV infection
    10. Do not engage in sexual acts with known herpes carriers when sores are present

4. List the people at high risk of HIV infection:

    1. Homosexual and bisexual men
    2. Intravenous drug abusers who share needles
    3. People who have had blood transfusions or pooled blood products between 1978 and 1985
    4. Sexually promiscuous men and women, especially prostitutes
    5. People with a history of hepatitis B, syphilis, or other sexually transmitted diseases

5. Describe the major characteristics of malaria.

    Malaria is caused by the injection into the blood stream of the *Plasmodium* protozoan usually through the bite of the *Anopheles* mosquito. Four species are involved: *P vivax, P. falciparum, P. malariae,* and *P. ovale.* The incubation period varies with the species involved, but generally is from 6 to 37 days. The symptoms are recurrent bouts of violent fever alternating with periods of good health. When the erythrocytes rupture, they release the protozoa, which causes the fever. Infected red blood cells adhere to each other and to walls of the capillaries. The vessels become plugged and the associated tissue is deprived of oxygen. The spleen enlarges due to its removal of large volumes of foreign material and abnormal blood cells from circulation. The treatment of malaria is related to the particular infectious species involved, but most of the medications are quinine derivatives. Prevention is dependent upon eradication of the reservoir mosquito and its habitat.

Review Questions (pg. 638)

1. Patients with infectious mononucleosis show an increase in the number of mononuclear leukocytes in the blood.

2. EB virus was first isolated from Burkitt's lymphoma.

3. EB virus infects B lymphocytes.

4. A heterophile antibody arises in the blood of a person or animal, and reacts with antigens of another animal.

5. Infectious mononucleosis is transmitted by saliva.

6. Infections by microorganisms of low virulence such as *P. carinii,* which can only cause disease in immunosuppressed people, were the first indications of AIDS.

7. The first symptoms of HIV disease are fever, muscle aches, headache, sore throat, enlarged lymph nodes, and rash; some people experience confusion, seizures, and paralysis.

8. Tuberculosis, pneumocystosis, toxoplasmosis, and generalized or localized infections by member of the Herpes-virus family commonly occur in AIDS.

9. Malignant tumors characteristic of AIDS are Kaposi's sarcoma, and central nervous system tumors arising from B lymphocytes.

10. People at risk of HIV infection are: promiscuous homosexual and bisexual men; intravenous drug abusers who share needles; people with a history of a sexually transmitted disease; people who received multiple blood transfusions or products prepared from pooled blood between 1978–1985; sexually promiscuous individuals; and people with sexual or blood contact with any of the above.

11. If you know you are HIV positive, you can take measures to avoid transmission to others and you can receive medical care to delay the onset of AIDS and the development of secondary infections.

Review Questions (pg. 642)

1. Jaundice is a symptom of severe cases, hence the name yellow fever.

2. Yellow fever is transmitted by the bite of *Aedes* mosquitoes.

3. Yellow fever can be complicated by DIC.

4. Yellow fever vaccination is important to travelers to parts of Central and South America, and Africa.

5. *Plasmodium falciparum* causes the most dangerous form of malaria.

6. Sporozoites infect liver cells.

7. In malaria contracted from a blood transfusion there are no sporozoites, and thus no liver infection develops.

8. Malaria is increasing because many *Plasmodium* strains are resistant to antimalarial medications; many mosquito vectors are resistant to insecticides; and control measures were abandoned too soon.

9. Current strategies for containing malaria include improving access to medical facilities for people living in malarious areas to foster prompt identification and earlier treatment, and faster control of epidemics.

Critical Thinking Questions (pg. 643)

1. Infectious mononucleosis is the manifestation of EB virus infection, and is most commonly recognized in adolescents and young adults. Most people with EB virus infection do not show the characteristic signs and symptoms of infectious mononucleosis.
2. AIDS represents the end stage of HIV disease and often occurs many years after HIV infection. Prompt identification of HIV infection helps prevent further transmission of the virus and helps assure optimum treatment for HIV-infected individuals.
3. If HIV-I had been introduced into the United States in 1928, it would have taken much longer to detect the epidemic because infectious disease surveillance was not as sophisticated then. The infectious agent

probably would not have been identified because of the absence of virological facilities of adequate capability.

The long incubation period and the lack of serological or virological diagnostic techniques would have made the epidemiology inscrutable.

In addition, the overall impact on humans would have been much greater.

Subcultures with strict sexual taboos might have escaped the epidemic in some cases, but rational strategies to avoid the disease would have been difficult to design.

4. The transmitting mosquito had to bite a yellow fever patient at a stage in the illness when there was significant viremia, otherwise the small sample of blood ingested by the mosquito might not have contained any yellow fever virus. The virus then had to replicate in the mosquito for a given time period (a minimum of 12 days, according to Commission findings) before there was a sufficient quantity of the virus in the mosquito saliva to infect Dr. Carroll.

CHAPTER 29

Introduction

The antimicrobial and antiviral medicines function to selectively kill or inhibit the growth of the invading microbes. Most effective medications interfere with the metabolism of rapidly growing cells; however, they are ineffective on the toxins produced by these disease-producing organisms. Resistance to many antimicrobial medicines has developed over the years, thus a greater selectivity of chemicals must be used to retard the progress of resistant strains. Laboratory sensitivity testing, which will target the most appropriate medication with its sensitive infectious agent, must be encouraged rather than the widespread use of broad-spectrum medications.

Since antiviral medications show significant damage to the human host, greater research into genetic engineering of interferon will continue until ideal therapeutic ratios can be reached.

Objectives

After studying this chapter, the student should be able to:

1. Describe the contributions of early scientists to our current knowledge of antimicrobial techniques.
2. Describe the principles of antimicrobial therapy.
3. Compare the differences among the antibacterial medicines.
4. Describe the modes of action of the antimicrobial medicines.
5. Describe bacterial sensitivity testing techniques.
6. Define the terms *sensitive* and *resistant* in relation to a microorganism to be treated.
7. Determine the limitations of the antibacterial agents.
8. Describe the uses and actions of the antifungal medicines.
9. List the antiprotozoal medications.
10. Describe the methods by which the antiviral medications function.

Important Terms

"- cidal" β-(beta) lacta ring
"- static" β-lactamase
antimicrobial chemotherapy   carboxypenems
therapeutic ratio   monobactams
antibiotic   suicide inhibitors
secondary metabolites   broad-spectrum antibiotics
semisynthetic   aplastic anemia
families of antibiotics   macrolide

pseudomembranous colitis
rifampin
sulfones
polyene antibiotics
quinolones
minimal inhibitory concentration (MIC)
minimal bactericidal concentration (MBC)
resistant variants
keratin
amphotericin B

flucytosine
ketoconazole
fluconazole
iodine compounds
griseofulvin
nystatin
clotrimazole
butoconazole
miconazole
interferon

Essay

1. Describe the antibacterial activity of the sulfa drugs.

    Sulfa drugs are purely synthetic antibacterial agents that are active against both Gram-positive and Gram-negative bacteria. They are bacteriostatic in their mode of action. Sulfa drugs are well absorbed from the gastrointestinal tract and are widely used to treat *E. coli* bladder infections. Side effects, which are infrequent, are primarily allergic reactions involving fever and skin rash.

2. Describe penicillin as an antibacterial agent.

    Several different penicillins have been identified in mold cultures. Of these, penicillin G was found to be most suitable for treating infections. It is still the drug of choice for susceptible organisms even with the drawback of its instability in acid solutions and possible degradation by stomach acid. Penicillin V is a relatively acid-stable compound and has replaced penicillin G in some cases. Penicillin acts to interfere with the cell wall synthesis of the bacteria and are almost completely nontoxic.

3. Describe the antibiotics that interfere with protein synthesis.

    The tetracyclines are produced by prokaryotes of the actinomycete genus *Streptomyces*. They are bacteriostatic and act by interfering with bacterial protein synthesis. Their favorable therapeutic value is due to their active transport into bacterial, but not mammalian, cells, so that their intracellular concentrations are several times greater than those of the surrounding medium. The tetracyclines are referred to as broad-spectrum antibiotics because they show activity against both Gram-positive and Gram-negative organisms. They are given orally but caution must be advised as they cannot be taken with milk, antacids, or iron-containing medications since the metal ions can inactivate their therapeutic value.

4. Describe some of the resistance mechanisms of the antimicrobial medications on the infectious agents.

    The aminoglycosides cause changes in the ribosomal structures of the organisms, which fail to attach and cannot interfere with protein synthesis.
    The β-lactam antibiotics activate cell-wall destroying enzymes that interfere with the peptidoglycan linkage of the cell wall formation.
    Rifampin causes changes in nucleic acid synthesis by binding to bacterial DNA-dependent RNA polymerase.
    Aminoglycosides stop bacterial enzyme activity by the addition of acetates or phosphates.

Review Questions (pg. 654)

1. Synthetic antibacterial medicines include sulfa drugs, trimethoprim, quinolones, isoniazid (INH), and metronidazole.

2. Antibiotics include penicillin G, tetracycline, cephalosporin, chloramphenicol, and erythromycin.

3. Chloramphenicol can cause aplastic anemia, which is often a fatal condition.

4. Bacterial growth continues for several cell divisions until the end products of folic acid metabolism already present are consumed.

5. Penicillin acts against peptidoglycan synthesis and activates a lysozymelike bacterial enzyme that weakens the cell wall, which eventually bursts.

6. Erythromycin, tetracycline, and chloramphenicol interfere with bacterial protein synthesis.

7. Quinolones and rifampin inhibit nucleic acid synthesis.

Review Questions (pg. 665)

1. "Sensitive" means that the bacterium is killed or inhibited by concentrations of an antimicrobial medicine reached in the blood of patients receiving usual doses of the medicine. "Resistant" means that the bacterium is not killed or inhibited at those concentrations.

2. Antimicrobial medicines are ineffective if the causative organism is resistant, enzymes inactivate them, and if there is a decrease in the cell wall permeability. In addition, the target substance may fail to bind the medication, the organism may not be actively multiplying, and the medicine may be inactivated by pus or acid in the urine.

3. Mechanisms of resistance include:
   1. Change in ribosomal structure so medication won't attach—aminoglycosides, tetracycline, erythromycin.
   2. Change in cytoplasmic membrane binding-proteins so medication won't attach—penicillins and other β-lactams.
   3. Medication is inactivated by enzymatic breakage of a ring structure—penicillins and other β-lactams, cephalosporins, erythromycin.
   4. Altered RNA polymerase structure so that medicine doesn't attach—rifampin.
   5. Medication is inactivated by addition of acetate or phosphate group—aminoglycosides, chloramphenicol.

4. Amphotericin B is effective in treating fungal infections.

5. Metronidazole is effective against bacteria and protozoa.

6. Amantadine helps prevent influenza A, and zidovudine is used in treating AIDS.

Critical Thinking Questions (pg. 666)

1. The bacteriostatic antibiotic would decrease the antibacterial effect because penicillin is only effective against growing organisms that are synthesizing cell wall components.
2. Selective toxicity implies a difference in structure or enzymes that allows a medication to attack one kind of cell and not another. Therapeutic ratio is the highest dose a patient can tolerate without toxic effects divided by the dose required to control a microbial infection.
3. Susceptibility of bacteria to antimicrobial medicines is determined by mixing a large quantity of the organisms with the medicine and a growth medium *in vitro* to determine whether they are killed or inhibited by concentrations of the medicine achieved *in vivo*. Since conditions in the test tube are similar to those in patients, susceptibility tests provide a reasonable estimate of a medicine's effectiveness in patients. However, many details relating to the nature of the bacterium and the disease process, and the status of the host's ability to absorb and excrete it, may need to be taken into account.
4. Scientists have dealt with antibacterial resistance by:
   1. constantly testing new antimicrobials for their effectiveness.
   2. learning the mechanisms of resistance and designing new forms of antimicrobial medicine to circumvent the mechanisms.
5. Decreasing the unnecessary prescriptions of the medicines should help limit the increasing number of resistant bacterial organisms.

CHAPTER 30

Introduction

Ecology is the study of the relationship of organisms to each other and to their environment. The ecology of the soil is extremely complex and subject to disturbances by the actions of humans. The world is completely dependent on maintaining the homeostasis of the soil and the biosphere. The earth is a closed system and all the essential ingredients for life must be recycled constantly to maintain the necessary ratios of carbon, hydrogen, oxygen, nitrogen, and water. Through the actions and reactions of the microorganisms of the soil these cycles are maintained in a state consistent with life. The study of these interactions is the basis of study in this chapter. Each species contributes its share to the balance of nature and as long as the balance is maintained, life will continue.

Objectives

After studying this chapter, the student should be able to:

1. Describe the contributions of history to the current knowledge of microbial ecology.
2. Describe and explain the principles of microbial ecology.
3. Describe the relationships of microbes and the soil with which they are associated.
4. Describe the environmental influences on the organisms present in the soil.
5. Understand the significance of biochemical cycling.
6. Discuss the effects of synthetic chemicals on the environment.

Important Terms

nitrogen fixation
pure cultures
community
ecosystem
ecological niche
ecology
biosphere
microenvironment
macroenvironment
oligotrophic
putrefaction
humus
actinomycetes

myxobacteria
geosmins
mesophiles
psychrophiles
thermophiles
rhizosphere
producers
consumers
decomposers
photosynthesis
fixed nitrogen
ammonification
denitrification

nitrification
lectins
infection thread
nodule

bacteroids
biological magnification
nitrogen fixation

Essay

1. Describe the meaning of the phrase that "microbial populations may both cause and adapt to environmental changes."

    Environmental changes often result in population changes. In an environment in which there are several populations with varying densities, the population with the greatest density is the dominant population. This dominant population utilizes the most abundant nutrients. As these nutrients are depleted, one of the lesser groups may be able to metabolize the products produced by the first group. This sequence of events continues when the products of one group become the nutrients of the next population, etc., until the environment changes drastically and the population changes with it. This series of events can be exemplified in the changes that take place in the nature of the change from fresh milk to completely soured milk by the populations of microbes present in the milk.

2. Distinguish between the microenvironment and the macroenvironment of the soil microorganisms.

    The environment immediately surrounding an individual cell is referred to as its microenvironment. To a bacterial cell, this small volume of soil is more important than that outside of it. The presence of moisture, oxygen, and organic matter immediately surrounding the cell supports its growth and metabolism. The macroenvironment of a microbial population may be more readily examined and measured but may be very different from its immediate surroundings. Not only is this true but also a variety of microenvironments may actually be found within a single crumb. For example, crevices can rapidly become anaerobic, while outer surfaces remain aerobic.

3. Describe the problems associated with the decomposition of synthetic chemicals.

    Most organic compounds of natural origin can be degraded by one or more species of soil or aquatic organisms. However, this is not the case with a large number of chemicals synthesized industrially. The chemical nature of nonbiodegradable synthetics covers an extremely broad range of organic compounds. Although certain of these chemically synthesized compounds disappear from an ecosystem within a reasonably short time, a few days or weeks, many remain undegraded for years. Chemically synthesized compounds are most likely to be biodegradable if they have a chemical composition similar to that of naturally occurring compounds.

    However, if a synthetic compound is totally different from any that occurs in nature, organisms will be less likely to have the enzymes necessary for degrading it rapidly, and the compound is likely to persist for long periods.

Review Questions (pg. 675)

1. The microenvironment is most important because the microorganism is in closest contact with it. The microenvironment could become anaerobic even if the macroenvironment is not, which would determine the availability of nutrients for survival.

2. When milk putrefies, the population of microorganisms goes through a succession of various organisms depending on the pH and nutrients available. *Streptococcus lactis* grows well at first, causing the milk to sour. In this highly acidic environment, *Lactobacillus casei* and *L. bulgaricus* multiply. Eventually yeasts and molds take over in the highly acid environment. The remaining milk protein (casein) is then broken down by *Bacillus*, which excretes proteolytic enzymes. The result is a clear but smelly fluid.

3. Yes, oligotrophic (low-nutrient) bacteria can grow in distilled water. In hospitals, the bacteria can grow in mist therapy units used to treat lung deseases. In laboratories, the bacteria can contaminate water used in experiments without producing a cloudy solution. Thus, this undiscovered growth can cause serious health problems and contaminate cultures.

4. Moisture affects the supply of oxygen available in the soil. Wet soggy soils have few air spaces and are favorable to anaerobes. Highly acid or alkaline soils inhibit the growth of common bacteria. Fungi are able to grow in a broader range of pH. Temperature governs the rate of biochemical processes through its effect on the rate of enzyme reactions. Most soil bacteria are mesophiles and grow best at temperatures between 20°C and 50°C. They are heterotrophic and thus need a supply of organic materials as a source of energy. Thus, the addition of organic material to soil increases the number of bacteria and fungi.

Review Questions (pg. 685)

1. Oxygen is cycled between respiration and photosynthesis. Green plants, algae, and cyanobacteria produce oxygen from water through the process of photosynthesis. Other plants and animals use the oxygen.
Carbon cycles between organic compounds and carbon dioxide. It is made into organic compounds from atmospheric carbon dioxide by the process of photosynthesis by green plants, cyanobacteria, and algae. Some chemoautotrophic bacteria also produce organic compounds. Then, consumer organisms use the carbon in the organic compounds as energy sources during respiration and produce carbon dioxide.
Nitrogen cycles between organic and inorganic compounds. Nitrogen is in the air but is unusable by plants and animals in this gaseous form. A few microorganisms, however, are able to fix nitrogen to a form that is then usable by the plant. Nitrogen fixation is accomplished by a few specialized bacteria in conjunction with the roots of special plants such as legumes. In addition, plants and microorganisms are able to convert nitrogen into ammonium ions and then into nitrate.
Phosphorus cycles between organic and inorganic forms. Bacteria and fungi are largely responsible for making organic phosphorus available to microorganisms and plants. The microorganisms synthesize the enzyme phosphatase, which separates the organic phosphate from a large variety of compounds.
Sulfur is present in soil as a part of proteins, and is taken up by plants as sulfate. Sulfate-containing proteins are degraded into amino acids by enzymes excreted by a variety of microorganisms. The sulfur

in amino acids is then converted to hydrogen sulfide by a variety of soil microorganisms. Under aerobic conditions, the hydrogen sulfide oxidizes to sulfur and is then converted into a more readily usable form, sulfate, by the sulfur bacteria.

3. Many synthetic chemicals are not biodegradable, thus they accumulate in the environment, producing toxic effects in plants, animals, and water systems in the ecosystem.

Critical Thinking Questions (pg. 686)

1. The fact that a compound is nonbiodegradable means that it lasts longer and is probably more effective over a longer time period. As a result, the compound accumulates in the environment and many organisms become resistant to it. Thus, it often becomes ineffective.

2. The leaching action of water depletes the soil of the nitrate fertilizer the farmer applied to his soil.

CHAPTER 31

Introduction

Water in one form or another covers a very large segment of the earth's surface. Its condition is extremely important to the maintenance of a healthy planet. The water levels in the lakes, oceans, and rivers are maintained by rainfall. When rainfall is acidic due to dissolved gases from fossil fuels and other sources, the long-term effects on the aquatic environment can be quite severe. Along with acid rain, the problem created by the nutrient enrichment of marine and fresh water due to the addition of sewage and industrial waste can seriously alter the ecological systems established by nature.

Objectives

After studying this chapter, the student should be able to:

1. Understand the contributions of history to our current knowledge of water as a reservoir.
2. Describe the nature of aquatic environments.
3. Compare the differences between the environments of freshwater and marinewater.
4. Describe the microbial flora of freshwater.
5. Describe the microbial life of marine environments.
6. Discuss the consequences of water pollution.
7. Discuss the effects of water pollution on the population of its inhabitants.

Important Terms

acid rain
eutrophication
eutrophic
oligotrophic
peat
photic zones
estuaries
indigenous population
cellulases
prosthecate

sulfur oxidizers
nitrifiers
Winogradsky column
epifluorescence counting
acridine orange
luciferin-luciferase enzyme
barophiles
algae
protozoa
accelerated eutrophication

Essay

1. What effect does acid rainfall have on ecosystems?

    Freshwater ecosystems are particularly sensitive to the increasing acidity that results from acid rain. They contain soft water (water with low concentrations of ions) that cannot neutralize the acidity created by rain with a pH of approximately 2.4. The result is acidification of lakes and streams as well as soils in the surrounding drainage areas. In such areas, animal, plant, and sensitive microbial populations can be eliminated, upsetting the delicate balance among the intermeshed ecosystems involved.

2. Describe the stratification that occurs in some slow moving lakes.

    Lakes become stratified during the warm weather when the water moves sluggishly. The colder, denser water is found at the bottom of the lake and is rapidly depleted of oxygen. The warmer water in the upper layers is replenished with oxygen from the air. Water taken from the various levels of this type (stratified) of lake contain quite different populations of organisms. Water systems that are high in nutrients will show a high population of anaerobic organisms in the deeper, cooler layers.

3. Describe some of the activities of freshwater bacteria.

    The activities of the freshwater bacteria are many and complex. In general, the bacteria perform three major functions:
    1. they decompose organic matter and liberate mineral nutrients
    2. they transform essential minerals, cycling them through forms other organisms can use
    3. they release dissolved organic compounds into the food webs, which supports the growth of other organisms.

4. Describe the environmental restrictions that regulate the microorganisms found in seawater.

    The numbers and kinds of microorganisms in seawater are restricted by several factors. Among them are:
    1. the salinity of the water
    2. the low levels of certain nutrients
    3. low temperature
    4. high pH (about 8.0)
    5. the intensity and wavelength of light that can penetrate the water to serve as an energy source.

5. Define eutrophication and describe its relation to pollution.

    When natural waters are nutrient enriched by either natural processes or accelerated by people, with the addition of pollutants, the result is referred to or defined as eutrophication. Nutrients can be added to a freshwater environment from a variety of sources, including sewage, agricultural runoff, and acid rain. These additions can lead to increased productivity that commonly upsets the normal ecological balance, and therefore creates various kinds of problems, including unsightly algal scums, obnoxious odors, and the death of fish resulting from oxygen depletion.

Review Questions (pg. 693)

1. Distilled water is not truly pure water because there are minute amounts of the components of glass or other containers dissolved in it.

2. Potable water means that it is safe to drink but is not pure in either a chemical or microbiological way.

3. Acid rain is harmful to the environment because it raises the pH of the freshwater ponds and lakes to the point where they are unable to support life. Freshwater has very few other chemicals with which to buffer the acid. In some areas, rain water has been measured to have a pH of 2.4, the equivalent of vinegar.

4. Freshwater lakes become stratified especially during warm weather because cold water is denser than warm water and sinks to the bottom of the lake. This layer is usually depleted of oxygen, whereas the warm surface waters are highly oxygenated from the air. Thus, different kinds of bacteria live in the different areas of the freshwater lakes.

5. 1) Seawater has a higher salt content (3.5% salt) compared to freshwater.
   2) Saltwater has fewer organic nutrients than freshwater.
   3) Open oceans have relatively stable environmental conditions compared to freshwater, which is more varied.
   4) Large areas of saltwater are unable to support the primary producers because they are below the photic zone.

Review Questions (pg. 701)

1. 1) Various species of bacteria degrade complex organic macromolecules into smaller molecules that are then used by other groups of bacteria.
   2) Bacteria are often the first step in the food web, providing the food for protozoa and small invertebrates.
   3) Bacteria decompose dead animals and plants into inorganic materials.

2. The availability of oxygen and the kind of nutrients available are the most important determinants of which type of microorganism can exist in freshwater. In seawater, factors such as salinity, pH, temperature, availability of light, and level of nutrients determine which microorganisms will grow and survive.

3. Eutrophication is caused by the addition of nutrients to a water source that is above the normal amount. This causes an unusual number of microorganisms to grow and multiply, often polluting the water source. Frequently, the source of nutrients is from sewage, agricultural runoff, and acid rain. Frequently, algae grows to such an extent that they deplete the supply of oxygen, causing other animals and plants to die. Eutrophication can be reversed by reorganizing the sewage disposal systems around a water source.

4. Epifluorescence counting involves staining a sample with the fluorescent dye acridine orange, which stains the DNA inside cells. When the sample is observed under the microscope with an ultraviolet light the cells are easy to see and appear fluorescent. The disadvantage is that not all microorganisms survive long enough to be counted. It is also a tedious process.

A more accurate method is the luciferin-luciferase enzyme system, which involves measuring the amount of ATP in the sample.

Also the use of nucleic acid probes gives an accurate count of the organisms in a sample.

Critical Thinking Questions (pg. 702)

1. This statement is not valid because even if a particular pollutant is diluted by large volumes of water, frequently it is not metabolized in a way that is beneficial to humans. For example, mercury salts are often present in industrial wastes. They are converted by bacterial action to methyl mercury, a powerful nerve toxin. Fish, which are not affected by it, can concentrate the methyl mercury in their tissues and when humans eat the fish, they are severely affected. Further, conditions in the ocean are frequently not conducive for any microbial action because it is too cold for microbes to grow or there are not enough microbes to degrade the pollutant. These pollutants can then accumulate and eventually become a problem.

2. Even small amounts of certain microbial toxins such as botulinum can be added to water supplies and seriously affect humans. Thus, protective measures such as patrolling, fencing, and regular testing for toxic substances should be done to protect the water supply against saboteurs.

CHAPTER 32

Introduction

The problem of delivering fresh water to concentrated populations and removing their accumulated waste has troubled community leaders since the evolvement of urban life. If the number of inhabitants is small and scattered, and the area covered is large the problem of human waste removal is far simpler. Septic systems, artificial wetlands, and lagooning are effective treatment methods. The passage of the Clean Water Restoration Act of 1972 eliminated the discharge of pollutants into navigable waters and forced cities to develop more effective means of safe waste disposal. Sanitary landfills have been a low-cost method of disposing of solid waste but since leaching of toxic materials into aquifers is still a major concern, recycling programs are now being implemented to reduce the volume and toxicity of landfills. In addition, backyard and commercial composting are being encouraged to lessen landfill problems and restore nutrients to the soil.

Objectives

After completing this chapter, the student should be able to:

1. Describe the historical contributions to our present understanding of sewage and waste treatment.
2. Describe the microbiology of waste treatment.
3. Describe how microorganisms can reduce the biological oxygen demand of sewage.
4. Explain how sewage can be treated on a small scale.
5. Explain how microbiology is involved in the treatment of solid waste.
6. Describe the positive results of recycling programs and their effect on landfills.

Important Terms

mineralization

stabilization

primary treatment

secondary treatment

tertiary treatment

digester

lagooning

activated sludge method

trickling filters

septic tanks

artificial wetlands

sanitary landfills

biochemical oxygen demand (BOD)

anaerobic stabilization

Essay

1. Describe the treatments of sewage in the primary, secondary, and tertiary stages.

    The primary treatment of sewage involves the removal of large objects and much of the particulate matter through the physical processes of screening and sedimentation. Secondary treatment involves the chemical and biological processes of converting the remaining materials in sewage into odorless inorganic substances that can be reused. Tertiary treatment involves the removal of phosphates and nitrogen compounds from sewage, which can cause eutrophication.

2. Describe the significance of BOD as it applies to sewage treatment.

    The term *biochemical oxygen demand* (BOD) describes the oxygen-consuming property of a waste-water sample. It is roughly proportional to the amount of degradable organic material present in the water sample. High BOD values reflect large amounts of degradable organic material in a sample of waste water or other substances. Effective sewage treatment should decrease the BOD of sewage as much as possible, and should also remove toxic and other questionable materials from the sewage.

3. Describe the advantages and disadvantages of using sanitary landfills for the disposal of solid waste.

    Landfills are used to dispose of solid wastes near towns and cities. When a layer of dirt is put over the wastes each day, the landfill is described as sanitary. When the landfill is completed, it can be used for recreation and eventually construction.
    There are several disadvantages to this type of waste management:
    1) the number of sites is limited
    2) the organic content undergoes slow anaerobic decomposition of at least 50 years
    3) methane gas is produced and must be removed to prevent dangerous explosions from occurring
    4) heavy metals and pesticides can leach from the landfill site into underground aquifers

4. Describe the organization and uses of a backyard compost pile.

    Composting is an attractive alternative to landfill disposal of organic waste. Garden debris along with kitchen organic waste can be layered with sand and soil to create an active decomposition site. In a short time, the inside of the pile heats up to 100°–150°F. At this temperature, the pathogens are killed but other thermophiles are not affected. When the temperature of the pile decreases, stirring it adds oxygen and allows the bacteria to resume growing and breaking down the organic materials. If a compost pile is turned frequently and aerobic digestion is good, the composting can be completed and ready for use in six weeks. The resulting black organic material can be used in gardens and flower beds. The material is rich in organic compounds and helps rejuvenate the soil for growing healthy plants.

Review Questions (pg. 713)

1. Primary treatment of sewage involves removing large materials and particulate matter through sedimentation or a series of screens. Sometimes aluminum sulfate or other chemicals are added to coagulate the particles so they will settle more quickly.

   Secondary treatment is designed to stabilize most of the organic materials and reduce the BOD of the sewage. Most secondary treatment plants use the activated sludge method of treatment. In this treatment, the sewage serves as a nutrient source for a mixed population of aerobic organisms. The sewage is aerated and the microbes are allowed to degrade the organic material to gases and other products. Later, the sewage is moved to the digester where anaerobic stabilization occurs. Water is removed from the sewage and solids are disposed of. Methane gas is produced during the anaerobic digestion.

   Tertiary treatment removes the materials left by secondary treatment. Lime is added to remove phosphate particles. Ammonia is removed by passing the water through a "stripping tower." A series of filters remove detergents, pesticides, and other toxic chemicals. The resulting water after treatment with chlorine is pure enough to use as drinking water.

2. BOD stands for biochemical oxygen demand, which designates the oxygen-consuming property of waste water. It is roughly proportional to the amount of degradable organic material present.

3. Most pathogenic bacteria are removed from sewage during secondary treatment, but some pathogenic viruses can remain. Any remaining microorganisms and viruses following tertiary treatment are killed by treating the water with chlorine.

Review Questions (pg. 714)

1. Sanitary landfill means that at the end of each day, a layer of dirt is put over the solid wastes to reduce the odor and the number of insects and rodents that are attracted to the garbage.

2. Composting and recycling are some alternatives to landfill disposal of solid waste.

3. Composting starts with a supply of organic materials, such as leaves and grass clippings. Then dirt and water are added and the compost pile is maintained in a moist condition, and heats up to 100°–150°F. The pile needs to be aerated regularly, and in about six weeks the organic materials are generally digested.

Critical Thinking Questions (pg. 716)

1. a. Septic tanks are fairly easy to install but often require large drain fields to work properly. Thus, a large amount of land is needed. Further, there is always the possibility of ground water contamination from a septic system.
   b. Trickling filters would be expensive to install for a small number of houses.
   c. Artificial wetlands are a good alternative but require expertise to develop and monitoring to maintain their effectiveness. Further, they require a large amount of land to work effectively.

CHAPTER 33

Introduction

Microorganisms are responsible for food spoilage, the fermentation of dairy products to produce various cheeses, and the fermentation of fruits and grains to produce alcoholic beverages.

Although food spoilage microorganisms are usually nonpathogenic for humans, several that invade food can be responsible for food poisonings.

Food preservation methods are designed to inhibit or destroy the microorganisms contaminating food and beverages. A variety of preservation methods are employed, depending on the nature of the product. Food is dried, irradiated, canned, or frozen. It can then be treated with chemicals or exposed to pasteurization techniques to destroy non-spore-forming pathogens.

Microbes can also be used in the production of food and beverages. Yogurt and cheese are made using lactic acid bacteria on specific milk product types. Alcoholic beverages and vinegars are the fermentation products of *Saccharomyces* or *Acetobacter* species.

Objectives

After completing this chapter, the student should be able to:

1. Understand the contributions of history to our current knowledge of food and beverage safety.
2. Describe the microorganisms responsible for food spoilage.
3. Describe the specific requirements for the growth of spoilage organisms.
4. Understand how some foods are protected by naturally occurring antimicrobial substances.
5. Describe some useful methods of food preservation.
6. Discuss the microbial hazards of food.
7. Describe the use of microbes in the production of food and beverages.

Important Terms

food spoilage

water activity

thermal death time (TDT)

high-temperature-short-time method (HTST)

low-temperature-long-time (LTLT) method

holding method

ultra-high-temperature method (UHT)

lyophilization

must

malt

wort

bottom yeast

Essay

1. List and briefly describe the growth requirements of food spoilage organisms.

    Several factors involved in determining the nature of the microorganisms involved in food spoilage are:

    a) availability of water—all microorganisms require water to grow; dried foods will inhibit the growth of bacteria but not fungi

    b) pH—neutral foods can support the growth of several types of organisms but highly acidic foods can be spoiled by molds

    c) available nutrients—if a certain food is deficient in a given nutrient, any organism that requires it will be inhibited. Since pseudomonads are capable of synthesizing essential nutrients and grow under a variety of conditions, they are capable of spoiling several different kinds of foods.

2. Describe the principles involved in the sterilization of foods.

    Several factors must be considered when evaluating the relationships of time and temperature involved in the sterilization of foods:

    a) the higher the sterilizing temperature, the shorter the time needed to preserve the food

    b) the higher the concentration of bacteria, the longer the heat treatment required if all other factors remain constant

    c) the growth phase of the vegetative cells or of the spores influences their heat resistance

    d) the environment or material in which microorganisms are heated influences their heat resistance:
    1) moist heat vs. dry heat
    2) pH
    3) concentrations of salt and sugar

3. Describe the four categories of food-borne illnesses.

    Food-borne illness still occurs in the United States in spite of the strict controls on the quality of the supply and the care taken in the preparation.
    The four general categories of food-borne illness are:

    1) food-borne idiosyncracies—include food hypersensitivities such as allergic reactions to wheat, eggs, milk, strawberries, and the like

    2) food-borne intoxications—occur when a microorganism growing in a food produces a toxin or converts a food to a toxic product, such as in botulism

    3) food-borne infections—occur when food contains microorganisms that are ingested and then colonize the person, such as salmonellosis

    4) chemical food poisonings—occur when people ingest foods that contain poisonous chemicals, such as those present in some mushrooms or in fish that have ingested toxic chemicals.

4. Describe the influence of microorganisms in the production of alcoholic beverages.

Many alcoholic beverages are produced using strains of the yeast *Saccharomyces*. Almost anything that can be fermented by yeasts has been used to make alcoholic beverages.

Grapes and other fruits are fermented to make wine. The yeasts needed to transform fruit juice into wine are often found naturally on the skin of the fruit.

Beer is made by the fermentation of starch, traditionally from germinated barley. The resulting fermented mash is sieved and filtered, and sometimes flavored with herbs or honey. Later, hops came into use, being added as a preservative and to give the characteristic bitter flavor to the beer.

Distilled spirits such as scotch, whiskey, and gin are manufactured by a process that is similar to the production of beer except that the wort is not boiled. When the fermentation is complete, the alcohol is separated by distillation. Different types of spirits are formed in different ways:
   a)   scotch whiskey is aged for several years in oak sherry casks
   b)   grain mash for sour-mash whiskey is fermented and distilled, resulting in whiskey with a characteristic flavor
   c)   tequila is traditionally made from the fermentation and distillation of juices of the agave plant.

Review Questions (pg. 721)

1. Since *Pseudomonas* sp. can metabolize a wide variety of compounds and are able to grow at refrigerator temperatures, they can spoil many different kinds of foods.

2. The water activity of a food is the amount of water that is available in foods, and provides an indication of microorganism growth.

3. You would expect a food with a high water activity to spoil faster because microorganisms thrive where water is available. If water is not available, microorganisms have a difficult time growing.

4. Since garlic in the juice inhibits the growth of bacteria, one would expect to recover fungi, which grow in acidic conditions.

Review Questions (pg. 727)

1. Care is taken in canning foods both at home and commercially to eliminate the possibility that anaerobic microorganisms are able to survive, grow, and produce toxins. *Corynebacterium botulinum* produces botulism through endospores whose toxin causes a serious nerve poison under anaerobic conditions.

2. Factors in sterilization of food include:
   1. The higher the temperature, the lower the time that is needed to kill the bacteria in food.
   2. The more bacteria contaminating the food, the longer the time it will take to reach the required killing.
   3. If bacteria are in their vegetative state rather than in their spore state, they are easier to kill with heat. Moist heat kills bacteria better than dry heat.

3. It is not safe to store foods in a refrigerator for long periods because some bacteria and molds can continue to grow at refrigerator temperatures, although their rate of growth will probably be slowed down.

4. When foods are salted, smoked, or sugared, the amount of water available for growth is greatly reduced. Thus without water, the microorganisms cannot grow.

5. The four categories of food-borne illness are:
   1) Food-borne idiosyncracies, such as allergies to milk and eggs.
   2) Food-borne infections are passed from one person to another through the food supply such as typhoid fever, salmonella, streptococcal infections, and hepatitis.
   3) Food-borne intoxications occur when microorganisms grow in food and produce a toxin or convert the food into a toxic product. *C. botulinum* is an example as well as the aflatoxins produced by *Aspergillus* sp.
   4) Food-borne chemical poisonings include eating poisonous mushrooms or the leaves of rhubarb, which contain oxalic acid.

Review Questions (pg. 735)

1. Foods produced by lactic acid fermentations include yogurt, cheese, and sauerkraut.

2. Sauerkraut is fermented by *Leuconostoc* sp., *Lactobacillus plantarum*, and *Lactobacillus brevis*. Yogurt is fermented by *Streptococcus thermophilus* and *Lactobacillus bulgaricus*. Cheese is fermented by *Lactobacillus bulgaricus*, *Streptococcus lactis*, *S. cremoris*, *S. thermophilus*, and many others.

3. *Saccharomyces* sp. are used to make bread and various alcoholic beverages.

4. Main steps in cheese production are:
   1) Milk is inoculated with a starter culture containing lactic acid bacteria.
   2) When there is sufficient lactic acid formed, the enzyme renin is added to coagulate the proteins.
   3) The curd (coagulated protein) is separated from the liquid (whey) and formed into blocks for ripening. Bacteria or yeasts may be added before and during the curing process to give cheeses their characteristic flavors.

5. Major steps in wine production are:
   1) Fruit juice is produced by crushing grapes.
   2) Sulfur dioxide is added to the juice to inhibit the growth of unwanted microorganisms.
   3) Carefully selected strains of *Saccharomyces cerevisiae* are added and fermentation is allowed to proceed at carefully controlled temperatures.
   4) When the final amount of alcohol has been achieved, the juice is removed from the pulp and the fermentation is allowed to proceed until it stops. Wines are then held for varying lengths of time to age.

Major steps in beer production are:

1) Ground malt, grains, and water are mixed and heated gradually to permit enzymes of the malt to degrade starch to sugars and dextrins. A mash results.
2) The spent grains of the mash are eliminated and sugar and hops are added. After brewing this mixture, the spent hops are eliminated leaving a hot wort mixture.
3) The hot wort mixture is allowed to settle and the coagulated solids are removed.
4) The hot wort mixture is passed over cooling coils.
5) The cooled wort mixture is fermented with brewer's yeast for about a week.
6) The beer is then allowed to settle, and is filtered to remove microorganisms and other particulates before aging in a lagering tank, carbonating, and bottling or canning.

Critical Thinking Questions (pg. 736)

1. A large food processing plant needs to monitor each stage of its operation with microbiological testing on a regular basis. Insisting that employees maintain safe food handling conditions at all times is also critical.
2. Refreezing thawed foods is not a good idea because thawing can activate bacteria in the food, which may then have a chance to grow increasing in number. Further, thawed foods generally have a breakdown of cell structure from the freezing process, making them more vulnerable to microbiological contamination.
3. Washing the hands, and contaminated surfaces and utensils between the preparation of meats, fish, and eggs, and other foods can help reduce the spread of disease organisms such as salmonella.

Chapter 1

Multiple Choice

1. The word "animalcule" was coined by _____ in _____.
   a. Jenner, 1794
   b. Pasteur, 1865
   c. Leeuwenhoek, 1674
   d. Needham, 1749

   ans. c.   diff. 6   obj. 1   p. 1

2. Organisms that can only be studied with the aid of a microscope are described as
   a. viruses
   b. animalcules
   c. invisible agents
   d. microorganisms

   ans. d.   diff. 6   obj. 1   p. 3

3. The microscopic technique in which faint objects appear brightly lit against a dark background is
   a. electron microscopy
   b. dark-field microscopy
   c. bright-field microscopy
   d. phase contrast microscopy

   ans. b.   diff. 4   obj. 1   p. 4

4. The founder of microbiology is
   a. Louis Pasteur
   b. Anton van Leeuwenhoek
   c. Edward Jenner
   d. Robert Koch

   ans. b.   diff. 4   obj. 2   p. 5

5. That life forms can arise spontaneously from nonliving matter is referred to as the theory of
   a. pure culture
   b. life's origin
   c. the cell
   d. spontaneous generation

   ans. d.   diff. 5   obj. 2   p. 6

6. Which of the following statements concerning Anton van Leeuwenhoek is not true?
   a. was a scientist
   b. described microorganisms
   c. used the term animalcules
   d. saw representatives of a wide variety of microbes

   ans. a.   diff. 4   obj. 1   p. 4

7. Which of the following scientists was not involved in the spontaneous generation controversy?
   a. Needham
   b. Pasteur
   c. Koch
   d. Spallanzani

   ans. c.   diff. 5   obj. 1   p. 8

8. Final settlement of spontaneous generation was accomplished by
   a. Pasteur
   b. Spallanzani
   c. Koch
   d. Tyndall

   ans. a.   diff. 4   obj. 2   p. 8

9. Pasteur disproved spontaneous generation by using _____ _____ in his experiments.
   a. cotton gauze
   b. cotton plugs
   c. hay infusions
   d. swan-necked flasks

   ans. d.   diff. 5   obj. 3   p. 9

10. Which of the following scientists, using hay infusions, had difficulty reproducing Pasteur's famous swan-necked flask experiment?
    a. Needham
    b. Redi
    c. Tyndall
    d. Koch

    ans. c.   diff. 6   obj. 3   p. 10

11. The structures present in the hay infusion that made it very difficult to sterilize were
    a. endospores
    b. solid material
    c. chlorophyll
    d. toxins

    ans. a.   diff. 5   obj. 3   p. 10

12. Which condition of valid experimentation was neglected by Tyndall in his attempt to duplicate Pasteur's swan-necked flask experiments?
    a. Needham was not invited to observe.
    b. The source of the nutrient infusion was different.
    c. The flasks were covered.
    d. The temperatures at which the experiment were done were not identical.

    ans. b.   diff. 7   obj. 3   p. 10

13. Most pathogenic bacteria were identified and early work on viruses was begun between the years
    a. 1920 and 1960
    b. 1675 and 1730
    c. 1727 and 1850
    d. 1875 and 1918

    ans. d.   diff. 6   obj. 4   p. 11

14. Approximately how many people have died of smallpox over the course of history?
    a. 10 million
    b. 1 million
    c. 500 thousand
    d. 5 million

    ans. a.   diff. 6   obj. 4   p. 11

15. Bubonic plague was a devastating disease of the Middle Ages in Europe. What percent of the population died during this period?
    a. 25%
    b. 13%
    c. 10%
    d. 37%

    ans. a.   diff. 5   obj. 4   p. 11

16. At the present time, the most important diseases in terms of frequency and death are
    a. viral diseases and gastrointestinal disorders
    b. heart disease and kidney failure
    c. respiratory infections and diarrheal diseases
    d. birth defects and geriatric diseases

    ans. c.   diff. 6   obj. 4   p. 12

17. Which of the following diseases is not considered one of the "new" diseases?
    a. AIDS
    b. toxic shock syndrome
    c. influenza
    d. Lyme disease

    ans. c.   diff. 7   obj. 4   p. 12

18. The use of vaginal tampons has been associated with
    a. toxic shock syndrome
    b. Legionnaires' disease
    c. candidiasis
    d. AIDS

    ans. a.   diff. 6   obj. 4   p. 13

19. It is possible that one of the agents that causes AIDS arose from a virus that at one time could only infect
    a. guinea pigs
    b. lemurs
    c. plants
    d. monkeys

    ans. d.   diff. 7   obj. 4   p. 14

20. Which of the following diseases is not being controlled by vaccination?
    a. measles
    b. hepatitis A
    c. polio
    d. whooping cough

    ans. b.   diff. 6   obj. 5   p. 14

21. For which of the following groups is TB not a serious threat?
    a. students
    b. the elderly
    c. those infected with HIV
    d. the poorly nourished

    ans. a.   diff. 4   obj. 5   p. 14

22. Approximately how many deaths have occurred in the U.S. due to AIDS?
    a. 100,000
    b. 300,000
    c. 500,000
    d. 200,000

    ans. d.　diff. 5　obj. 5　p. 15

23. One of the most important and beneficial contributions of microorganisms to human life is
    a. the nitrogen cycle
    b. the production of antibiotics
    c. the process of decay
    d. the action of primary producers

    ans. c.　diff. 5　obj. 6　p. 15

24. Which group of ancient populations first used yeast to produce leavened bread?
    a. Romans
    b. Hebrews
    c. Egyptians
    d. Phoenicians

    ans. c.　diff. 6　obj. 6　p. 15

25. Which of the following foods is not a product of fermented milk?
    a. buttermilk
    b. yogurt
    c. cheese
    d. butter

    ans. d.　diff. 5　obj. 6　p. 16

26. Biotechnology is revolutionizing all aspects of life. Its cornerstones are
    a. physics and immunology
    b. genetics and chemistry
    c. microbiology and virology
    d. microbiology and immunology

    ans. d.　diff. 6　obj. 6　p. 16

27. The process of using bacteria to chew up and detoxify pollutants is termed
    a. recycling
    b. bioremediation
    c. biofeedback
    d. detoxification

    ans. b.　diff. 7　obj. 6　p. 18

28. The fundamental unit of all living matter is the
    a. cell
    b. atom
    c. molecule
    d. organelle
    e. DNA

    ans. a.　diff. 4　obj. 7　p. 20

29. Before the 1930s, the fundamental unit of structure was thought to be the
    a. cell
    b. organism
    c. atom
    d. molecule

    ans. b.　diff. 6　obj. 7　p. 20

30. The cell theory was proposed in 1838 by
    a. Watson and Crick
    b. Schleiden and Schwann
    c. Pasteur and Tyndall
    d. Hooke and Jenner

    ans. b.　diff. 4　obj. 7　p. 20

31. The cells have one major function in life. It is to
    a. ingest nutrients
    b. degrade organic material
    c. metabolize nutrients
    d. reproduce themselves exactly

    ans. d.　diff. 6　obj. 8　p. 21

32. Most cells get their energy by
    a. reducing CO_2 to C and O
    b. recycling organic material
    c. decay
    d. degrading foodstuffs

    ans. d.　diff. 4　obj. 8　p. 21

33. The biological catalysts that are responsible for regulating the rate of chemical reactions of cells are
    a. hormones
    b. enzymes
    c. vitamins
    d. minerals

    ans. b.   diff. 5   obj. 8   p. 21

34. The molecule that stores genetic information in cells is
    a. RNA
    b. DNA
    c. ATP
    d. CP

    ans. b.   diff. 5   obj. 8   p. 22

35. The storage form of energy in cells is
    a. RNA
    b. DNA
    c. ATP
    d. CP

    ans. c.   diff. 5   obj. 8   p. 22

36. The cell types that lack a membrane-bound nucleus are the
    a. prokaryotes
    b. eukaryotes
    c. fungi
    d. yeasts

    ans. a.   diff. 5   obj. 8   p. 22

37. Bacteria are described structurally as
    a. Protista
    b. eukaryotes
    c. plants
    d. prokaryotes

    ans. d.   diff. 6   obj. 8   p. 23

38. The cells that have a true nucleus are
    a. eukaryotes
    b. prokaryotes
    c. blue-green algae
    d. viruses

    ans. a.   diff. 5   obj. 8   p. 23

39. A cell type that seems to have properties of both the prokaryotes and the eukaryotes is termed the
    a. *Archezoa*
    b. Protista
    c. monad
    d. virus

    ans. a.   diff. 6   obj. 9   p. 25

40. The property that all bacteria have in common is that they are all
    a. heterotrophs
    b. autotrophs
    c. prokaryotes
    d. aerobic

    ans. c.   diff. 7   obj. 9   p. 26

41. The differences between the eubacteria and the archaebacteria relate to fundamental differences in their
    a. actions
    b. physical appearance
    c. chemical make-up
    d. staining properties

    ans. c.   diff. 7   obj. 9   p. 27

42. Which of the following groups of organisms is thought to be more important in causing disease and more common in the environment?
    a. archaebacteria
    b. eubacteria
    c. viruses
    d. fungi
    e. protozoa

    ans. b.   diff. 8   obj. 9   p. 28

43. Which of the following is not considered a distinguishing feature of true bacteria?
    a. divide by binary fission
    b. rigid cell wall
    c. unicellular
    d. membrane-bound nucleus

    ans. d.   diff. 6   obj. 9   p. 29

44. A distinguishing characteristic of the archaebacteria is the ability to grow in
    a. humans only
    b. living cells
    c. extreme environmental conditions
    d. freezing temperatures

    ans. c.   diff. 7   obj. 10   p. 30

45. Some archaebacteria are commonly found in
    a. fossil fecal matter of early humanoids
    b. the ocean bottom mud
    c. the ionosphere
    d. boiling hot springs

    ans. d.   diff. 8   obj. 10   p. 31

46. Which of the following single-celled organisms are not considered part of the microbial world?
    a. worms
    b. yeasts
    c. bacteria
    d. protozoa

    ans. a.   diff. 5   obj. 10   p. 32

47. Which of the following members of the microbial world are considered as nonliving?
    a. bacteria
    b. viruses
    c. algae
    d. fungi

    ans. b.   diff. 5   obj. 10   p. 33

48. Viruses are often referred to as
    a. archaebacteria
    b. eubacteria
    c. infectious agents
    d. cellular agents

    ans. c.   diff. 5   obj. 10   p. 34

49. Viroids are
    a. smaller and simpler than viruses
    b. enclosed in an envelope
    c. more complex than viruses
    d. extracellular parasites

    ans. a.   diff. 6   obj. 10   p. 34

50. Both viruses and viroids are
    a. macroscopic particles
    b. capable of independent reproduction
    c. interdependent for metabolism
    d. obligate intracellular parasites

    ans. d.   diff. 8   obj. 10   p. 35

51. Infectious agents that appear to be proteins without any nucleic acid are
    a. viruses
    b. viroids
    c. prions
    d. archaebacteria

    ans. c.   diff. 8   obj. 10   p. 35

52. The system by which organisms are named is referred to as
    a. taxonomy
    b. system
    c. families
    d. nomenclature

    ans. d.   diff. 5   obj. 1   p. 36

53. The binomial system of nomenclature was devised by
    a. Carl von Linné
    b. James Watson
    c. Ernst Haeckel
    d. Aristotle

    ans. a.   diff. 5   obj. 11   p. 36

54. The scientific name of an organism includes its
    a. family and genus
    b. genus and species
    c. species and variety
    d. kingdom and family

    ans. b.   diff. 5   obj. 11   p. 37

55. A slight distinction in members of the same species is designated its
    a. genus
    b. variety
    c. family
    d. group

    ans. b.   diff. 6   obj. 11   p. 37

56. The reference text for bacteria is
    a. *Bergey's Manual of Systematic Bacteriology*
    b. Linnaeus Journal
    c. Group Systemic
    d. Organismic Listing

    ans. a.   diff. 5   obj. 11   p. 37

57. Rod-shaped bacteria are
    a. bacilli
    b. cocci
    c. spirals
    d. coccobacilli

    ans. a.   diff. 4   obj. 11   p. 38

58. The suffix "coccus" refers to organisms that are
    a. helical
    b. ovoid
    c. spherical
    d. curved

    ans. c.   diff. 4   obj. 11   p. 38

True/False

1. The theory of spontaneous generation was disproved about one hundred and fifty years ago by Tyndall.
   false (100 years by Pasteur)

2. The number of deaths from bubonic plague in the present era is approximately one thousand.
   false (100 deaths)

3. Bioremediation was used to clean up oil spills on the oil-soaked beaches of Alaska in 1989.
   true

4. The cell theory was proposed by Schleiden and Schwann in 1838.
   true

5. The molecule of genetics in all cells is RNA.
   false (DNA)

6. Proteins are composed of amino acids and vitamins.
   false (amino acids only)

7. Generally, eukaryotic cells are much larger than prokaryotic cells.
   true

8. Prokaryotic cells have mitochondria and chloroplasts.
   false (prokaryotes lack these)

9. The true bacteria are the archaebacteria.
   false (eubacteria)

10. The primitive bacteria are referred to as the archaebacteria.
    true

11. A virus is described as a piece of RNA without a protein coat.
    false (genetic material with a protein coat)

12. *Proteus vulgaris* is the scientific name of a Gram-negative rod.
    true

13. The scientific name of an organism is made up of its genus and strain.
    false (genus and species)

14. Binomial nomenclature was developed by Ernst Haeckel.
    false (Carl von Linné)

15. Spherical bacteria are commonly referred to as cocci.
    true

Chapter 2

Multiple Choice

1. Louis Pasteur, who was an important contributor to the development of microbiology, started his scientific career as a(n)
   a. astronomer
   b. physicist
   c. chemist
   d. biologist

   ans. c.   diff. 4   obj. 3   p. 2

2. Pasteur's observations and experiments with polarized light led him to conclude that
   a. molecules synthesized by living organisms have a preferred handedness
   b. mold growing in a solution destroyed some of the components in the solution
   c. isomers will not polarize light
   d. molecules synthesized by chemical means are right-handed

   ans. a.   diff. 7   obj. 3   p. 4

3. The chemistry of life is termed
   a. organic chemistry
   b. vivachemistry
   c. chemo-bioscience
   d. biochemistry

   ans. d.   diff. 5   obj. 1   p. 5

4. The pure substance that consists of a single type of atom is a(n)
   a. molecule
   b. element
   c. compound
   d. neutron
   e. enzyme

   ans. b.   diff. 4   obj. 1   p. 5

5. Four elements make up over 98% of all living material by weight. They are
   a. sulfur, phosphorus, carbon, oxygen
   b. carbon, hydrogen, oxygen, nitrogen
   c. carbon, nitrogen, oxygen, sulfur
   d. hydrogen, nitrogen, chlorine, iodine

   ans. b.   diff. 5   obj. 1   p. 5

6. The remaining 10% of all living material is made up of
   a. iron and zinc
   b. chlorine and bromine
   c. sulfur and phosphorus
   d. silicon and calcium

   ans. c.   diff. 5   obj. 1   p. 5

7. The percent of hydrogen in cells is approximately
   a. 49%
   b. 25%
   c. 0.5%
   d. 1%

   ans. a.   diff. 6   obj. 1   p. 6

8. Cells in bone have a high calcium content. Diatoms have a high _____ content.
   a. silicon
   b. sulfur
   c. helium
   d. phosphorus

   ans. a.   diff. 6   obj. 1   p. 6

9. The positively charged atomic particle is the
   a. proton
   b. electron
   c. quark
   d. neutron

   ans. a.   diff. 5   obj. 1   p. 7

10. The negatively charged atomic particle is the
    a. neutron
    b. proton
    c. isotope
    d. electron

    ans. d.   diff. 5   obj. 1   p. 7

11. The uncharged atomic particle is the
    a. electron
    b. positron
    c. neutron
    d. proton

    ans. c.   diff. 5   obj. 1   p. 7

12. The atom as a whole is uncharged because the number of
    a. electrons in the nucleus are too light to carry a charge
    b. neutrons in the nucleus are neutral
    c. protons equals the number of electrons
    d. nuclear particles are uncharged

    ans. c.   diff. 6   obj. 1   p. 7

13. If an atom has 11 electrons, they would be arranged
    a. 2, 7, 2
    b. 3, 8, 0
    c. 2, 6, 3
    d. 2, 8, 1

    ans. d.   diff. 4   obj. 1   p. 7

14. Molecules are formed when
    a. electrons change orbits
    b. electrons bond with neutrons
    c. compounds disassociate
    d. atoms bond with each other

    ans. d.   diff. 5   obj. 1   p. 7

15. The atomic number of an atom is equal to the number of its
    a. electrons
    b. neutrons
    c. electron orbits
    d. protons + neutrons

    ans. a.   diff. 4   obj. 1   p. 7

16. The atomic weight of an atom is equal to the
    a. sum of the protons and neutrons
    b. difference between the number of neutrons and protons
    c. number of electrons
    d. charges of the electrons

    ans. a.   diff. 5   obj. 1   p. 8

17. Different forms of the same elements that have different atomic weights are termed
    a. isomers
    b. isotopes
    c. polymers
    d. nucleics

    ans. b.   diff. 5   obj. 1   p. 8

18. Two or more atoms held together by chemical bonds are termed a(n)
    a. molecule
    b. unit
    c. bonding
    d. isotope

    ans. a.   diff. 5   obj. 1   p. 10

19. The molecular weight, in grams, of a molecule is termed a(n)
    a. sum of atomic numbers
    b. compound weight
    c. mole
    d. unit

    ans. c.   diff. 5   obj. 1   p. 10

20. The sum or the number of protons and neutrons that comprise a molecule is termed its
    a. atomic weight
    b. molecular weight
    c. radiation differential weight
    d. gram weight

    ans. b.   diff. 6   obj. 1   p. 10

21. Bonds that are formed when atoms share electrons are termed
    a. unstable
    b. ionic
    c. electrovalent
    d. covalent

    ans. d.   diff. 5   obj. 2   p. 11

22. Protein catalysts that are capable of breaking covalent bonds are
    a. hormones
    b. enzymes
    c. vitamins
    d. nucleic acids

    ans. b.   diff. 6   obj. 2   p. 12

23. If one atom has a much greater attraction for electrons than the other, the electrons are shared unequally. The resulting covalent bond is described as being
    a. polar
    b. nonpolar
    c. ionic
    d. molecular

    ans. a.   diff. 6   obj. 2   p. 13

24. If electrons are gained or lost in the formation of a bond, the bond is termed
    a. ionic
    b. covalent
    c. shared
    d. nonpolar

    ans. a.   diff. 6   obj. 2   p. 13

25. Ionic bonds are formed when electrons are exchanged between atoms. This is an example of
    a. atomic negativity
    b. sharing in a balanced formula
    c. extreme polarity
    d. organic molecule formation

    ans. c.   diff. 7   obj. 2   p. 13

26. Atoms that gain electrons are
    a. positively charged
    b. negatively charged
    c. heavier
    d. lighter

    ans. b.   diff. 6   obj. 2   p. 13

27. Atoms that give up electrons are
    a. heavier
    b. negatively charged
    c. positively charged
    d. lighter

    ans. c.   diff. 6   obj. 2   p. 13

28. Charged atoms are termed
    a. electrified
    b. ions
    c. anions
    d. active

    ans. b.   diff. 5   obj. 2   p. 14

29. The most common type of bond in inorganic molecules is
    a. neutral
    b. covalent
    c. hydrogen
    d. ionic

    ans. d.   diff. 6   obj. 2   p. 14

30. In aqueous solutions, which type of bond is weakened by the activity of water molecules?
    a. ionic
    b. covalent
    c. hydrogen
    d. neutral

    ans. a.   diff. 6   obj. 2   p. 14

31. The bonds responsible for the strength of marble are
    a. hydrogen
    b. ionic
    c. covalent
    d. neutral

    ans. b.   diff. 7   obj. 2   p. 14

32. What type of bonding is responsible for holding the two strands of DNA together?
    a. ionic
    b. hydrogen
    c. covalent
    d. neutral

    ans. b.   diff. 7   obj. 2   p. 14

33. Since molecules will always assume their most stable form, which type of bond will be formed if the opportunity arises?
    a. neutral
    b. hydrogen
    c. ionic
    d. covalent

    ans. d.   diff. 7   obj. 2   p. 16

34. The lock-and-key relationship describes hydrogen bonds that are closely held together. This occurs when the atoms of two molecules form
    a. complementary structures
    b. an inorganic molecule
    c. an enzyme
    d. a macromolecule

    ans. a.   diff. 8   obj. 2   p. 17

35. The percent of water in a living cell is approximately ___%.
    a. 40
    b. 50
    c. 60
    d. 90

    ans. d.   diff. 7   obj. 2   p. 18

36. Ions with a positive charge are termed
    a. reduced
    b. anions
    c. charged
    d. cations

    ans. d.   diff. 6   obj. 2   p. 19

37. Ions with a negative charge are termed
    a. anions
    b. cations
    c. charged
    d. oxidized

    ans. a.   diff. 6   obj. 2   p. 19

38. Carbon atoms covalently bonded to another carbon or hydrogen atom are termed
    a. organic
    b. inorganic
    c. polar
    d. nonpolar
    e. carboxyl
    ans. a.   diff. 5   obj. 2   p. 20

39. Covalent bonds are not found in which of the following biologically important groups?
    a. amino
    b. hydroxyl
    c. phosphate
    d. carboxyl

    ans. c.   diff. 7   obj. 3   p. 23

40. Pyruvate and acetate are common terms used interchangeably with
    a. the names of organic salts
    b. pyrimidine and aspartine
    c. pyruvic acid and acetic acid
    d. inorganic molecules with similar structures

    ans. c.   diff. 7   obj. 3   p. 23

41. Which of the following molecules is not included among the classes of biologically important macromolecules?
    a. nucleic acids
    b. proteins
    c. lipids
    d. polysaccharides

    ans. c.   diff. 7   obj. 3   p. 23

42. Macromolecules formed by the joining together of the same small molecules are termed
    a. biosynthetics
    b. subunits
    c. supraunits
    d. polymers

    ans. d.   diff. 6   obj. 3   p. 24

43. The subunits of proteins are
    a. fatty acids
    b. glycerols
    c. amino acids
    d. monomers enzymes
    e. nucleic acids
    ans. c.   diff. 4   obj. 4   p. 24

44. How many different amino acids exist in nature?
    a. 20
    b. 30
    c. 15
    d. unknown number

    ans. a.   diff. 5   obj. 4   p. 24

45. The process of joining subunits of a macromolecule involves a chemical reaction termed
    a. dehydration synthesis
    b. a hydrolytic reaction
    c. polymerization
    d. bioremediation

    ans. a.   diff. 6   obj. 4   p. 24

46. When a macromolecule is broken down, the chemical reaction is termed a
    a. dehydration reaction
    b. hydrolytic reaction
    c. reduction of molecules
    d. bioremediation

    ans. b.   diff. 6   obj. 4   p. 25

47. In either the synthesis or degradation of a large number of molecules, specific enzymes and what other molecule are always involved?
    a. hormones
    b. water
    c. vitamins
    d. proteins

    ans. b.   diff. 5   obj. 3   p. 25

48. If the side chains of amino acids contain carboxyl or amino groups, they readily form ions that
    a. give negative electric charges to the amino acids
    b. are termed basic amino acids
    c. will readily bond to lipids to form lipoproteins
    d. will readily be reduced to their components

    ans. a.   diff. 9   obj. 4   p. 26

49. If the side chains of amino acids contain the ammonium ion, they readily form ions that
    a. will be reduced to their substructures
    b. are described as acidic amino acids
    c. will readily bond to lipids to form lipoproteins
    d. give positive electric charges to the amino acid

    ans. d.   diff. 9   obj. 4   p. 26

50. If acidic amino acids contain carboxyl groups, then basic amino acids contain
    a. H^+
    b. COO^-
    c. OH^-
    d. NH_3^+

    ans. d.   diff. 8   obj. 4   p. 26

51. Amino acids that contain many methyl groups carry
    a. a positive charge
    b. no charge
    c. a negative charge
    d. reversible ends to the molecule

    ans. b.   diff. 9   obj. 4   p. 27

52. Molecules that are attracted to water (water loving) are termed
    a. hydrophilic
    b. acidic
    c. basic
    d. hydrophobic

    ans. a.   diff. 8   obj. 4   p. 27

53. L-amino acids occur in proteins and are designated
    a. right-handed
    b. unnatural
    c. rare
    c. natural

    ans. d.   diff. 8   obj. 4   p. 27

54. D-amino acids are associated with
    a. human proteins
    b. bacterial cell walls
    c. plant proteins
    d. radioactive isotopes

    ans. b.   diff. 8   obj. 4   p. 27

55. Amino acids are held together by peptide bonds between the
    a. carbon atoms of two adjacent amino acids
    b. carboxyl group of one amino acid and the amino group of another
    c. methyl group of one amino acid and an adjacent lipid
    d. nitrogen atom and the carboxyl ion

    ans. b.   diff. 8   obj. 4   p. 27

56. The average-sized protein consists of about _____ amino acids.
    a. 100
    b. 20
    c. 400
    d. 80

    ans. c.   diff. 6   obj. 4   p. 28

57. The most important feature of a protein is termed its
    a. primary structure
    b. shape
    c. bonding type
    d. electric charge

    ans. a.   diff. 6   obj. 4   p. 28

58. Which of the following is considered unimportant in maintaining the three-dimensional shape of a protein molecule?
    a. acidity of the solution
    b. temperature of the solution
    c. substrate acted upon
    d. binding of ions to the protein

    ans. c.   diff. 7   obj. 4   p. 29

59. If a protein molecule is exposed to an elevated temperature or the solution becomes very acidic or basic, then the protein may be
    a. activated
    b. denatured
    c. bonded to albumin
    d. capable of reversing the reaction

    ans. b.   diff. 7   obj. 4   p. 30

60. Proteins that consist of more than one polypeptide chain held together by weak bonds assume a specific shape termed its _____ structure.
    a. quaternary
    b. primary
    c. activated
    d. tertiary

    ans. a.   diff. 6   obj. 4   p. 30

61. Biologically important proteins covalently bonded to amino acids are called
    a. substituted proteins
    b. activated molecules
    c. enzymes
    d. organelles

    ans. a.   diff. 7   obj. 4   p. 31

62. Polymers of monosaccharide subunits are termed
    a. complexes
    b. polypeptides
    c. glycoproteins
    d. polysaccharides

    ans. d.   diff. 6   obj. 5   p. 31

63. Carbohydrates are compounds containing carbon, hydrogen, and oxygen in the ratio of
    a. 1:1:3
    b. 2:1:1
    c. 1:2:1
    d. 1:3:1

    ans. c.   diff. 5   obj. 5   p. 31

64. Monosaccharides are classified by
    a. their sweetness
    b. the number of carbon atoms they contain
    c. their dissolving factor
    d. the proportion of hydrogen to oxygen atoms

    ans. b.   diff. 5   obj. 5   p. 32

65. The sugars of the nucleic acids have ___ carbon atoms.
    a. 5
    b. 6
    c. 4
    d. 5 to 6

    ans. a.   diff. 4   obj. 5   p. 32

66. Which of the following sugars is not a hexose?
    a. sucrose
    b. glucose
    c. galactose
    d. mannose

    ans. a.   diff. 4   obj. 5   p. 32

67. Glucose and mannose contain aldehyde groups; fructose has a(n) _____ group.
    a. hydroxyl
    b. acid
    c. basic
    d. keto

    ans. d.   diff. 4   obj. 5   p. 33

68. Most monosaccharides in living organisms are of the
    a. a mixture of D- and L-configurations with L being more plentiful
    b. L-configuration
    c. D-configuration
    d. S-configuration

    ans. c.   diff. 5   obj. 5   p. 33

69. The most common disaccharides are
    a. maltose and sucrose
    b. fructose and hexose
    c. maltose and ribose
    d. lactose and sucrose

    ans. d.   diff. 6   obj. 5   p. 34

70. The linkage joining disaccharide monosaccharide sugars is termed a
    a. glycosidic linkage
    b. glycogen bond
    c. formed from a reaction between carboxyl groups
    d. hydrolysis reaction

    ans. a.   diff. 6   obj. 5   p. 34

71. The principal glucose polymer of plant cell walls is
    a. glycogen
    b. cellulose
    c. dextran
    d. chitin

    ans. b.   diff. 4   obj. 5   p. 34

72. The carbohydrate storage product of animals is
    a. glycogen
    b. cellulose
    c. dextran
    d. chitin

    ans. a.   diff. 4   obj. 5   p. 35

73. The subunits of nucleic acids are termed
    a. phosphates
    b. nucleotides
    c. deoxyriboses
    d. nitrogen bases

    ans. b.   diff. 4   obj. 6   p. 35

74. The purines of deoxyribonucleic acid are
    a. thymine and adenine
    b. cytosine and uracil
    c. adenine and guanine
    d. cytosine and guanine

    ans. c.   diff. 4   obj. 6   p. 36

75. The covalent bond between an alcohol and an acid is termed a(n) _____ linkage.
    a. keto
    b. aldehyde
    c. ester
    d. peptide

    ans. c.   diff. 6   obj. 6   p. 36

76. The DNA of a typical bacterium is arranged as a(n)
    a. single molecule
    b. series of plasmids
    c. complex of fragments
    d. single-stranded helix

    ans. a.   diff. 6   obj. 6   p. 37

77. Which of the following nitrogen bases is found in RNA but not DNA?
    a. guanine
    b. thymine
    c. adenine
    d. uracil

    ans. d.   diff. 5   obj. 6   p. 37

78. The characteristic common to all lipids is their
    a. solubility in organic solvents
    b. hydrophilic nature
    c. extraordinary size
    d. homogenicity

    ans. a.   diff. 6   obj. 7   p. 38

79. Fats are a combination of
    a. phosphorus and fatty acids
    b. glycerol and fatty acids
    c. sterol and lipid
    d. cholesterol and fatty acids

    ans. b.   diff. 5   obj. 7   p. 39

154

80. Sterols can rarely be formed in which of the following cell types?
    a. virus
    b. plant
    c. protozoa
    d. prokaryotic

    ans. d.   diff. 5   obj. 7   p. 40

81. Phospholipids are commonly referred to as
    a. compound lipids
    b. sterols
    c. inorganic lipids
    d. simple compounds

    ans. a.   diff. 4   obj. 7   p. 40

True/False

1. Louis Pasteur is known as the Father of Bacteriology.
   true

2. Compounds are pure substances that consist of a single type of molecule.
   true

3. Five elements make up over 98% of all living material by weight.
   false (four)

4. The basic chemical composition of all living cells is extremely dissimilar.
   false (similar)

5. The heaviest part of the atom is made up of the protons and neutrons.
   true

6. The isotope of hydrogen ^2H is called hydronium.
   false (deuterium)

7. Covalent bonds are highly unstable and can be broken at physiological temperatures.
   false (stable, only broken at very high temperatures)

8. When electrons are shared equally in covalent bonds, the bonds are termed polar.
   false (nonpolar)

9. Atoms that exchange electrons are bonded ionically.
   true

10. Covalent bonds are stronger in aqueous solutions than are ionic bonds.
    true

11. The most abundant molecule on earth is carbon dioxide.
    false (water)

12. Since hydrogen bonds are strongest at room temperature, very strong catalysts are needed to break them.
    false (bonds are weak and need no catalyst)

13. Complementary structures are responsible for the bonding between RNA and amino acids.
    false (between strands of DNA)

14. Lipids are considered among the smallest of the biologically important macromolecules.
    true

15. Amino acids are bonded together by protein bonds.
    false (peptide)

16. L-amino acids are termed natural and are found in many proteins.
    true

17. The specific shape of the protein is determined by the bonds that form between the amino acids of that protein.
    true

18. The primary structure of the protein determines its angle of curvature.
    false (three-dimensional shape)

19. Disaccharides are the macromolecular form of carbohydrates.
    false (polysaccharides)

20. The least common disaccharide is maltose.
    true

21. Lactose consists of glucose and galactose.
    true

22. The most abundant organic compound in the world is water.
    false (cellulose)

23. An ester linkage forms between an acid and a phosphate.
    false (acid and an alcohol)

24. The pyrimidine uracil is found in RNA but not DNA.
    true

25. Polyunsaturated fats tend to be liquid and are called oils.
    false (unsaturated)

26. In bacteria, phospholipids occur in the cell wall.
    false (cytoplasmic membrane)

Matching

1. bonds formed when electrons are exchanged
2. bonds between sodium and chloride in salt
3. bonds between C—C and C—H bonds
4. bonds between the carbon atoms in a sugar
5. bonds between nitrogen bases in DNA

a. covalent
b. ionic
c. hydrogen

1 - b; 2 - b; 3 - a; 4 - a; 5 - c

Match the molecule with its subunit(s):

1. nucleic acid
2. protein
3. carbohydrate
4. lipids

a. monosaccharide
b. amino acid
c. nucleotide
d. glycerol and fatty acids
e. minerals

1 - c; 2 - b; 3 - a; 4 - d

Match the molecule with its class:

1. enzymes
2. polymers of repetitive nitrogen bases, phosphate, and hexose sugar
3. polymers of monosaccharides
4. fats, steroids, and phospholipids

a. proteins
b. nucleic acid
c. lipids
d. carbohydrate

1 - a; 2 - b; 3 - d; 4 - c

Chapter 3

Multiple Choice

1. The antibiotic discovered by Alexander Fleming in 1929 is
   a. triple sulfa
   b. tetracycline
   c. penicillin
   d. ampicillin

   ans. c.   diff. 4   obj. 1   p. 2

2. The scientists that purified penicillin so that it could be tested were
   a. Fleming and Scott
   b. Florey and Chain
   c. Todaro and Huebner
   d. Watson and Crick

   ans. b.   diff. 5   obj. 1   p. 3

3. Which of the scientists listed here did not share in the Nobel Prize in 1945 for the pioneering studies of penicillin?
   a. Florey
   b. Fleming
   c. Todaro
   d. Chain

   ans. c.   diff. 5   obj. 1   p. 3

4. Definitive studies of prokaryotic cell structure are possible by studying them with
   a. the electron microscope
   b. biochemistry
   c. electrophoresis
   d. the light microscope

   ans. a.   diff. 5   obj. 1   p. 4

5. The microscope developed by van Leeuwenhoek is usually described as a(n) _____ microscope.
   a. compound
   b. simple
   c. electron
   d. objective

   ans. b.   diff. 4   obj. 1   p. 5

6. The two sets of magnifying lenses of the light microscope are
   a. basic and revolving
   b. eyepiece and high power
   c. simple and compound
   d. objective and ocular

   ans. d.   diff. 4   obj. 2   p. 5

7. Resolving power of a microscope is described as the ability of the microscope to
   a. see structures at various depths in a tissue
   b. magnify an object without distortion
   c. separate the colors of the organism's organelles
   d. separate clearly two objects that are very close together

   ans. d.   diff. 5   obj. 2   p. 6

8. The maximum resolving power of the best light microscope under optimal operating conditions is
   a. $0.2 \mu m$
   b. $1.0 \mu m$
   c. $2.0 \mu m$
   d. $0.8 \mu m$

   ans. a.   diff. 6   obj. 2   p. 6

9. The minimum distance that can exist between two points so that the points are observed as separate entities is
   a. working distance
   b. resolving power
   c. magnification
   d. aperture

   ans. b.   diff. 5   obj. 2   p. 6

10. The resolving power of the light microscope can be increased by using _____ between the specimen being viewed and the objective lens.
    a. water
    b. oil
    c. air
    d. a cover slide

    ans. b.   diff. 5   obj. 2   p. 7

157

11. The microscope most commonly used for observing living microorganisms is the _____ microscope.
   a. electron
   b. fluorescence
   c. compound
   d. phase contrast

   ans. d.   diff. 5   obj. 2   p. 7

12. The microscope that allows the specimen to appear three-dimensional is the _____ microscope.
   a. fluorescence
   b. phase contrast
   c. interference
   d. dark-field

   ans. c.   diff. 6   obj. 2   p. 8

13. Which of the following microscopes can be used for viewing *Treponema pallidum?*
   a. light
   b. fluorescence
   c. electron
   d. dark-field

   ans. d.   diff. 6   obj. 2   p. 9

14. Which of the following microscope types is not ordinarily used to view living cells?
   a. light
   b. dark-field
   c. phase contrast
   d. interference

   ans. a.   diff. 6   obj. 2   p. 9

15. The microscope constructed by Knoll and Hruska in Berlin in 1931 is the _____ microscope.
   a. scanning electron
   b. transmission electron
   c. phase contrast
   d. dark-field

   ans. b.   diff. 6   obj. 2   p. 10

16. Fluorescence microscopy involves the use of _____ and _____ to view objects.
   a. electrons, vacuum
   b. fluorochrome, ultraviolet light
   c. stain, visible light
   d. light waves, oil

   ans. b.   diff. 6   obj. 2   p. 10

17. The electron microscope allows both magnification and resolving power to be greater than with the light microscope. The resolving power is increased _____ times.
   a. 50
   b. 100
   c. 1,000
   d. 5,000

   ans. c.   diff. 7   obj. 2   p. 10

18. The resolving power of the electron microscope is about
   a. 0.3 nm
   b. 0.8 μm
   c. 1.0 nm
   d. 0.5 μm

   ans. a.   diff. 7   obj. 2   p. 10

19. Specimens to be viewed by the transmission electron microscope are sliced in very thin sections by
   a. a glass knife
   b. freeze drying
   c. staining and smashing
   d. drying

   ans. a.   diff. 7   obj. 2   p. 11

20. An excellent method of observing both external and internal structures in a specimen with the TEM is by using a preparation procedure termed
   a. freeze-etching
   b. a glass knife
   c. freeze drying
   d. fluorescence drying

   ans. a.   diff. 7   obj. 2   p. 11

21. A three-dimensional view of an object with excellent viewing of surface features is allowed by the
   a. light microscope
   b. transmission electron microscope
   c. scanning electron microscope
   d. phase contrast microscope

   ans. c.   diff. 6   obj. 2   p. 12

22. Individual atoms on the surface of prepared samples can be observed by using the
   a. transmission electron microscope
   b. scanning electron microscope
   c. scanning tunneling microscope
   d. phase contrast microscope

   ans. c.   diff. 6   obj. 2   p. 13

23. Which two of the following scientists received the Nobel Prize in Physics for creation of the scanning tunneling microscope?
   a. Todaro and Fleming
   b. Florey and Chain
   c. Binning and Rohrer
   d. Scott and Hueber

   ans. c.   diff. 6   obj. 2   p. 13

24. How are organisms to be stained fixed to the glass slide?
   a. ice
   b. heat
   c. glue
   d. tape

   ans. b.   diff. 4   obj. 3   p. 13

25. Stains that have a strong attraction for cell components and stain them on the microscope slide are referred to as _____ stains.
   a. negative
   b. positive
   c. capsule
   d. acid-fast

   ans. b.   diff. 6   obj. 3   p. 14

26. The simple staining procedure frequently utilizes _____ as the dye.
   a. methylene blue
   b. Sudan black
   c. iodine
   d. nigrosin

   ans. a.   diff. 6   obj. 3   p. 15

27. Basic dyes have _____ charges.
   a. electrically neutral
   b. negative
   c. positive
   d. either

   ans. c.   diff. 6   obj. 3   p. 15

28. The most frequently used differential stains are the
   a. acid-fast and capsule stains
   b. positive and negative stains
   c. capsule and negative stains
   d. Gram and acid-fast stains

   ans. d.   diff. 5   obj. 3   p. 16

29. The order of reagents in the Gram stain reaction are
   a. crystal violet, iodine, alcohol, safranin
   b. safranin, alcohol, methylene blue, iodine
   c. methylene blue, alcohol, safranin
   d. crystal violet, alcohol, iodine, safranin

   ans. a   diff. 6   obj. 3   p. 16

30. Organisms that absorb crystal violet and iodine, and are not easily decolorized by alcohol are referred to as
   a. pathogenic
   b. Gram-negative
   c. acid-fast
   d. Gram-positive

   ans. d.   diff. 6   obj. 3   p. 17

31. The major criteria used in placing bacteria into different groups is based on differences in
   a. cell wall structure
   b. cell membrane permeability
   c. presence or absence of flagella
   d. detergent susceptibility

   ans. a.   diff. 6   obj. 3   p. 17

32. Organisms are acid-fast because of a unique lipid in their cell walls that binds to the basic fuchsin. This lipid is
    a. teichoic acid
    b. mycolic acid
    c. lipopolysaccharide
    d. cholesterol

    ans. b.   diff. 6   obj. 3   p. 18

33. Most bacteria have three shapes. They are
    a. rods, spheres, and spirals
    b. squares, coccobacilli, and spirochetes
    c. cocci, spheres, and bacilli
    d. rods, bacilli, and spirals

    ans. a.   diff. 4   obj. 4   p. 20

34. The process by which prokaryotic cells divide into two identical cells is
    a. mitosis
    b. binary fission
    c. fusion
    d. splitting

    ans. b.   diff. 4   obj. 4   p. 20

35. A typical prokaryotic cell lacks which of the following structures?
    a. cell material
    b. rigid cell wall
    c. flagella
    d. nuclear membrane

    ans. d.   diff. 5   obj. 4   p. 22

36. Which of the following cell outer structures is not essential to the life of the cell?
    a. plasma membrane
    b. cell wall
    c. cytoplasmic membrane
    d. capsule

    ans. d.   diff. 6   obj. 4   p. 23

37. The cytoplasmic membrane, the cell wall, and possibly the capsule collectively comprise the cell's
    a. transduction apparatus
    b. inner structures
    c. genetic transfer membranes
    d. envelope

    ans. d.   diff. 6   obj. 4   p. 23

38. Structures produced only by certain bacteria and often only when grown with special nutrients are termed the
    a. glycocalyx
    b. cell wall
    c. envelope
    d. membrane

    ans. a.   diff. 6   obj. 4   p. 24

39. Which of the following is not a function of capsules?
    a. alter Gram stain reaction
    b. attach to plant roots
    c. avoid phagocytosis
    d. increase virulence

    ans. a.   diff. 7   obj. 4   p. 25

40. The macromolecule found in the cell walls of all eubacteria is
    a. glycocalyx
    b. peptidoglycan
    c. peptides
    d. diaminopimelic acid

    ans. b.   diff. 8   obj. 5   p.27

41. What makes the cell walls of Gram-negative bacteria weaker than those of Gram-positive bacteria?
    a. the periplasmic space
    b. cellulose
    c. the amino acid bridges
    d. the single peptidoglycan layer

    ans. d.   diff. 7   obj. 5   p. 28

42. Which of the following molecules is lacking in Gram-negative bacterial cell walls?
    a. D-amino acids
    b. lipoprotein
    c. lipopolysaccharide
    d. phospholipid

    ans. a.   diff. 6   obj. 5   p. 29

43. The archaebacteria that lack peptidoglycan in their cell wall are the
    a. *Mycoplasma*
    b. thermophiles
    c. *Chlamydia*
    d. halophiles

    ans. d.   diff. 7   obj. 5   p. 30

44. Which of the following eubacteria lack a cell wall?
    a. *Mycobacterium leprae*
    b. *Treponema pallidum*
    c. *Mycoplasma pneumoniae*
    d. *Klebsiella pneumoniae*

    ans. c.   diff. 7   obj. 5   p. 31

45. Which of the following chemicals damage the cell walls of old cultures, causing Gram-positive organisms to stain Gram-negative?
    a. autolysins
    b. lipids
    c. carbohydrates
    d. phospholipids

    ans. a.   diff. 8   obj. 5   p. 33

46. The primary component of the cell walls of eukaryotic organisms is
    a. glycocalyx
    b. peptidoglycan
    c. cellulose
    d. lipoprotein

    ans. c.   diff. 3   obj. 6   p. 34

47. The cytoplasmic membrane of both prokaryotic and eukaryotic cells functions to
    a. form endoplasmic reticulum
    b. produce energy
    c. regulate movement of molecules that enter and leave the cell
    d. form lysosomes and Golgi apparatus

    ans. c.   diff. 7   obj. 6   p. 36

48. Which of the following mechanisms require cellular energy?
    a. osmosis
    b. active transport
    c. diffusion
    d. facilitated diffusion

    ans. b.   diff. 5   obj. 5   p. 38

49. Bacteria use membrane transport proteins to carry small molecules into the cell. These molecules are called
    a. permeases
    b. lipoproteins
    c. endases
    d. vitamins

    ans. a.   diff. 4   obj. 5   p. 39

50. The system bacteria use for the rapid transport of nutrients into the cell is
    a. active transport
    b. facilitated diffusion
    c. exocytosis
    d. osmosis

    ans. a.   diff. 5   obj. 5   p. 40

51. Which of the following is not found in all prokaryotic cells?
    a. vacuole
    b. cell membrane
    c. genome
    d. cell wall

    ans. d.   diff. 5   obj. 5   p. 43

52. The chemicals frequently found in the plasma membrane of eukaryotes that contribute to the ability of these membranes to resist osmotic lysis are
    a. vitamins
    b. enzymes
    c. sterols
    d. receptors

    ans. c.   diff. 6   obj. 6   p. 44

53. Which of the following structures of eukaryotic cells is responsible for the addition of lipids to proteins to form lipoproteins?
    a. mitochondria
    b. endoplasmic reticulum
    c. ribosomes
    d. Golgi apparatus

    ans. d.   diff. 5   obj. 6   p. 45

54. The eukaryotic cellular organelles responsible for the generation of energy are the
   a. Golgi apparatus
   b. mitochondria
   c. ribosomes
   d. lysosomes

   ans. b.   diff. 5   obj. 6   p. 45

55. The eukaryotic cellular organelle concerned with the generation of energy in organisms that use light in the process of photosynthesis is the
   a. chloroplast
   b. mitochondria
   c. ribosomes
   d. endoplasmic reticulum

   ans. a.   diff. 5   obj. 6   p. 46

56. The structures responsible for motility in both prokaryotes and eukaryotes are
   a. cilia
   b. flagella
   c. pili
   d. pseudopods

   ans. b.   diff. 5   obj. 6   p. 47

57. Basal body, hook, and filament are all parts of the cellular extensions referred to as the
   a. pili
   b. cilia
   c. flagella
   d. pseudopods

   ans. c.   diff. 7   obj. 6   p. 47

58. The bacteria that have tufts of flagella around the cell are referred to as
   a. anatrichous
   b. peritrichous
   c. monotrichous
   d. lophotrichous

   ans. d.   diff. 6   obj. 6   p. 48

59. The process by which bacteria are attracted to nutrients in their environment is referred to as
   a. phototaxis
   b. negative chemotaxis
   c. positive chemotaxis
   d. aerotaxis

   ans. c.   diff. 5   obj. 6   p. 49

60. The mechanism bacteria used to move toward or away from signals in the environment is
   a. related to oxygen needs as related to metabolism
   b. positive chemotaxis
   c. a primitive form of the nervous system in animals
   d. a well-developed system of territorialism

   ans. c.   diff. 7   obj. 6   p. 51

61. Protein appendages of attachment are
   a. adhesions
   b. pili
   c. glycocalyces
   d. flagella

   ans. b.   diff. 6   obj. 6   p. 53

62. Which of the following statements cannot be attributed as related to the structure and/or function of pili?
   a. have adhesins attached at the tip or along the length
   b. attach to a wide variety of surfaces
   c. can be removed by shaking the cells in a blender
   d. sex pili are longer than other attached pili

   ans. b.   diff. 8   obj. 6   p. 53

63. Typical prokaryotic cells have their genetic material in the form of
   a. a single double-stranded molecule of DNA
   b. a single strand of DNA
   c. pairs (23) of chromosomes
   d. several DNA fragments called the genome

   ans. a.   diff. 5   obj. 8   p. 55

64. Extra chromosomal DNA in prokaryotes is found in
   a. mitochondria
   b. RNA
   c. plasmids
   d. histones

   ans. c.   diff. 6   obj. 8   p. 56

65. When the eukaryotic cells are not actively dividing, the DNA occurs in long, threadlike strands. Each strand is wound around basic proteins called
    a. microtubules
    b. chromatin
    c. nucleoli
    d. histones

    ans. d.   diff. 6   obj. 8   p. 59

66. The dense, spherical structure found within the nucleus that is important in the synthesis of ribosomes is the
    a. nucleolus
    b. histone
    c. chromatin
    d. lysosome

    ans. a.   diff. 5   obj. 8   p. 60

67. Ribosomes in both prokaryotic and eukaryotic cells are responsible for
    a. energy production
    c. protein synthesis
    c. replication of genetic material
    d. internal transportation

    ans. b.   diff. 5   obj. 8   p. 61

68. Bacteria store glycogen granules as sources of both
    a. carbon and energy
    b. enzymes and hormones
    c. sulfur and nucleotides
    d. enzymes and nucleotides

    ans. a.   diff. 6   obj. 9   p. 64

69. Two genera of bacteria, the *Bacillus* and the *Clostridium,* are capable of surviving in environments low in essential nutrients due to the production of
    a. capsules
    b. endospores
    c. plasmids
    d. volutin

    ans. b.   diff. 7   obj. 9   p. 64

70. Phosphates are stored as cellular inclusions in several genera of bacteria as
    a. chromatin
    b. endospores
    c. plasmids
    d. volutin

    ans. d.   diff. 6   obj. 9   p. 65

71. Which of the following environmental factors will not stimulate endospore formation?
    a. freezing
    b. mesophilic temperatures
    c. aridity
    d. boiling

    ans. b.   diff. 5   obj. 9   p. 67

72. An important component of the endospore that aids in its resistance to destruction is
    a. chromatin
    b. dipicolinic acid
    c. microtubules
    d. histone

    ans. b.   diff. 8   obj. 9   p. 68

Matching

1. streptococci
2. diplococci
3. staphylococci
4. tetrads

   a. chains of cocci
   b. clusters of cocci
   c. two cells joined together
   d. packets of four or cubes

1 - a; 2 - c; 3 - b; 4 - d

1. mitochondria
2. ribosomes
3. lysosomes
4. cell membrane

   a. regulates movement of molecules into and out of cell
   b. protein synthesis
   c. energy production
   d. internal digestion

1 - c; 2 - b; 3 - d; 4 - a

True/False

1. The surface structures of bacteria determine the staining properties when the simple stain is used.
   false (Gram or acid-fast stain)

2. Organisms that are stained with a basic red dye and not easily decolorized with an acidic solution of alcohol are said to be acid-fast.
   true

3. Starlike bacteria have flagella on their surface.
   false (prosthecae)

4. The major cause of dental caries is *Streptococcus pyogenes*.
   false (*Streptococcus mutans*)

5. Cell membranes are found only in prokaryotes, not in eukaryotes.
   false (found in all cells)

6. The cell wall functions to hold the cell together; without the wall, the cell would burst under the environmental conditions in which bacteria grow.
   true

7. The cell membranes of *Mycoplasma pneumoniae* contain sterols, which strengthen the cell membrane.
   true

8. Penicillin is toxic to bacteria because it affects the synthesis of the lipopolysaccharides portion of the cell wall.
   false (peptidoglycan)

9. Penicillin is an extremely effective antibiotic to Gram-negative bacteria.
   false (Gram-positive)

10. The small molecules that can pass through the cell membrane freely are hydrophobic.
    true

11. When using the active transport mechanism, the cell moves molecules until equilibrium is reached.
    false (until they are concentrated inside the cell)

12. The enzyme concerned with generating energy in eukaryotic cells is bound to the cytoplasmic membrane.
    false (prokaryotic cells)

13. Flagella are responsible for movement in prokaryotes, and cilia are responsible for movement in eukaryotes.
    false (flagella in both)

14. When bacteria flagella rotate counterclockwise, they fly apart and swimming time increases.
    false (flagella are in a tight bundle)

15. Motile bacteria that react to the earth's magnetic field contain magnetic particles such that cells line up in a north-south direction.
    true

16. Pili attach to specific receptors for transfer of genetic material.
    true

17. The chromosomal material of the prokaryote makes up about 25 percent of the total cell volume.
    false (10 percent)

18. The nucleolus of the eukaryotic cell is responsible for the synthesis of chromatin.
    false (ribosomes)

19. The structures responsible for protein synthesis are the mitochondria.
    false (ribosomes)

20. Endospores are exceptionally efficient reproductive structures of the bacterial genera *Bacillus* and *Clostridium*.
    false (protective)

21. Botulism food poisoning is due to ingestion of the endospores of *Clostridium botulinum*.
    false (toxins)

Chapter 4

Multiple Choice

1. The first person to demonstrate that bacteria could be responsible for changing one substance into another was
   a. Jenner
   b. Koch
   c. Leeuwenhoek
   d. Pasteur

   ans. d.   diff. 4   obj. 1   p. 2

2. If all the bacterial cells that result from the replication of a single organism are identical, the culture in which these organisms are found is said to be a(n)
   a. pure culture
   b. population
   c. lag stage
   d. mutation

   ans. a.   diff. 4   obj. 1   p. 3

3. Which of the following represents the most significant bacteriological advance of the events given here?
   a. pure culture technique
   b. polio vaccine
   c. Gram stain
   d. electron microscope

   ans. a.   diff. 6   obj. 1   p. 4

4. The physician who contributed the most to pure culture methods was
   a. Louis Pasteur
   b. Robert Koch
   c. Edward Jenner
   d. Alexander Fleming

   ans. b.   diff. 5   obj. 1   p. 4

5. In order for bacteria to be visible to the human eye, about how many cells must be in a colony?
   a. 10×10^2
   b. 1×10^4
   c. 1×10^6
   d. 1×10^3

   ans. c.   diff. 5   obj. 1   p. 4

6. The simplest technique for isolating bacteria in a growth media is referred to as the
   a. streak-plate method
   b. pour-plate method
   c. divided plate method
   d. serial dilution method

   ans. a.   diff. 6   obj. 1   p. 5

7. The solidifying agent used most successfully in bacterial nutrient media is gelatin .
   a. heat
   b. gelatin
   c. agar
   d. cold

   ans. c.   diff. 5   obj. 1   pl 6

8. Julius Petri invented a dish that bears his name. It functions well to allow the organisms to grow yet
   a. prevents the organisms from spreading
   b. keeps the population low
   c. keeps the media free of contaminants
   d. provides for easy distribution of the colonies

   ans. c.   diff. 6   obj. 1   p. 6

9. Development of the pure culture techniques provided the means for
   a. disproving spontaneous generation
   b. accurate identification of organisms by their colony shape
   c. differentiation of organisms based on nutrition
   d. rapid isolation of pathogens

   ans. d.   diff. 6   obj. 1   p. 6

10. The orderly increase in the quantity of bacterial cell components is referred to as
    a. reproduction
    b. growth
    c. fission
    d. isolation

    ans. b.   diff. 5   obj. 2   p. 7

11. Prokaryotic cells divide by a process known as
    a. mitosis
    b. binary fission
    c. fusion
    d. conjugation

    ans. b.   diff. 4   obj. 2   p. 7

12. A synonym for generation time is
    a. streak time
    b. fission time
    c. growth time
    d. doubling time

    ans. d.   diff. 4   obj. 2   p. 7

13. In a rapidly multiplying bacterial population, cell numbers increase exponentially. The phase of the growth cycle is referred to as the _____ phase.
    a. log
    b. lag
    c. stationary
    d. decline

    ans. a.   diff. 5   obj. 2   p. 8

14. If a given microorganism divides every thirty minutes, a single cell will result in a population of about _____ cells in ten hours.
    a. 10,000
    b. 10×10^6
    c. 1×10^8
    d. 1×10^6

    ans. d.   diff. 6   obj. 2   p. 9

15. In a rapidly multiplying bacterial population, cell numbers increase
    a. indirectly
    b. arithmetically
    c. logarithmically
    d. as related to size

    ans. c.   diff. 5   obj. 2   p. 9

16. Which of the following methods cannot be successfully employed to measure bacterial cell number?
    a. plate count
    b. turbidity
    c. direct microscopic count
    d. dilution and plating

    ans. b.   diff. 6   obj. 2   p. 10

17. The number of colonies that appear after a period of incubation represents
    a. the number of living cells that result from the original inoculation
    b. the surviving mutants in the inoculation
    c. the dilution factor used in the original inoculation
    d. the living and dead organisms in the culture

    ans. a.   diff. 6   obj. 2   p. 10

18. Species of organisms that do not live under very harsh environmental conditions are referred to as
    a. psychrophiles
    b. mesomorphs
    c. extremophiles
    d. pathogens

    ans. c.   diff. 5   obj. 3   p. 12

19. Which of the following conditions would not be considered a harsh environment for the growth of extremophiles?
    a. Lake Tahoe
    b. Dead Sea
    c. Yellowstone hot springs
    d. Great Salt Lake

    ans. a.   diff. 6   obj. 3   p. 12

20. The two factors that must be considered when determining the microbial growth conditions are
    a. nutrition and growth
    b. genetics and mutants
    c. environment and nutrition
    d. lag and log phases of growth

    ans. c.   diff. 6   obj. 3   p. 13

21. The temperature range of the majority of bacteria is approximately
    a. 10°C to 20°C
    b. 35°C to 40°C
    c. 20°C to 30°C
    d. 20°C to 50°C

    ans. d.   diff. 6   obj. 3   p. 13

22. The speed of enzymatic reaction in the cell approximately doubles for each _____ rise in temperature.
    a. 12°C
    b. 5°C
    c. 10°C
    d. 6°C

    ans. c.   diff. 5   obj. 3   p. 14

23. Organisms that have their optimum growth temperatures between −5° and 20°C are
    a. psychrophiles
    b. extreme psychrophiles
    c. mesophiles
    d. range requirements

    ans. a.   diff. 5   obj. 3   p. 14

24. Mesophiles have their optimum growth range of
    a. 0°C to 10°C
    b. 20°C to 50°C
    c. 10°C to 35°C
    d. 50°C to 80°C

    ans. b.   diff. 6   obj. 3   p. 14

25. Organisms that have their optimum growth temperatures between _____ are thermophiles.
    a. 50°C and 80°C
    b. 35°C and 40°C
    c. -5°C and 20°C
    d. 20°C and 50°C

    ans. a.   diff. 6   obj. 3   p. 14

26. Extreme thermophiles have an optimum growth temperature of above
    a. 100°C
    b. 90°C
    c. 80°C
    d. 70°C

    ans. c.   diff. 6   obj. 3   p. 14

27. The optimal temperature for most human pathogens is between
    a. 30°C and 45°C
    b. 20°C and 40°c
    c. 35°C and 55°C
    d. 35°C and 40°C

    ans. d.   diff. 7   obj. 3   p. 14

28. Generally speaking, the protein molecules of thermophiles are not denatured at high temperatures. This property is described as
    a. endurability
    b. chemostasicity
    c. thermostability
    d. duricity

    ans. c.   diff. 7   obj. 3   p. 15

29. The growth of microorganisms that cause food spoilage can generally be retarded by storing food at a temperature of
    a. 15°C
    b. 25°C
    c. 20°C
    d. 30°C

    ans. a.   diff. 6   obj. 3   p. 16

30. *Mycobacterium leprae* preferentially involves the ears, toes, and fingers due to its
    a. need for cooler body temperature in order to grow
    b. easier access to these parts
    c. need for blood available in these parts
    d. long incubation period

    ans. a.   diff. 5   obj. 3   p. 17

31. Organisms that require gaseous oxygen for metabolism are referred to as
    a. microaerophiles
    b. facultative aerobes
    c. facultative anaerobes
    d. obligate aerobes

    ans. d.   diff. 5   obj. 3   p. 17

32. Members of the obligate aerobe group include the genera
    a. *Staphylococcus* and *Streptococcus*
    b. *Escherichia* and *Brucella*
    c. *Enterobacter* and *Bacillus*
    d. *Bacillus* and *Pseudomonas*

    ans. d.   diff. 6   obj. 3   p. 17

33. Organisms that become metabolically inactive in the presence of oxygen are
    a. microaerophiles
    b. facultative anaerobes
    c. aerobes
    d. obligate anaerobes

    ans. d.   diff. 7   obj. 3   p. 17

34. Which of the following genera of organisms is an anaerobe?
    a. *Clostridium*
    b. *Escherichia*
    c. *Staphylococcus*
    d. *Halobacter*

    ans. a.   diff. 6   obj. 3   p. 18

35. Organisms that are said to be indifferent to the presence of oxygen are known as
    a. aerobes
    b. anaerobes
    c. aerotolerant
    d. microaerophiles

    ans. c.   diff. 6   obj. 3   p. 18

36. Which of the following pathogens is classified as aerotolerant?
    a. *Staphylococcus aureus*
    b. *Bacillus subtilis*
    c. *Clostridium rubrum*
    d. *Streptococcus pyogenes*

    ans. d.   diff. 6   obj. 3   p. 18

37. The optimum pH for most species of bacteria to grow is
    a. 6
    b. 7
    c. 8
    d. 6.5

    ans. b.   diff. 5   obj. 4   p. 21

38. Yeasts and mold grow most rapidly in a(n) _____ medium.
    a. acid
    b. basic
    c. neutral
    d. warm

    ans. a.   diff. 4   obj. 4   p. 21

39. Products that limit pH changes by combining chemically with H^+ ions of strong acids and the OH^- ions of strong bases are referred to as
    a. hormones
    b. enzymes
    c. buffers
    d. cytoplasmic granules

    ans. c.   diff. 5   obj. 4   p. 22

40. The pressure required to prevent the net flow of water across a semipermeable membrane is defined as the _____ pressure.
    a. hydrostatic
    b. osmotic
    c. permeability
    d. diffusion

    ans. b.   diff. 5   obj. 4   p. 23

41. If the concentration of NaCl increases in the medium on the outside of the cell, the osmotic pressure increases and the net volume of the water in the cell
    a. decreases
    b. increases
    c. stays the same
    d. varies from high to low

    ans. a.   diff. 6   obj. 4   p. 24

42. Using high salt or sugar concentrations in food preservation inhibits the growth of bacteria due to the _____ osmotic pressure surrounding the cells.
    a. high
    b. low
    c. variable
    d. equal

    ans. a.   diff. 5   obj. 4   p. 24

43. The prefix "photo" indicates that organisms use the energy of _____ to synthesize nutrients.
    a. chemicals
    b. light
    c. organics
    d. synthetics

    ans. b.   diff. 3   obj. 4   p. 26

44. Organisms that use organic molecules as their source of carbon are called
    a. autotrophs
    b. heterotrophs
    c. organotrophs
    d. chemoheterotrophs

    ans. b.   diff. 6   obj. 4   p. 26

45. Some organisms have the word "_____" appended to their nutritional description, indicating that they are flexible in their nutritional capabilities.
    a. micro
    b. limited
    c. obligate
    d. facultative

    ans. d.   diff. 5   obj. 4   p. 27

46. Small organic molecules that bacteria must be provided with in order to grow are referred to as
    a. vitamins
    b. minerals
    c. water
    d. growth factors

    ans. d.   diff. 5   obj. 4   p. 29

47. Examples of organisms that lack some of the enzymes as structures necessary for independent existence and can only multiply inside living cells are represented by which two genera?
    a. *Staphylococcus* and *Streptococcus*
    b. *Bacillus* and *Clostridium*
    c. *Chlamydia* and *Rickettsia*
    d. *Escherichia* and *Enterobacter*

    ans. c.   diff. 6   obj. 5   p. 29

48. Microbiological media that contain measured amounts of pure chemical compounds are referred to as _____.
    a. complex
    b. synthetic
    c. organic
    d. inorganic

    ans. b.   diff. 6   obj. 5   p. 31

49. Organisms that can synthesize all of their cellular constituents from glucose and inorganic salts are said to grow on _____ media.
    a. enriched
    b. synthetic
    c. complex
    d. simple

    ans. b.   diff. 7   obj. 5   p. 31

50. Organisms that require specific growth factors can be provided by the host to the detriment of that host are described as
    a. tolerant
    b. fastidious
    c. complex
    d. pathogens

    ans. d.   diff. 6   obj. 5   p. 34

51. Only one of the following organisms has never been cultivated in the laboratory in the absence of living cells. It is
    a. *Neisseria gonorrhoeae*
    b. *Halobacterium halobium*
    c. *Rickettsia rickettsii*
    d. *Micrococcus lutea*

    ans. c.   diff. 6   obj. 6   p. 36

52. A medium that enhances the growth of one particular organism in a mixed population is known as a(n)
    a. complex medium
    b. specific culture
    c. enrichment medium
    d. synthetic medium

    ans. c.   diff. 6   obj. 5   p. 38

53. A selective medium is used to
    a. enrich another medium to enhance growth of pathogens
    b. isolate a nutritionally demanding organism
    c. differentiate pathogens from nonpathogens
    d. change a synthetic medium to a complex medium

    ans. b.   diff. 7   obj. 6   p. 39

54. A bacterial medium able to differentiate between two different species based on differences in their growth patterns is described as a(n) _____ medium.
    a. complex
    b. selective
    c. enrichment
    d. indicator

    ans. d.   diff. 6   obj. 6   p. 40

55. Which of the following is mismatched?
    a. blood agar - selective medium
    b. thioglycolate media - obligate anaerobes
    c. chocolate agar - enriched medium
    d. living cells - viruses

    ans. b.   diff. 7   obj. 6   p. 41

56. In which phase of the bacterial growth curve do the number of viable cells not increase significantly?
    a. lag
    b. log
    c. stationary
    d. death

    ans. a.   diff. 6   obj. 7   p. 43

57. The phase of the bacterial growth curve in which the total number of viable cells decreases is the _____ phase.
    a. log
    b. lag
    c. death
    d. stationary

    ans. c.   diff. 6   obj. 7   p. 43

58. Which of the following is not mismatched?
    a. lag phase - low multiplication, adaptation
    b. stationary phase - exponential growth
    c. death phase - number die equals number produced
    d. log phase - insignificant increase in population

    ans. a.   diff. 6   obj. 7   p. 43

59. The growth phase during which the organisms are most susceptible to antibiotics is the _____ phase.
    a. decline
    b. lag
    c. stationary
    d. exponential

    ans. d.   diff. 6   obj. 7   p. 44

60. The growth of a bacterial colony results from
    a. an increase in cell number toward the center of the colony
    b. cells multiplying at the outer edges of the colony
    c. cells increasing under previously formed cells
    d. an increase in cell size within the colony

    ans. b.   diff. 7   obj. 7   p. 47

61. When bacteria are grown in the laboratory, the system is referred to as a _____ system.
    a. open
    b. closed
    c. artificial
    d. manufactured

    ans. b.   diff. 6   obj. 7   p. 48

62. The devise used to simulate the open system found in nature is called a
    a. selector
    b. closure
    c. chemostat
    d. generator

    ans. c.   diff. 6   obj. 7   p. 49

True/False

1. It is impossible for bacteria to grow in temperatures above 100°C.
   false (it's possible)

2. The person who contributed most to pure culture methods was Louis Pasteur.
   false (Robert Koch)

3. Robert Koch's wife suggested the use of agar to solidify nutrient media for bacteria.
   false (Frau Hesse suggested agar)

4. Binary fission is the process by which prokaryotic cells ordinarily divide.
true

5. The special glass slides that hold a specific volume of liquid and are used for counting bacteria are called concave slides.
false (counting chambers)

6. The two broad categories that must be considered when discussing microbial growth factors are nutrition and genetics.
false (nutrition and environment)

7. The optimum temperature for bacterial growth is close to the lower limit of its range.
false (upper limit)

8. Thermophiles are usually members of the archaebacteria group.
true

9. Most species of bacteria grow best in a growth medium with a pH of 7.
true

10. Yeasts and bacteria grow best in an acidic medium.
true

11. Some organisms can grow in a broad range of pH because their cytoplasm can withstand this wide variety.
false (because of internal buffers)

12. Organisms that use organic molecules as their source of energy are called autotrophs.
false (heterotrophs)

Matching

1. Match the group of bacteria with their optimum growth temperatures.

   1. psychrophile
   2. mesophile
   3. thermophile
   4. extreme thermophile

   a. > 80°C
   b. 50°C to 80°C
   c. -5°C to 20°C
   d. 35°C to 40°C
   e. 20°C to 50°C

   1 - c; 2 - e; 3 - b; 4 - a

2. Match the class of bacteria with its source of carbon.

   1. photoautotroph
   2. photoheterotroph
   3. chemoautotroph
   4. chemoheterotroph
   5. lithoheterotrophs

   a. various organic compounds
   b. CO_2

   1 - b; 2 - a; 3 - a; 4 - b; 5 - a

171

Chapter 5

Multiple Choice

1. The decomposition of natural products in the absence of air is described as the process of
   a. anabolism
   b. metabolism
   c. respiration
   d. fermentation

   ans. d.   diff. 5   obj. 1   p. 2

2. Experiments designed by _____ _____ proved that living cells caused the fermentation of sugar to produce alcohol.
   a. Louis Pasteur
   b. Robert Koch
   c. Justus von Liebig
   d. Friedrich Wohler

   ans. a.   diff. 6   obj. 1   p. 3

3. Studies done in 1897 demonstrated that compounds could be transformed to other compounds by the cytoplasmic materials from disrupted cells. These active ingredients of the ground up cells were _____.
   a. enzymes
   b. DNA
   c. substrates
   d. antibodies

   ans. a.   diff. 7   obj. 1   p. 3

4. The man awarded the Nobel Prize in 1907 for his pioneering studies of biochemical pathways was
   a. Robert Koch
   b. Louis Pasteur
   c. Eduard Buchner
   d. Edward Jenner

   ans. c.   diff. 6   obj. 1   p. 3

5. Biosynthetic reactions that require energy for the conversion of molecular subunits into larger molecules are called
   a. precursor metabolites
   b. catabolic reactions
   c. anabolic reactions
   d. kinetic energy

   ans. c.   diff. 6   obj. 2   p. 4

6. The general term used to describe the anabolic and catabolic reactions in the cells is
   a. metabolism
   b. enzymatic
   c. reactionism
   d. anabolism

   ans. a.   diff. 5   obj. 2   p. 4

7. The key proteins in the cell that determine its properties are
   a. hormones
   b. enzymes
   c. vitamins
   d. minerals

   ans. b.   diff. 5   obj. 2   p. 5

8. Enzymes function to convert the substrate into a product in a short time. Enzymes act as _____.
   a. products
   b. substrates
   c. catalysts
   d. catabolites

   ans. c.   diff. 5   obj. 2   p. 5

9. Enzymes act on _____ to produce _____.
   a. substrates, anabolites
   b. products, catabolites
   c. lactose, lactic acid
   d. substrates, products

   ans. d.   diff. 4   obj. 2   p. 6

10. Enzymes speed up reactions by
    a. breaking the bonds between hydrogen and oxygen
    b. converting products into substrates
    c. lowering their energy of activation
    d. producing heat

    ans. c.   diff. 6   obj. 2   p. 5

11. Which of the following equations is an example of the chain reaction that results in energy release?
    a. $2H + O \rightarrow H_2O + energy$
    b. $2H—H + O—O \rightarrow 2H—O—H + energy$
    c. $H_2O + CO_2 \rightarrow H_2CO_3 + energy$
    d. $C_6H_{12}O_6 + CO_2 \rightarrow H_2O + CO_2 + energy$

    ans. b.   diff. 5   obj. 3   p. 6

12. Energy is defined as
    a. the capacity to do work
    b. a snowball rolling up hill
    c. the use of high-level phosphate bonds
    d. the potential to exercise

    ans. a.   diff. 6   obj. 2   p. 6

13. Energy in motion is referred to as
    a. metabolism
    b. potential energy
    c. heat
    d. kinetic energy

    ans. d.   diff. 5   obj. 2   p. 6

14. Substances that lower the activation energy hurdle are
    a. products
    b. substrates
    c. fuels
    d. enzymes

    ans. d.   diff. 5   obj. 3   p. 7

15. Since the shape of the substrate must fit the shape of the enzyme that acts on it, this is commonly referred to as the
    a. enzyme substrate
    b. lock-and-key arrangement
    c. catalytic site
    d. energy of activation

    ans. b.   diff. 6   obj. 2   p. 7

16. The features of the environment that are most important in influencing enzyme activities are
    a. temperature, pH, and salt concentration
    b. concentration, substrate, and pH
    c. age, heat, and concentration
    d. salt, sugar, and acid

    ans. a.   diff. 6   obj. 4   p. 8

17. Most enzymes function at a _____ salt concentration and at a pH near _____.
    a. low, 5
    b. high, 6
    c. low, 7
    d. moderate, 8

    ans. c.   diff. 5   obj. 4   p. 8

18. Some toxic substances that combine with and prevent the functioning of enzymes necessary for life are
    a. cyanide, arsenic, and mercury
    b. hydrogen, helium, and carbon
    b. lead, mercury, and fluorine
    d. calcium, lead, and strychnine

    ans. a.   diff. 5   obj. 4   p. 8

19. A substance that competes with a normal substrate for the active site of the enzyme is termed a(n)
    a. competitive inhibitor
    b. active site
    c. product activator
    d. substrate

    ans. a.   diff. 6   obj. 4   p. 10

20. The ingestion of ethylene glycol can be fatal due to the conversion of this product by enzymes to _____.
    a. acetic acid
    b. lactic acid
    c. oxalic acid
    d. carbolic acid

    ans. c.   diff. 6   obj. 4   p. 10

21. Mercury in mercurochrome inhibits microorganisms because it oxidizes the S—H groups of the amino acid _____ and converts it to _____.
    a. valine, glutamic acid
    b. adenosine, adenine
    c. threonine, tyrosine
    d. cysteine, cystine

    ans. d.   diff. 6   obj. 4   p. 10

22. The readily usable energy currency of cells is
    a. electron transport system
    b. ADP
    c. electricity
    d. ATP

    ans. d.   diff. 4   obj. 5   p. 13

23. The microorganisms that depend on organic nutrients for energy are referred to as
    a. fermentors
    b. photoorganotrophs
    c. fungi
    d. chemoheterotrophs

    ans. d.   diff. 5   obj. 5   p. 13

24. When an organic product such as pyruvic acid is the final electron or hydrogen acceptor in the metabolic pathway, this pathway is termed
    a. electron transport
    b. respiration
    c. digestion
    d. fermentation

    ans. d.   diff. 6   obj. 5   p. 14

25. In the pathway referred to as _____, oxygen serves as the final acceptor of electrons or hydrogen atoms.
    a. oxidation
    b. fermentation
    c. respiration
    d. reduction

    ans. c.   diff. 6   obj. 5   p. 14

26. The most common pathway for degrading sugars is
    a. glycolysis
    b. respiration
    c. fermentation
    d. oxidation

    ans. a.   diff. 4   obj. 6   p. 16

27. Another term used to describe glycolysis is
    a. respiration
    b. the Embden-Meyerhof pathway
    c. biosynthetic reaction
    d. oxidation

    ans. b.   diff. 5   obj. 6   p. 16

28. The name given to the reaction that involves the removal of electrons or hydrogen atoms from a compound is termed
    a. metabolism
    b. reduction
    c. glycolysis
    d. oxidation

    ans. d.   diff. 5   obj. 6   p. 17

29. The name given to the reaction that involves the addition of electrons or hydrogen atoms to a compound is termed
    a. reduction
    b. oxidation
    c. glycolysis
    d. metabolism

    ans. a.   diff. 5   obj. 6   p. 17

30. The small, nonprotein molecules that can be readily separated from an enzyme and are responsible for the transfer of atoms from one molecule to another are referred to as
    a. coenzymes
    b. enzymes
    c. hormones
    d. vitamins

    ans. a.   diff. 5   obj. 6   p. 18

31. Coenzymes are specific in that they
    a. transfer only one type of small molecule or atom
    b. take part in only a single reaction
    c. are proteins that catalyze reactions
    d. are involved in attaching molecules to others permanently

    ans. a.   diff. 7   obj. 6   p. 18

32. All coenzymes are derivatives of
    a. minerals
    b. vitamins
    c. proteins
    d. enzymes

    ans. b.   diff. 5   obj. 6   p. 18

33. NAD in its oxidized form is abbreviated NAD$^+$; in its reduced form it is symbolized as
    a. FAD
    b. NAD$^-$
    c. NAD
    d. NADH

    ans. d.   diff. 5   obj. 6   p. 20

34. Cells degrade sugar largely to
    a. convert fat to energy
    b. use energy
    c. gain energy
    d. utilize coenzymes

    ans. c.   diff. 6   obj. 6   p. 21

35. The amount of heat required to raise the temperature of 1 gram of water 1°C is described as a(n)
    a. Calorie
    b. calorie
    c. Kilocalorie
    d. energy in motion

    ans. b.   diff. 5   obj. 6   p. 22

36. Each mole of glucose contains _____ calories of bond energy.
    a. 52,000
    b. 634,000
    c. 686,000
    d. 376,000

    ans. c.   diff. 6   obj. 6   p. 22

37. The amount of energy produced by glycolysis can be measured as
    a. 52,000 calories
    b. 686,000 calories
    c. 634,000 calories
    d. 482 calories

    ans. a.   diff. 6   obj. 6   p. 22

38. One compound that can accept hydrogen atoms from NADH is
    a. pyruvic acid
    b. lactic acid
    c. hydrogen
    d. water

    ans. a.   diff. 6   obj. 6   p. 23

39. Pyruvic acid can be metabolized along two major routes:
    a. metabolism and anabolism
    b. oxidation and reduction
    c. fermentation and respiration
    d. anabolism and catabolism

    ans. c.   diff. 6   obj. 6   p. 23

40. The major product responsible for the formation of cheese is
    a. tricarboxylic acid
    b. pyruvic acid
    c. acetic acid
    d. lactic acid

    ans. d.   diff. 6   obj. 6   p. 24

41. The most common lactic acid-producing bacteria belong to the genera *Lactobacillus* and
    a. *Staphylococcus*
    b. *Streptococcus*
    c. *Pseudomonas*
    d. *Saccharomyces*

    ans. b.   diff. 5   obj. 6   p. 25

42. When several members of different genera produce only a single product they are called
    a. single producers
    b. homofermenters
    c. unit enzymes
    d. heterofermenters

    ans. b.   diff. 6   obj. 6   p. 25

43. Lactic acid fermenters, referred to as heterofermenters, readily produce
    a. H_2O and CO_2
    b. CO_2 and ethanol
    c. alcohol and O_2
    d. salt rising bread

    ans. b.   diff. 5   obj. 6   p. 25

44. The group of molecules to which the vitamins are converted are
    a. coenzymes
    b. hormones
    c. enzymes
    d. metabolites

    ans. a.   diff. 6   obj. 5   p. 27

45. The TCA cycle oxidizes pyruvic acid to
    a. $O_2 + CO_2 + ATP$
    b. $H_2O + O_2 + ADP$
    c. $CO_2 + H_2O + ATP$
    d. $H_2O + CO + ADP$

    ans. c.   diff. 6   obj. 7   p. 27

46. Another name for the TCA cycle is the
    a. pyruvate cycle
    b. Hans cycle
    c. energy cycle
    d. Krebs cycle

    ans. d.   diff. 6   obj. 7   p. 27

47. The coenzyme FAD is synthesized from the vitamin
    a. riboflavin
    b. niacin
    c. pantothenic acid
    d. ascorbic acid

    ans. a.   diff. 5   obj. 3   p. 29

48. The large amounts of energy gained in respiration result in the transport of electrons of the _____ atoms through the electron transport system.
    a. hydrogen
    b. oxygen
    c. carbon
    d. nitrogen

    ans. a.   diff. 5   obj. 6   p. 30

49. The final electron acceptor in the oxidative phosphorylation process is
    a. carbon
    b. hydrogen
    c. oxygen
    d. nitrogen

    ans. c.   diff. 5   obj. 6   p. 30

50. In what part of the prokaryotic cell does ATP synthesis occur?
    a. within the ER
    b. on the inner mitochondrial membrane
    c. on the pili extensions
    d. within the cytoplasmic membrane

    ans. d.   diff. 7   obj. 6   p. 31

51. Which of the following processes generates the greatest amount of energy?
    a. oxidative phosphorylation
    b. fermentation
    c. glycolysis
    d. anaerobic respiration

    ans. a.   diff. 6   obj. 7   p. 32

52. Under anaerobic conditions, which of the following molecules is the final electron acceptor?
    a. oxygen
    b. pyruvic acid
    c. carbon dioxide
    d. hydrogen

    ans. b.   diff. 6   obj. 7   p. 33

53. Which of the following is the most common starting material for the ethanol that has been widely used in combination with gasoline to make "gasohol"?
    a. molasses
    b. corn
    c. starch
    d. sugarcane

    ans. b.   diff. 7   obj. 6   p. 34

176

54. Another name for the Entner-Doudoroff pathway, which produces pyruvic acid in the bacterial energy cycle, is
    a. aerobic shunt
    b. pentose phosphate pathway
    c. glycolytic pathway
    d. hexose monophosphate shunt

    ans. d.   diff. 7   obj. 7   p. 35

55. In the TCA cycle, an important reaction is one in which the cell incorporates ammonia into an organic molecule. One such example occurs when alpha-ketoglutaric acid reacts with NH_4 to form
    a. glutamic acid
    b. protein
    c. glucose
    d. valine

    ans. a.   diff. 6   obj. 7   p. 41

56. Bacteria that can absorb light energy and convert it into ATP are commonly called
    a. chemotrophs
    b. phototrophs
    c. organotrophs
    d. cytochromes

    ans. b.   diff. 6   obj. 8   p. 43

57. The overall equation of photosynthesis is
    a. $C_{12}H_{22}O_{11} + 6O_2 \rightarrow 2C_6H_{12}O_6 + 6CO_2$
    b. $6O_2 + 6H_2O \rightarrow C_6H_{12}O_6 + 6CO_2$
    c. $6CO_2 + 6H_2O \rightarrow C_6H_{12}O_6 + 6O_2$
    d. $CO_2 + H_2O \rightarrow 2C_6H_{12}O_6 + O_2$

    ans. c.   diff. 5   obj. 8   p. 44

58. The reaction by which glucose is completely metabolized via glycolysis and the TCA cycle is
    a. water + oxygen → sugar + carbon dioxide
    b. $C_{12}H_{22}O_{11} + CO_2 \rightarrow 6CO_2 + 6O_2 + energy$
    c. $6H_2O + 6CO_2 \rightarrow C_6H_{12}O_6 + 6O_2$
    d. $C_6H_{12}O_6 + 6O_2 \rightarrow 6CO_2 + 6H_2O + energy$

    ans. d.   diff. 5   obj. 8   p. 44

59. In the phototrophic production of energy, the O_2 originates from
    a. CO_2
    b. H_2O
    c. $C_6H_{12}O_6$
    d. NH_4OH

    ans. b.   diff. 7   obj. 8   p. 44

60. In the type of photosynthesis carried on by the purple and green bacteria, instead of oxygen being produced _____ is produced.
    a. hydrogen
    b. sulfur
    c. hydrogen sulfide
    d. carbon dioxide

    ans. b.   diff. 7   obj. 8   p. 45

True/False

1. Pasteur and other scientists observed that bacteria were always present in the fermentation of fruit juices to alcohol and carbon dioxide.
   false (yeast)

2. Pasteur, von Liebig, and Wohler believed that the breakdown of sugar in fruit juice was purely chemical and not biological.
   false (Pasteur disagreed with this premise)

3. Buchner performed experiments that showed that ground yeast cells would convert sugar to ethanol and carbon dioxide.
   true

4. Precursor metabolites are the end products of the synthesis of cellular products.
   false (beginning materials)

5. Anabolism is the general term for all chemical reactions in the body.
   false (metabolism)

6. Biosynthetic reactions that require energy are termed catabolic reactions.
   false (anabolic reactions)

7. The substance on which the enzyme acts is termed the substrate.
   true

8. Energy of activation is commonly known as kinetic energy.
   true

9. Energy required to lift molecules to their activation level is described as kinetic energy.
   false (energy of activation)

10. In an enzyme-controlled reaction, the substrate binds to a specific portion of the enzyme, the active site, to form an enzyme-substrate complex.
    true

11. Most enzymes function best at low-salt concentrations and at pH values near 5.
    false (7)

12. The readily usable energy available in all cells is ADP.
    false (ATP)

13. Reduced compounds contain very little energy; oxidized compounds contain a great deal.
    false (opposite is true)

14. Coenzymes take part in single reactions very much like enzymes.
    false (coenzymes take part in many reactions)

15. All coenzymes are derivatives of proteins.
    false (vitamins)

16. The amount of energy present in a mole of glucose is 686,000 calories.
    true

17. Oxidative phosphorylation uses the energy of the electron transport system to stimulate cell growth.
    false (produces energy by the electron transport system)

18. Photosynthesis performed in the absence of oxygen is termed anoxygenic.
    true

Chapter 6

Multiple Choice

1. The phrase "one gene - one enzyme," which means that one gene is responsible for the synthesis of one enzyme, was coined by
   a. Beadle and Tatum
   b. Watson and Crick
   c. Jacob and Esther Lederberg
   d. Gregor Mendel

   ans. a.   diff. 5   obj. 1   p.2

2. One of the earliest scientists connected with genetic experiments is
   a. Edward Tatum
   b. George Beadle
   c. Gregor Mendel
   d. James Watson

   ans. c.   diff. 4   obj. 1   p. 2

3. Beadle and Tatum studied common molds to determine the functions of genes. These early experiments involved the production of mutants by treating the organisms with
   a. X rays
   b. sodium salts
   c. UV light
   d. benzene

   ans. a.   diff. 6   obj. 1   p. 2

4. That the structure of enzymes is determined by genes was reported in 1940 by
   a. Gregor Mendel
   b. Beadle and Tatum
   c. Watson and Crick
   d. Wilhelm Johannsen

   ans. b.   diff. 4   obj. 1   p. 2

5. DNA and RNA share many features but differ in a number of ways. Which of the following is true of DNA but not RNA?
   a. contains the nitrogen bases adenine, uracil, guanine and cytosine
   b. double-stranded molecule
   c. functions in the cytoplasm
   d. contains ribose sugar

   ans. b.   diff. 5   obj. 2   p. 4

6. The bonds that hold the DNA strands together are _____ bonds.
   a. sulfur
   b. oxygen
   c. carbon
   d. hydrogen

   ans. d.   diff. 4   obj. 2   p. 4

7. In DNA, the pairing between the nitrogen bases is complementary. This means that
   a. cytosine always base pairs with thymine
   b. adenine always base pairs with thymine
   c. guanine always base pairs with adenine
   d. uracil base pairs with thymine

   ans. b.   diff. 4   obj. 2   p. 4

8. When two new strands of DNA are synthesized prior to cell division
   a. the old DNA is denatured and all new strands are formed
   b. DNA and RNA are formed
   c. two identical helices are formed
   d. uracil replaces thymine in the new DNA

   ans. c.   diff. 5   obj. 2   p. 5

9. When the two strands of DNA wind around each other their orientation is in opposite directions. The two strands are described as being
   a. similar
   b. identical
   c. antiparallel
   d. mutated
   e. semi-conservative

   ans. c.   diff. 5   obj. 2   p. 6

10. Which of the following facts is true of RNA but not DNA?
    a. contains uracil
    b. functions in nucleus
    c. double-stranded
    d. contains deoxyribose

    ans. a.   diff. 6   obj. 2   p. 6

11. The information stored in DNA is in the
    a. number of nucleotides it contains
    b. sugar-phosphate bonds
    c. hydrogen bonding of nitrogen bases
    d. sequence of its nucleotide subunits

    ans. d.   diff. 6   obj. 2   p. 6

12. The process that ensures that after cell division both cells will have the same genetic information is termed
    a. DNA replication
    b. RNA transcription
    c. protein synthesis
    d. ribosome formation

    ans. a.   diff. 5   obj. 3   p. 6

13. The transfer of information from DNA into the sequence of amino acids that make up a protein is referred to as the
    a. central dogma of molecular biology
    b. Mendelian genetics principle
    c. one gene - one enzyme principle
    d. operon theory

    ans. a.   diff. 5   obj. 3   p. 6

14. The sequence of nucleotides in DNA that determines the pattern of the nucleotides in RNA is referred to as a(n)
    a. copy
    b. template
    c. mentor
    d. parallel
    e. teacher

    ans. b.   diff. 5   obj. 3   p. 7

15. Synthesis of new strands of DNA begins at a specific site referred to as the
    a. central dogma
    b. replication fork
    c. antiparallel
    d. origin of replication

    ans. d.   diff. 5   obj. 3   p. 8

16. When DNA replicates, each new cell contains one of the original strands and one newly synthesized strand. This type of replication is termed
    a. identical
    b. conservative
    c. semiconservative
    d. chain reference
    e. independent

    ans. c.   diff. 6   obj. 3   p. 9

17. The bacterial chromosome is divided into _____ genes.
    a. 4,000
    b. 2,000
    c. 1,000
    d. 7,000 plus

    ans. a.   diff. 3   obj. 4   p. 9

18. Each gene on a chromosome has 2 regions. The first region codes for a polypeptide chain. This region is commonly called the
    a. regulatory region
    b. coding region
    c. template region
    d. forking region

    ans. b.   diff. 4   obj. 4   p. 10

19. Each gene on a chromosome has 2 regions. The second region contains nucleotide sequences to which various proteins bind that controls the expression of the gene. This region is called the
    a. coding region
    b. regulatory region
    c. template region
    d. message region

    ans. b.   diff. 4   obj. 4   p. 10

20. A specific amino acid is coded for by a sequence of _____ nucleotides.
    a. 4
    b. 6
    c. 2
    d. 3

    ans. d.   diff. 3   obj. 4   p. 10

21. The DNA code is copied into a specific sequence of amino acids in mRNA by the process of
    a. replication
    b. parallelization
    c. templating
    d. transcription

    ans. d.   diff. 3   obj. 4   p. 11

22. The nucleotide sequence for each of the twenty amino acids is known. The sequence of each is printed into the mRNA from the DNA in a series of _____ nucleotides in the form of a(n) _____.
    a. 3, codon
    b. 6, template
    c. 2, regulator
    d. 4, transcriber

    ans. a.   diff. 5   obj. 4   p. 11

23. There are several mRNA sequences of nitrogen bases that do not code for any of the twenty amino acids. These are referred to as
    a. nontemplate codons
    b. anticodons
    c. nonsense codons
    d. extra codons

    ans. c.   diff. 5   obj. 3   p. 11

24. The codon on mRNA that signals the start of a protein is recognized as
    a. AUG
    b. UUA
    c. UAG
    d. TAG

    ans. a.   diff. 6   obj. 3   p. 11

25. The scientists responsible for "cracking" the genetic code of DNA are
    a. Nirenberg and Khorana
    b. Esther and Jacob Lederberg
    c. Watson and Crick
    d. Habner and Tobaro

    ans. a.   diff. 5   obj. 1   p. 11

26. The specific sequence of nucleotides in the DNA to which the RNA polymerase attaches is termed the
    a. core region
    b. regulatory region
    c. sigma region
    d. promoter region

    ans. d.   diff. 6   obj. 4   p. 12

27. The RNA polymerase enzyme necessary for the transcription of a gene is termed the
    a. sigma
    b. holoenzyme
    c. core
    d. promoter

    ans. b.   diff. 4   obj. 4   p. 12

28. The strand of DNA from which the RNA is copied is termed the
    a. core
    b. master strand
    c. copy strand
    d. template strand

    ans. d.   diff. 5   obj. 4   p. 13

29. Although genes vary in size, their average length is about _____ nucleotides.
    a. 500
    b. 3,000
    c. 1,000
    d. 5,000

    ans. c.   diff. 5   obj. 4   p. 13

181

30. If the transcribed mRNA represents more than one gene, it is termed
    a. multiple
    b. aggregate
    c. polygenic
    d. a template

    ans. c.   diff. 4   obj. 4   p. 13

31. Which of the following types of RNA is responsible for *directly* coding of proteins?
    a. rRNA
    b. tRNA
    c. mRNA
    d. sRNA

    ans. c.   diff. 4   obj. 4   p. 14

32. tRNA functions to
    a. terminate the proteins on the mRNA
    b. transcribe the mRNA
    c. transfer an amino acid to mRNA
    d. translate the genetic code on DNA

    ans. c.   diff. 6   obj. 4   p. 14

33. The placement of the amino acid transferred from tRNA to mRNA is determined by the
    a. sequence of nucleotides on tRNA
    b. need of the protein in the organism
    c. DNA, which transcribes both molecules
    d. complementarity of nucleotides on the two molecules

    ans. d.   diff. 7   obj. 4   p. 14

34. The tRNA is said to be "charged" when it
    a. carries an amino acid
    b. is in contact with mRNA
    c. has completed the transfer of its amino acid
    d. carries unsatisfied hydrogen bonds

    ans. a.   diff. 7   obj. 4   p. 14

35. The first amino acid in virtually all proteins in bacteria is
    a. guanine
    b. methionine
    c. threonine
    d. lysine

    ans. b.   diff. 5   obj. 2   p. 15

36. The three nucleotides that are the start signal for translation of proteins in bacteria is the code for
    a. leucine
    b. lysine
    c. methionine
    d. valine

    ans. c.   diff. 5   obj. 2   p. 15

37. In the process of translation, the _____ serves as the structure that brings the amino acids into a favorable position to be joined by a _____ bond.
    a. translation, saccharide
    b. DNA, hydrogen
    c. RNA, covalent
    d. ribosome, peptide

    ans. d.   diff. 7   obj. 2   p. 16

38. In which type of cell is one end of a mRNA molecule being translated at the same time the other end is being synthesized from the DNA template?
    a. animal
    b. eukaryote
    c. virus
    d. prokaryote

    ans. d.   diff. 6   obj. 5   p. 19

39. In which type of cell is the mRNA transported out of the nucleus before it can be translated in the cytoplasm?
    a. prokaryote
    b. eukaryote
    c. virus
    d. bacteria

    ans. b.   diff. 5   obj. 5   p. 20

40. Some segments of mRNA are coded by segments of DNA that do not code for amino acids. These are removed from the useful mRNA before it leaves the nucleus and is translated on the ribosomes. This removed piece of RNA is termed an
    a. intron
    b. acceptor site
    c. peptide site
    d. template

    ans. a.   diff. 8   obj. 5   p. 20

41. The scientists responsible for observing and documenting the discovery that RNA molecules can act as enzymes were
    a. Altman and Cech
    b. Beadle and Tatum
    c. Watson and Crick
    d. Lederberg and Mendel

    ans. a.   diff. 5   obj. 7   p. 21

42. When is it possible for cells to synthesize degradative enzymes?
    a. when the cell is eukaryotic
    b. when the presence of the identical enzyme is present in the environment
    c. only if DNA is active and transcribing RNA
    d. when the substrates are present in the environment

    ans. d.   diff. 7   obj. 7   p. 23

43. The control of gene expression in a biosynthetic pathway is termed
    a. allosteric inhibition
    b. negative feedback
    c. end product repression
    d. homeostasis

    ans. c.   diff. 6   obj. 7   p. 25

44. End product repression is the term used to describe the _____ in biosynthetic pathways.
    a. environment
    b. homeostasis
    c. allosteric inhibition
    d. control of gene expression

    ans. d.   diff. 6   obj. 7   p. 25

45. The mechanism that involves the inhibition of only the first enzyme of the biosynthetic pathway by the end product of the pathway is
    a. environmental trigger
    b. homeostasis
    c. allosteric inhibition
    d. acceptor site

    ans. c.   diff. 5   obj. 7   p. 25

46. An example of a polygenic message is the mRNA coding for _____ biosynthesis in *E. coli.*
    a. tryptophan
    b. tyrosine
    c. methionine
    d. leucine

    ans. a.   diff. 6   obj. 8   p. 26

47. A gene that codes for a protein that is not an enzyme is termed a(n)
    a. regulatory gene
    b. operator gene
    c. polygenic activator
    d. end product

    ans. a.   diff. 5   obj. 8   p. 26

48. The DNA site to which the repressor protein binds is the _____ region, and is part of the regulatory region of the gene.
    a. repressor
    b. regulator
    c. operator
    d. acceptor

    ans. c.   diff. 6   obj. 8   p. 26

49. The set of genes that are linked together and transcribed as a single unit is referred to as a(n) _____.
    a. regulator
    b. operon
    c. operator
    d. repressor

    ans. b.   diff. 6   obj. 8   p. 27

50. Biosynthetic pathways controlled by end product repression are referred to as _____ pathways.
    a. allosteric
    b. repressible
    c. intron
    d. ribosome

    ans. b.   diff. 6   obj. 8   p. 28

51. In biosynthetic pathways, the end product turns off gene expression and inhibits the first enzyme of the pathway. This sequence of events describes
    a. introns
    b. repression
    c. negative feedback
    d. ribosomes

    ans. c.   diff. 7   obj. 8   p. 29

52. The metabolites that activate gene transcription are called
    a. introns
    b. activators
    c. repressors
    d. inducers

    ans. d.   diff. 7   obj. 8   p. 30

53. *E. coli* breaks down glucose with enzymes that are not subject to regulation by induction or repression. Such enzymes are termed
    a. sigma
    b. template
    c. constitutive
    d. holoenzyme

    ans. c.   diff. 7   obj. 8   p. 31

True/False

1. Beadle and Tatum were responsible for determining that genes hold the key to inherited traits.
   false (Gregor Mendel)

2. DNA is a double-stranded helix that contains the nitrogen bases adenine, cytosine, guanine, and uracil.
   false (A, C, G, and thymine)

3. In DNA, the number of molecules of cytosine always equals the number of molecules of guanine, which is true also for adenine and thymine.
   true

4. When the two strands of DNA wind around each other to form a double helix, the orientation is antiparallel.
   true

5. The incorporation of nucleotides into newly synthesized strands requires the enzyme DNA polymerase for both DNA and RNA.
   false (DNA polymerase for DNA, RNA polymerase for RNA)

6. A single codon on mRNA may specify for more than one amino acid, although there may be several that specify for the same amino acid.
   false (several codons may specify for one)

7. Several codons do not code for any of the twenty amino acids; these are called anticodons.
   false (nonsense, or stop, codons)

8. The start signal for proteins in all eukaryotic organisms is methionine.
   false (in bacteria only)

9. tRNA molecules are "charged" when they are carrying an amino acid.
   true

10. Each ribosome is involved with the synthesis of only a single series of proteins at any one time.
    false (only one protein at a time)

11. In eukaryotic cells, the mRNA is transcribed and translated at the same time.
    false (prokaryotic)

12. End product repression involves the control of gene expression in a biosynthetic pathway.
    true

13. In bacteria, genes involved in a biochemical pathway are located very close to one another on the chromosome.
    true

14. *Escherichia coli* breaks down glucose with enzymes that are not subject to regulation by induction or repression by enzymes termed inducive.
    false (constitutive)

Chapter 7

Multiple Choice

1. Frederick Griffith was one of the earliest bacteriologists to suggest that genetic traits could be _____
   a. evaluated by specific biochemical tests.
   b. transferred from killed organisms to living ones.
   c. used to develop antibiotics.
   d. discarded as important in the disease process.

   ans. b.   diff. 5   obj. 1   p. 2

2. The scientists first responsible for identifying DNA as being the most important molecule in the cell were
   a. Avery, MacLeod, and McCarty
   b. Griffith, Dawson, and Fleming
   c. Koch, Jenner, and Lister
   d. Watson and Crick

   ans. a.   diff. 5   obj. 1   p. 2

3. The "transforming principle" was discovered by
   a. Dawson
   b. Griffith
   c. Avery
   d. Crick

   ans. b.   diff. 5   obj. 1   p. 2

4. The "transforming principle" first discovered by Griffith was later identified as
   a. RNA
   b. cellular protein
   c. DNA
   d. the nucleus

   ans. c.   diff. 4   obj. 1   p. 2

5. That capsule formation, a genetic trait, could be transferred from a filtrate of dead organisms to living organisms was proved by
   a. Watson
   b. Griffith
   c. Dawson
   d. McCarty

   ans. c.   diff. 4   obj. 1   p. 2

6. The studies of Avery, MacLeod, and McCarty pointed out DNA as an important molecule in the scheme of life and led to the determination of its structure by
   a. Fielding
   b. Griffith
   c. Dawson
   d. Watson and Crick

   ans. d.   diff. 4   obj. 1   p. 3

7. Sulfa drugs were the first chemotherapeutic agent to be successful against
   a. pneumonia
   b. syphilis
   c. bacterial dysentery
   d. gonorrhea

   ans. c.   diff. 5   obj. 2   p. 4

8. Organisms resistant to antibiotics were discovered in the 1950s. It seemed that sensitive organisms developed resistance after a period of time. How was this possible?
   a. Antibiotic resistance could be transferred from one organism to another.
   b. Organisms develop resistance after long-term exposure to antibiotics.
   c. Viruses, living in bacteria, are resistant to antibiotics.
   d. The patient becomes immune to the antibiotic.

   ans. a.   diff. 5   obj. 2   p. 4

9. The source of variation among microorganisms that were once identical is
   a. antibiotic resistance
   b. a mutation
   c. capsule formation
   d. new classification methods

   ans. b.   diff. 6   obj. 3   p. 5

10. The total genetic constitution of an organism is its
    a. genome
    b. phenotype
    c. molecular metabolism
    d. genotype

    ans. d.   diff. 5   obj. 3   p. 6

11. Together, the chromosomes and plasmids of an organism make up its
    a. genotype
    b. genome
    c. phenotype
    d. mutagens

    ans. b.   diff. 4   obj. 3   p. 6

12. The characteristics displayed by an organism in any given environment comprise the _____ of the organism.
    a. genotype
    b. phenotype
    c. mutation
    d. genome

    ans. b.   diff. 5   obj. 3   p. 6

13. The interaction of the genome of the cell with its environment describes its
    a. phenotype
    b. genotype
    c. mutation
    d. organization

    ans. a.   diff. 6   obj. 3   p. 6

14. Changes in the nucleotide sequence of the DNA of a few organisms within a large population of organisms are described as
    a. cellular organization
    b. phenotypic expression
    c. genotypic changes
    d. specificity

    ans. c.   diff. 6   obj. 3   p. 6

15. When a great number of cells in a population appear to be altered in their environment, the change is described as
    a. numerical taxonomy
    b. genotypic alteration
    c. phenotypic alteration
    d. mutation statistics

    ans. c.   diff. 6   obj. 3   p. 6

16. A change in the nucleotide sequence of DNA that results in a recognizable change in the organism is called a(n)
    a. mutation
    b. substitution
    c. alteration
    d. growth

    ans. a.   diff. 5   obj. 4   p. 7

17. The most frequent (although rare) type of DNA mutation involves errors in
    a. protein synthesis
    b. lactose fermentation
    c. pigment production
    d. base substitution

    ans. d.   diff. 6   obj. 4   p. 8

18. In bacterial DNA, the nitrogen base bonding occurs very specifically. Which of the following bondings is incorrect?
    a. A-U
    b. C-G
    c. G-C
    d. T-A

    ans. a.   diff. 3   obj. 4   p. 8

19. A DNA codon for an amino acid always involves _____ nitrogen bases.
    a. nine
    b. four
    c. two
    d. three

    ans. d.   diff. 3   obj. 4   p. 8

20. The deletion or addition of a purine or pyrimidine nucleotide will alter the amino acid sequence in a protein. This type of mutation is called a(n)
    a. phenotypic alteration
    b. codon sequence
    c. frame shift
    d. environmental trigger

    ans. c.   diff. 5   obj. 4   p. 8

21. More amino acid sequences are changed by _____ _____ rather than by _____ _____.
    a. nucleotide addition, base substitution
    b. nucleotide substitution, base addition
    c. base addition, base deletion
    d. mutation frequency, mutation duration

    ans. a.   diff. 6   obj. 4   p. 8

22. The process by which a mutation is produced is an infrequent occurrence. The range is
    a. 1 in 10^2 to 1 in 10^6
    b. 1 in 10^{-4} to 1 in 10^{-12}
    c. 1 in 10^{-12} to 1 in 10^{-2}
    d. 1 in 10 to 1 in 1000

    ans. b.   diff. 5   obj. 4   p. 9

23. An agent that can increase the rate of mutation is a mutagen. If it is deliberately exposed to a group of organisms it may bring about a(n)
    a. spontaneous change
    b. induced change
    c. triggered frequency
    d. chemical element

    ans. b.   diff. 6   obj. 4   p. 9

24. Chemical mutagens often act by altering the
    a. nitrogen base sequences
    b. alkyl groups of the nitrogen bases
    c. hydrogen-bonding properties of the nitrogen bases
    d. number of bonding sites on the nitrogen bases

    ans. c.   diff. 5   obj. 4   p. 9

25. The chemical mutagen most frequently used to induce DNA changes is
    a. hydrogen
    b. nitrous acid
    c. ketone forming agents
    d. radiation

    ans. b.   diff. 6   obj. 4   p. 9

26. Nitrous acid most frequently causes mutations by
    a. converting cytosine to uracil
    b. converting keto groups to amino groups
    c. altering the bonding ratios of nitrogen bases
    d. substituting oxygen for hydrogen bonds

    ans. a.   diff. 6   obj. 4   p. 9

27. An alkylating agent used in cancer therapy to alter bonding is (are)
    a. nitrous acid
    b. nitrogen mustard
    c. base analogs
    d. guanine

    ans. b.   diff. 6   obj. 4   p. 10

28. Chemical mutagens that resemble the chemical structure of the naturally occurring bases and are incorporated into the DNA in place of the natural bases during DNA replication are called
    a. alkylating agents
    b. nitrogen mustard
    c. base analogs
    d. nitrous oxide

    ans. c.   diff. 5   obj. 4   p. 10

29. Three-ringed molecules that become inserted between base pairs of double-stranded DNA and allow additional bases to be inserted into the spaces are called
    a. alkylating agents
    b. nitrous oxide
    c. base analogs
    d. intercalating agents

    ans. d.   diff. 6   obj. 5   p. 11

187

30. Segments of DNA found in virtually all species of bacteria that can move from one site to another are called
    a. insertion sequences
    b. base analogs
    c. transposons
    d. palindromic sequences

    ans. c.   diff. 5   obj. 5   p. 12

31. The scientist who received the Nobel Prize in Medicine for her discovery of transposons was
    a. James Watson
    b. Barbara McClintock
    c. Edmund McLeary
    d. Ernst Haeckel

    ans. b.   diff. 6   obj.   p. 14

32. The formation of a covalent bond between adjacent thymine molecules on DNA that results in the formation of a thymine dimer may be caused by
    a. X rays and UV light
    b. chemical mutagens
    c. mustard gas
    d. transposons

    ans. a.   diff. 6   obj. 7   p. 15

33. DNA repair mechanisms are present in
    a. both prokaryotic and eukaryotic organisms
    b. only prokaryotes
    c. only eukaryotes
    d. neither prokaryotes nor eukaryotes

    ans. a.   diff. 4   obj. 8   p. 17

34. Thymine dimers caused by exposure to UV light can be repaired only by a phenomenon called _____ repair.
    a. dark
    b. SOS
    c. dimer
    d. light

    ans. d.   diff. 4   obj. 7   p. 18

35. Bacterial DNA can be repaired by removal of the damaged segment and replacement of an undamaged segment by a process termed
    a. SOS repair
    b. light repair
    c. dark repair
    d. dimer repair

    ans. c.   diff. 4   obj. 7   p. 18

36. If the mutation rate for a given gene in a given cell is said to be one chance in a million, it is expressed as
    a. 1×10^{-6}
    b. 1×10^{-4}
    c. 1×10^{-12}
    d. 1×10^{-7}

    ans. a.   diff. 4   obj. 4   p. 20

37. Occasionally a nucleotide will change back to its original state through mutation. The process is termed
    a. mutagen
    b. excision
    c. reversion
    d. transposon

    ans. c.   diff. 6   obj. 9   p. 21

38. Prokaryotic cell mutations can be observed right after the mutation occurs because the prokaryotic chromosome is
    a. polyploid
    b. diploid
    c. haploid
    d. repetitive

    ans. c.   diff. 4   obj. 4   p. 32

39. The diploid character of eukaryotic cells may mask the appearance of a mutation since the
    a. nature of the mutation may be palindromic
    b. mutation may be reversible
    c. matching chromosome may carry the dominant trait
    d. mutation may be a frame shift

    ans. c.   diff. 5   obj. 4   p. 22

40. Direct selection technique involves inoculating cells onto a medium on which
   a. the mutagen is present
   b. the nutrients will reverse the mutation
   c. the nutrients necessary for mutation occur
   d. the mutant but not the parent will grow

   ans. d.   diff. 6   obj. 10   p. 23

41. Among the easiest of the mutations to isolate are those that
   a. involve antibiotic resistance
   b. involve diploid chromosomes
   c. allow populations to be measured
   d. use the indirect technique for measuring

   ans. a.   diff. 6   obj. 10   p. 23

42. A term that has been coined to describe a bacterium that requires a growth factor in order to grow is
   a. auxotroph
   b. palindromic
   c. replica plating
   d. direct selection

   ans. a.   diff. 6   obj. 10   p. 24

43. A prototroph is a bacterium that does not require a growth factor to grow. This term is often confused with
   a. auxotroph
   b. wild type
   c. mutagen
   d. replica

   ans. b.   diff. 5   obj. 10   p. 24

44. A clever technique that reduces the tedious and time consuming nature of isolating and identifying auxotrophs is
   a. replica plating
   b. mass chromatography
   c. direct selection
   d. reversion

   ans. a.   diff. 5   obj. 10   p. 24

45. The technique of replica plating was devised by
   a. Frederick Griffith
   b. Watson and Crick
   c. Joshua and Esther Lederberg
   d. Alexander Fleming

   ans. c.   diff. 6   obj. 10   p. 24

46. To increase the proportion of mutants in a population of bacteria, a technique termed _____ can be used.
   a. individual transfer
   b. replica plating
   c. direct selection
   d. penicillin enrichment

   ans. d.   diff. 6   obj. 10   p. 26

47. One class of conditionally lethal mutants will grow and function at the low temperature (25°C) but not the higher (37°C) temperature. This class is termed _____ mutants.
   a. replication
   b. translation-sensitive
   c. master plate
   d. temperature-sensitive

   ans. d.   diff. 6   obj. 9   p. 27

48. The commercial use of mutant bacteria is well illustrated in the production of _____ by the fermentation of molasses.
   a. lysine
   b. arginine
   c. tyrosine
   d. threonine

   ans. a.   diff. 6   obj. 9   p. 28

49. A quick microbiological test for potential carcinogens was developed in the 1960s by
   a. Bruce Ames
   b. Esther Lederberg
   c. Alexander Fleming
   d. George Carcin

   ans. a.   diff. 5   obj. 11   p. 29

50. The Ames test is not a certain detector of a carcinogen. Some substances are not carcinogenic in themselves, but can be converted to active carcinogens by a(n)
    a. replica plating with penicillin
    b. low temperature synthesis of the carcinogen
    c. metabolic reaction occurring in animals but not in bacteria
    d. reverse transcriptase use

    ans. c.   diff. 7   obj. 1   p. 30

51. The Ames test is useful as a rapid screening test to identify those compounds that
    a. will respond to chemical agents
    b. have a high probability of being carcinogenic
    c. respond to deletion enzymes of DNA
    d. will protect an organism from cancer

    ans. b.   diff. 8   obj. 1   p. 32

52. Bacteria that have properties of the donor cell and the recipient cell are the results of
    a. genetic recombination
    b. ultraviolet light
    c. SOS repair
    d. frame shift mutation

    ans. a.   diff. 7   obj. 1   p. 33

53. Recombinant DNA molecules are termed homologous when there is
    a. deletion of carcinogenic nucleotides
    b. addition of nucleotides in exact areas
    c. reversal of the nucleotide order when necessary
    d. exact substitution of donor genes for recipient genes

    ans. d.   diff. 6   obj. 1   p. 35

54. Competent cells that are able to take up large DNA molecules do so by _____ in an unknown manner.
    a. passing through the cytoplasmic membrane
    b. rupturing the cell wall and then repairing it
    c. active transport of sodium ions that allow absorption
    d. spontaneous mutation

    ans. a.   diff. 8   obj. 12   p. 36

55. Homologous recombination of DNA can occur when an electric current is used to treat a mixture of released DNA and live bacteria. The technique is termed
    a. light repair
    b. spontaneous recombination
    c. ellipsis
    d. electroporation

    ans. d.   diff. 7   obj. 12   p. 37

56. The mechanism by which a bacterial gene is transferred from one bacterium to another by a bacteriophage is termed
    a. transduction
    b. ellipsis   translation
    c. light repair   transformation
    d. posturing   transcription

    ans. a.   diff. 7   obj 12   p. 38

57. The common type of gene transfer by Gram-negative bacteria in which there must be cell-to-cell contact is
    a. transduction
    b. conjugation
    c. transformation
    d. competency
    e. comingling

    ans. b.   diff. 5   obj. 12   p. 38

58. For conjugation to take place between adjacent cells, one cell is the donor and the other the recipient. The structure that identifies the donor cell is the
    a. host plasmid gene
    b. Y chromosome
    c. diploid gene
    d. F plasmid
    e. R plasmid

    ans. d.   diff. 6   obj. 12   p. 40

59. The F plasmid of the donor cell carries the genetic information for the synthesis of the
    a. host gene
    b. diploid character
    c. sex pilus
    d. Y chromosome

    ans. c.   diff. 6   obj. 12   p. 40

60. When the F plasmid replicates as part of the chromosome, rather than independently, the progeny are
    a. undetermined without further study
    b. F-cells
    c. F plasmid cells F+
    d. Hfr cells
    e. F' cells

    ans. d.   diff. 6   obj. 12   p. 42

61. In conjugation, the time required for the entire transfer of the donor chromosome is about _____.
    a. 60 seconds
    b. 60 minutes
    c. 30 minutes
    d. 90 minutes

    ans. d.   diff. 6   obj. 12   p. 43

62. The two scientists most responsible for the research on the F plasmid and Hfr strains were
    a. Jacob and Wollman
    b. Esther and Jacob Lederberg
    c. Frederick Griffith
    d. Ernest Hackle

    ans. a.   diff. 5   obj. 12   p. 44

63. The fact that some DNA may be extrachromosomal or exist as an episome was discovered by
    a. Jacob
    b. Monas
    c. Wollman
    d. Lederman

    ans. a.   diff. 6   obj. 12   p. 44

64. Among the best studied groups of plasmids are those that confer resistance to antibiotics and heavy metals. Many of these plasmids are composed of two parts:
    a. ABC and D genes
    b. RTF and R genes
    c. F$^+$ and F$^-$ plasmids
    d. Hfr and F$^+$ plasmids

    ans. b.   diff. 6   obj. 12   p. 48

65. Antibiotic-resistant organisms are most frequently found in
    a. schools
    b. hospitals
    c. research laboratories
    d. rural areas
    e. microbial classrooms

    ans. b.   diff. 6   obj. 12   p. 49

66. Crown gall tumors are common in plants. Study of this bacterial disease showed that the altered tissue synthesized opine, which
    a. neither plant nor bacteria could synthesize
    b. the plant could synthesize but bacteria could not
    c. the bacteria could synthesize but the plant could not
    d. was penicillin sensitive

    ans. a.   diff. 6   obj. 12   p. 50

67. The study of the bacterial contributions to the formation of crown gall tumors show that
    a. opine is a carcinogen
    b. plant cells can be resistant to penicillin
    c. *Agrobacterium* can cause other diseases
    d. DNA could be transferred across kingdoms

    ans. d.   diff. 7   obj. 12   p.51

68. Proteins that can code for antibiotic resistance as well as genes that code for toxin synthesis have been shown to be manufactured by
    a. episomes
    b. transducted bacteria
    c. viral particles
    d. transposable genes

    ans. d.   diff. 8   obj. 12   p. 52

69. Recipient bacteria will take up DNA if the donor DNA is from the same species of bacteria by
    a. replica plating
    b. viral transfer and conjugation
    c. transformation, transduction, and conjugation
    d. transposons

    ans. c.   diff. 9   obj. 12   p. 53

70. The DNA enzyme that reacts with specific short sequences of nucleotides in foreign DNA and breaks the DNA at these sites is termed
    a. ribonuclease
    b. restriction endonuclease
    c. transferase
    d. DNAse

    ans. b.   diff. 7   obj. 12   p. 53

71. Restriction enzymes have been found in all species of
    a. viruses but not prokaryotes
    b. yeasts but not in viruses
    c. prokaryotes but not in any eukaryotes
    d. eukaryotes but not prokaryotes

    ans. c.   diff. 8   obj. 12   p. 54

72. Conjugation occurs most frequently between Gram-negative organisms, although some Gram-positive organisms have been known to conjugate. Among these are
    a. streptococci and staphylococci
    b. pseudomonads and enterobacteria
    c. *Neisseria* and *Micrococcus*
    d. *Salmonella* and *Shigella*

    ans. a.   diff. 5   obj. 12   p. 55

True/False

1. Avery, MacLeod, and McCarty earned a Nobel Prize for their discovery that DNA was the "transforming" principle.
   false (published, but no award)

2. Antibiotic resistance among microorganisms develops as microbes mutate in response to increasingly sophisticated drug therapies.
   true

3. The genome includes the genes, chromosomes, and plasmids of an organism.
   true

4. Genotypic alterations are reversible when the environment changes.
   false (phenotypic alterations are reversible)

5. Genotypic changes are fixed and relatively stable in the population's environment.
   true

6. Change in the nucleotide sequence of DNA can only occur by deletion of one or several purines.
   false (deletion or addition of purine or pyrimidine)

7. Mutations occur infrequently. The rate is often stated as 1 in 10^9.
   true

8. Mutations that occur at an infrequent rate without any known trigger are known as induced mutations.
   false (spontaneous)

9. A transposon whose sequence in double-stranded DNA is inserted in opposite directions on either strand is termed a base analog.
   false (inverted sequence or palindromic sequence)

10. The major mutagenic action of UV light does not damage DNA directly, but rather from the cell's attempt to repair the damage by a mechanism termed a frame shift.
    false (SOS repair)

11. Once a DNA molecule is broken, it cannot be repaired and the organism dies.
    false (SOS repair is possible)

12. Prokaryotic cells show their mutations immediately because the chromosome complement is haploid.
    true

13. Replica plating is an easier technique for identifying auxotrophs. It was devised by Frederick Griffith.
    false (Joshua and Esther Lederberg)

14. The Ames test is a simple, inexpensive test to detect potential carcinogens.
    true

15. Another name for the F plasmid is the sex pilus.
    false (conjugative plasmid)

16. In conjugation, the entire donor chromosome is usually transferred into the recipient cell in a short time period.
    false (only part or takes 90 minutes)

17. The DNA that exists interchangeably as part of the chromosome or as an extrachromosomal fragment is termed a plasmid.
    false (episome)

18. Plasmids replicate independently of the chromosome.
    true

19. If DNA is methylated before it is degraded by the restriction endonuclease, it can be recognized easily by the restriction enzyme and can be modified.

false (it cannot be recognized once it is methylated)

Matching

1. Match the term in Column A with its closest definition from Column B.

Column A	Column B
1. genome	a. characteristics of an organism
2. genotype	b. permanent genetic change
3. phenotype	c. genetic constitution of an organism
4. mutation	d. DNA and plasmids of an organism

1 - d; 2 - c; 3 - a; 4 - b

2. Match the chemical in Column A with its mutagenic effect in Column B.

Column A	Column B
1. 5-bromouracil	a. add alkyl group to nitrogenous base
2. nitrous acid	b. forms thymine dimer
3. nitrogen mustard	c. AT base pair becomes GC base pair
4. UV light	d. converts amino group to keto group

1 - c; 2 - d; 3 - a; 4 - b

3. Match the DNA repair mechanism in Column A with its result in Column B.

Column A	Column B
1. light repair	a. excise thymine dimer and synthesis of damaged DNA by DNA polymerase
2. dark repair	b. uses an inducible replication
3. SOS repair	c. forms covalent bonds between the thymine
	d. breaks covalent bond joining thymine dimers

1 - d; 2 - a; 3 - b; no ans. for c

4. Choose the most appropriate answer in describing the mutagenic ability of the chemical listed.

a) mutagenic   b) carcinogenic in animals   c) nonmutagenic

1. dieldrin (insecticide)		4. phenobarbital (drug)
2. sevin (insecticide)		5. hydrogen peroxide
3. nicotine (tobacco)		

1 - b; 2 - c; 3 - c; 4 - b; 5 - a

5.   Match the plasmid traits in Column A with the organism from Column B in which it is found.

Column A   Column B

1.   antibiotic synthesis   a.   *Rhizobium*
2.   oil degradation   b.   *Escherichia coli*
3.   nitrogen fixation   c.   *Yersinia enterocolitica*
4.   antibiotic resistance   d.   *Streptomyces*
5.   increased virulence   e.   *Pseudomonas*

1 - d; 2 - e; 3 - a; 4 - b; 5 - c

Chapter 8

Multiple Choice

1. The technique of gene cloning was perfected by
    a. Maxam and Gilbert
    b. Smith and Nathans
    c. Arber
    d. Cohen and Boyer

    ans. d.   diff. 5   obj. 1   p. 2

2. Bacteria may protect themselves from viral invasion by the manufacture of restriction enzymes that
    a. degrade the viral DNA
    b. cover the viral receptor sites
    c. alter the viral genetic code
    d. reverse the invasion process

    ans. a.   diff. 6   obj. 1   p. 2

3. The arrangement of DNA nucleotides is accomplished by a process known as
    a. gene sequencing
    b. plasmid orientation
    c. bacterial transduction
    d. viral genome analysis

    ans. a.   diff. 4   obj. 1   p. 3

4. The modern biochemical technique to solve practical problems is defined as
    a. recombinant DNA technology
    b. biotechnology
    c. gene cloning
    d. genetic engineering

    ans. b.   diff. 4   obj. 1   p. 4

5. DNA molecules that contain pieces of DNA not normally found together is defined as
    a. biotechnology
    b. gene cloning
    c. recombinant DNA
    d. genetic engineering

    ans. c.   diff. 5   obj. 2   p. 4

6. The process of removing genes from their normal DNA sequence and propagating the genes in a new host is described as
    a. genetic engineering
    b. gene cloning
    c. biotechnology
    d. laboratory interference

    ans. b.   diff. 6   obj. 2   p. 4

7. When breaking up the DNA in a human cell, how many fragments can one expect to find?
    a. 100,000
    b. 100
    c. 7,000
    d. 700,000

    ans. a.   diff. 4   obj. 3   p. 6

8. The most common vectors used for cloning genes are
    a. plasmids and viruses
    b. viruses and nucleotides
    c. molecular transcriptase and reverse transcriptase
    d. bacteria and plasmids

    ans. a.   diff. 5   obj. 3   p. 7

9. Restrictive endonucleases are commonly called
    a. sticky ends
    b. blunt ends
    c. "scissors"
    d. knives

    ans. c.   diff. 5   obj. 3   p. 8

10. Restriction enzymes are used to cut DNA molecules, which will then be mixed in a test tube. These are joined permanently by a second type of enzyme, which is commonly called
    a. DNA reverse transcriptase
    b. RNA transcriptase
    c. DNA ligase
    d. RNA ligase

    ans. c.   diff. 5   obj. 3   p. 9

11. What is the approximate number of plasmid molecules needed to clone a human genome?
    a. 10×10^7
    b. 10×10^6
    c. 1×10^7
    d. 1×10^6

    ans. d.   diff. 6   obj. -   p. 9

12. The entire set of cloned fragments of the completed human genome is termed the
    a. recombinant genome
    b. gene library
    c. book of genes
    d. restructured genome

    ans. b.   diff. 5   obj. 3   p. 9

13. DNA fragments can be separated according to size by
    a. agarose probes
    b. ethidium bromide
    c. DNA ligase
    d. agarose gel electrophoresis

    ans. d.   diff. 5   obj. 3   p. 10

14. When using agarose gel electrophoresis to separate DNA fragments, the energy for the operation is supplied by
    a. separative phoresis
    b. active transport
    c. electricity
    d. agarosis

    ans. c.   diff. 6   obj. 3   p. 10

15. The danger of cloning a human DNA fragment into *E. coli* is that
    a. the outer membrane is poisonous to humans
    b. the fragments may disassociate in the cell
    c. the human may reject the insertion
    d. gel may be incorporated into the clone

    ans. a.   diff. 6   obj. 3   p. 12

16. Host cells containing recombinant DNA can be selected on the basis of the properties conferred by the
    a. vector
    b. host
    c. enzymes
    d. virus

    ans. a.   diff. 7   obj. 3   p. 12

17. A common way to identify the *E. coli* that carries the desired recombinant DNA is by using a
    a. host
    b. vector
    c. probe
    d. plasmid

    ans. c.   diff. 6   obj. 3   p. 13

18. The term *Southern hybridization* is used to describe the technique of using
    a. a metal probe for isolation of the hybrid
    b. single-stranded DNA to identify its complement
    c. electrophoresis with silicon gel
    d. a method of gene selection

    ans. b.   diff. 7   obj. 3   p. 13

19. In many cases of gene cloning, it is the _____ of the cloning rather than the _____ _____ that is the object of the investigation.
    a. transcription, process
    b. enzyme, DNA hybrid
    c. cloning, hybrid process
    d. product, cloned gene

    ans. d.   diff. 7   obj. 4   p. 15

20. As a general rule in hybridization,
    a. the ribosomes and the RNA polymers used must be closely related
    b. the more closely the two organisms are related the greater the success
    c. large quantities of the gene are inserted into the chromosome used
    d. *S. aureus* rather than *E. coli* are used

    ans. b.   diff. 5   obj. 5   p. 15

21. Some important human proteins produced by biotechnology are
    a. pancreatic enzymes and blood clotting factors
    b. insulin, growth hormone, and interferons
    c. oxytocin, insulin, and pancreatic enzymes
    d. testosterone, oxytocin, and growth hormone

    ans. b.   diff. 6   obj. 5   p. 16

196

22. The great demand for bacterial production of human insulin is based on
    a. ease of manufacture and quality
    b. its cost and speed of reaction
    c. its safety and nonallergenic qualities
    d. hypoallergenic qualities and effectiveness

    ans. c.   diff. 5   obj. 5   p. 16

23. Human growth hormone is used to treat
    a. hypopituitary dwarfism
    b. hyperpituitary giantism
    c. Simmonds disease
    d. diabetes

    ans. a.   diff. 5   obj. 5   p. 17

24. Interferon has been successfully produced after cloning in
    a. viral plasmids
    b. *S. aureus* and *E. coli*
    c. *Candida albicans* and *S. aureus*
    d. *E. coli* and *Saccharomyces cerevisiae*

    ans. d.   diff. 6   obj. 5   p. 18

25. Genetically engineered microorganisms are now being used to produce
    a. vaccines
    b. hybrid microorganisms
    c. digestive enzymes
    d. antibiotics

    ans. a.   diff. 5   obj. 5   p. 19

26. A large-scale bioremediation experiment was carried out in Alaska in an attempt to
    a. adapt new species to the environment
    b. promote the evaluation of new plant forms
    c. clean up a large oil spill
    d. manufacture nitrogen fertilizers

    ans. c.   diff. 6   obj. 5   p. 20

27. Natural insect toxins are rapidly replacing chlorinated hydrocarbons to control insects on plants. One of the most widely used toxins is produced by
    a. *Saccharomyces cerevisiae*
    b. *Escherichia coli*
    c. *Bacillus thuringiensis*
    d. *Staphylococcus aureus*

    ans. c.   diff. 7   obj. 5   p. 21

28. An effective method of using the insect toxin of *B. thuringiensis* is by
    a. using a viral plasmid to carry the gene to the plant
    b. mixing the toxin with DDT and spraying over the affected area
    c. cloning the genes for its production in *Pseudomonas*
    d. spraying it directly on the insect population

    ans. c.   diff. 7   obj. 5   p. 22

29. Which of the following genera of plant pathogens has been frequently used to transfer selected genes to specific plants?
    a. *Escherichia*
    b. *Bacillus*
    c. *Pseudomonas*
    d. *Agrobacterium*

    ans. d.   diff. 6   obj. 5   p. 24

30. The polymerase chain reaction technology is a very important breakthrough in genetic engineering. It allows researchers to
    a. duplicate specific pieces of DNA
    b. split DNA into single genetic operons
    c. plug the specific gene into an organism
    d. remove undesirable genes from molecules of DNA

    ans. a.   diff. 7   obj. 5   p. 26

31. In the polymerase chain reaction of DNA synthesis, the third and final cycle shows the production of
    a. 1×10^6 genes/hour of the combination of old and new molecules
    b. 1×10^9 molecules/hour of the newly synthesized chain
    c. an untold number of new polymers of DNA
    d. 1×10^7 template combinations of DNA and RNA

    ans. b.   diff. 6   obj. 5   p. 27

32. The polymerase chain reaction is a new biotechnology used to amplify small segments of
    a. proteins
    b. RNA
    c. DNA
    d. nucleotides

    ans. c.   diff. 5   obj. 5   p. 27

33. Forensic medicine can now use biotechnology to compare body samples at a crime scene with those of a suspect by using the technique of
   a. RNA typing
   b. DNA typing
   c. semen matching
   d. blood analyses

   ans. b.   diff. 5   obj. 5   p. 28

34. When human cells are lysed, about _____ fragments of _____ may be released.
   a. 100, genetic material
   b. 10,000, single-stranded RNA
   c. 1×10^6, protein particles
   d. 100,000, double-stranded DNA

   ans. d.   diff. 5   obj. 3   p. 6

True/False

1. The use of modern biochemical techniques to solve practical problems is defined as recombinant DNA technology.
   false (biotechnology)

2. The DNA molecule that carries the new DNA in shotgun cloning is called a vector.
   true

3. The "scissors" used to cut DNA into specific pieces are described as cloning vehicles.
   false (restriction endonuclease)

4. The plasmid used for cloning DNA into a bacterium must be sensitive to the antibiotic to which the bacterium is sensitive.
   false (plasmid must be resistant)

5. The most commonly used antibiotic for hybrid DNA transfer into *E. coli* is kanamycin.
   true

6. As a general rule in hybridization, the more closely the two organisms are related, the less likely the genes of one will be expressed in the cells of another.
   false (more likely to be expressed)

7. Products of genetic engineering like insulin are used to treat human diabetes.
   true

8. Bacterial production of insulin is safer but less economical than using animal glands.
   false (more economical)

9. When hormone extracts from cadaver glands are used in treatment, a common danger is infection.
   true

10. *Bacillus thuringiensis* produces an insecticide that can be used to protect plants against harmful insects. It is transferred into the plant by using an RNA virus and a promoter plasmid.
    false (transferred via *Agrobacterium*)

11. By using the polymerase chain reaction of DNA synthesis, very small segments of DNA can be amplified even in an impure state.
    true

12. One of the current areas of biotechnology is the improvement of the quantities of plants through selective breeding and genetic engineering.
    false (quality not quantity)

Matching

Match the statement in Column A with its most appropriate response from Column B.

Genetically Engineered Products

1. insulin
2. growth hormone
3. streptokinase
4. alpha interferon
5. interferon alpha-2b

a. dissolve blood clots
b. treat hepatitis B infection
c. treat dwarfism
d. treat diabetes
e. treat cancer

1 - d; 2 - c; 3 - a; 4 - e; 5 - b

Chapter 9

Multiple Choice

1. Lister developed his ideas on asepsis after studying the work of
    a. Pasteur
    b. Koch
    c. Jenner
    d. Watson

    ans. a.   diff. 4   obj. 1   p. 2

2. One of the first chemicals used by Lister to prevent surgical sepsis was
    a. carbolic acid
    b. iodine
    c. mercurochrome
    d. alcohol

    ans. a.   diff. 4   obj. 2   p. 2

3. The process of removing or killing all microorganisms on or in a material is termed
    a. sanitation
    b. disinfection
    c. sterilization
    d. cleanliness

    ans. c.   diff. 3   obj. 3   p. 3

4. If a material is treated to reduce the population of microorganisms, it is said to be
    a. clean
    b. dried
    c. sterilized
    d. disinfected

    ans. d.   diff. 4   obj. 3   p. 3

5. A term often used interchangeably with disinfection is
    a. sterilization
    b. sanitization
    c. decontamination
    d. cleanliness

    ans. c.   diff. 3   obj. 3   p. 4

6. If a chemical substance is primarily inhibitory rather than killing in its action, the suffix used to describe it is
    a. -cidal
    b. -static
    c. -cillin
    d. -mysin

    ans. b.   diff. 4   obj. 3   p. 4

7. Which of the following microorganism/microbe structures are most resistant to destruction?
    a. fungi
    b. viruses
    c. endospores
    d. capsules

    ans. c.   diff. 3   obj. 4   p. 6

8. In order for sterilization to proceed successfully, which of the following must precede it?
    a. mechanical removal of organics
    b. chemical soaking of substance
    c. nothing
    d. drying

    ans. a.   diff. 5   obj. 4   p. 6

9. Effectiveness of microcidal agents is greatest during which phase of the growth curve?
    a. decline
    b. stationary
    c. lag
    d. log

    ans. d.   diff. 4   obj. 5   p. 6

10. Heat kills microorganisms by
    a. dissolving the capsule
    b. destruction of the carbohydrates in the cell wall
    c. denaturing of the DNA
    d. coagulation of proteins

    ans. d.   diff. 4   obj. 6   p. 6

11. A common application of dry heat in the laboratory is used to
    a. prepare specimens for study
    b. sterilize media
    c. disinfect equipment
    d. sterilize the inoculating loop

    ans. d.   diff. 3   obj. 6   p. 8

12. Which of the following comparisons between dry heat and wet heat to kill microorganisms is most correct?
    a. dry heat 121°C/1.5 hrs.; wet heat 200°C/.25 hrs
    b. dry heat 160°C/1.0 hrs.; wet heat 200°C/.5 hrs
    c. dry heat 200°C/1.5 hrs.; wet heat 121°C/.25 hrs
    d. dry heat 100°C/2.0 hrs.; wet heat 100°C/.5 hrs

    ans. c.   diff. 6   obj. 6   p. 8

13. Which of the following is not a sterilization method?
    a. filtration
    b. autoclaving
    c. hot air oven
    d. pasteurization

    ans. d.   diff. 5   obj. 6   p. 9

14. Boiling is not a reliable sterilizing technique because
    a. water boils at higher temperatures at higher altitudes
    b. heat-sensitive instruments may be damaged
    c. heat-resistant spores are unaffected
    d. viruses are more sensitive to heat than bacteria

    ans. c.   diff. 3   obj. 6   p. 10

15. Under conditions in which rapid sterilization is necessary, flash autoclaving is employed. The temperature and pressure of this technique are
    a. 121°C at 10 pounds pressure
    b. 135°C at 15 pounds pressure
    c. 100°C at 5 pounds pressure
    d. 121°C at 15 pounds pressure

    ans. b.   diff. 6   obj. 6   p. 12

16. Coagulation of proteins is achieved most readily
    a. by moist heat under pressure
    b. by dry heat at lowered pressure
    c. in a hot air oven
    d. in the presence of antagonists

    ans. a.   diff. 5   obj. 6   p. 13

17. Pasteurization methods are used to protect milk consumers from diseases such as
    a. botulism, typhoid fever, and amoebiasis
    b. syphilis, pneumonia, and botulism
    c. tuberculosis, pneumonia, and salmonellosis
    d. tuberculosis, typhoid fever, and brucellosis

    ans. d.   diff. 6   obj. 7   p. 16

18. Heat-sensitive products can be sterilized by a process involving intermittent heating (to 60°C) and cooling (to 37°C) to destroy both vegetative and endospores. The process is called
    a. sterilization
    b. tyndallization
    c. pasteurization
    d. accommodation

    ans. b.   diff. 4   obj. 8   p. 17

19. Fluids that are delicate and cannot be sterilized by even moderate amounts of heat may successfully be sterilized by
    a. filtration
    b. ultra violet light (3000 Å)
    c. freezing
    d. drying
    e. Pasteurization

    ans. a.   diff. 4   obj. 9   p. 18

20. Generally speaking, filtration methods of sterilization cannot be used to remove
    a. viruses from blood products
    b. bacteria from protein-rich media
    c. infectious agents from alcoholic beverages
    d. bacterial enzymes from filtrate

    ans. a.   diff. 5   obj. 9   p. 21

21. Alcohols are most effective in disinfecting materials when the final concentration is ____ %.
    a. 85
    b. 65
    c. 75
    d. 45

    ans. c.   diff. 3   obj. 9   p. 24

22. Alcohols are ineffective in the destruction of
    a. lipid-based enzymes
    b. bacterial capsules
    c. bacterial endospores
    d. viral capsids

    ans. c.   diff. 4   obj. 9   p. 24

23. Which of the following is not used as a gas in the destruction of microorganisms?
    a. chlorine
    b. phenol
    c. ethylene oxide
    d. formaldehyde
    e. Hydrogen peroxide
    ans. b.   diff. 4   obj. 9   p. 25

24. Tincture of iodine and iodophores are widely used as
    a. antiseptics
    b. disinfectants
    c. sterilizing agents
    d. emulsifiers
    e. Sanitizers
    ans. a.   diff. 4   obj. 9   p. 26

25. Formaldehyde has been used as a highly effective disinfectant. It has been replaced recently due to its suspected carcinogenicity by
    a. glutaraldehyde
    b. iodophors
    c. aldehydes since they are gentler on the skin
    d. iodine tinctures since they are more dependable
    e. Quats
    ans. a.   diff. 6   obj. 9   p. 27

26. Which of the following chemical compounds are related to phenol and are used to kill vegetative bacteria?
    1) Lysol 2) cresol 3) hexachlorophene 4) aldehyde
    5) ammonia 6) xylenol
    a. 5, 6, 2, 3
    b. 1, 2, 3, 6
    c. 3, 1, 2, 5
    d. 1, 3, 5, 7

    ans. b.   diff. 7   obj. 7   p. 28

27. Burn patients are particularly susceptible to the Gram-negative rod _____ _____.
    a. *Pseudomonas aeruginosa*
    b. *Staphylococcus aureus*
    c. *Escherichia coli*
    d. *Micrococcus lutea* Proteus vulgaris
    e. *Salmonella paratyphi*

    ans. a.   diff. 6   obj. 4   p. 29

28. The leading cause of wound infections commonly destroyed by hand washing with hexachlorophene is
    a. *Enterobacter aerogenes*
    b. *Micrococcus aureus*
    c. *Escherichia coli*
    d. *Staphylococcus aureus*

    ans. d.   diff. 6   obj. 4   p. 30

29. The disinfection activity of _____ is (are) easily inactivated by soaps and detergents.
    a. phenol compounds
    b. quaternary ammonia
    c. aldehydes
    d. formaldehyde

    ans. b.   diff. 5   obj. 9   p. 31

30. Currently, 1% silver nitrate is out of favor as a treatment for pathogens in the eyes of newborns. It has been replaced by
    a. antibiotics
    b. alcohol drops
    c. saline flushing
    d. a stronger than 1% solution of silver nitrate

    ans. a.   diff. 5   obj. 10   p. 32

31. The eyes of newborns are presently treated with _____ to protect them from infection with _____.
    a. antibiotics, *Neisseria gonorrhoeae*
    b. 1% silver nitrate, herpes simplex
    c. hydrogen peroxide, *S. aureus*
    d. 70% alcohol, *S. pyogenes*

    ans. a.   diff. 7   obj. 10   p. 33

201

32. Ethylene oxide is an effective sterilizing agent in a closed chamber for
    a. 24 hours
    b. 1 to 3 hours
    c. 3 to 12 hours
    d. 20 minutes at 15 pounds pressure

    ans. c.   diff. 4   obj. 10.   p. 34

33. Germicides that can kill large numbers of bacteria as well as their endospores can be labelled
    a. antimicrobics
    b. disinfectants
    c. sterilants
    d. germicides

    ans. c.   diff. 6   obj. 10   p. 36

34. The most resistant of the vegetative organisms to the action of germicides is
    a. *Streptococcus pneumoniae*
    b. *Salmonella typhi*
    c. *Bacillus cereus*
    d. *Mycobacterium tuberculosis*
    e. S. aureus
    ans. d.   diff. 5   obj. 11   p. 37

35. The greatest hindrance to the germicidal activity of a chemical agent is the
    a. temperature at which it is used
    b. presence of organic material
    c. quality of the container in which it is used
    d. personnel performing the duty
    e. lg. initial # target microorgs,
    ans. b.   diff. 6   obj. 4   p. 37

36. In a medical setting, the term "critical item" is used to describe an instrument used to
    a. invade the tissues
    b. come in contact with skin
    c. take samples from mucous membranes
    d. perform therapeutic treatments

    ans. a   diff. 7   obj. 1   p. 38

37. In a medical setting, ordinary work surfaces can be safely disinfected with substances labelled as _____ by the EPA.
    a. "tuberculocidal agents"
    b. "semicritical"
    c. "critical"
    d. "noncritical"

    ans. d.   diff. 6   obj. 1   p. 38

38. The phenol coefficient is frequently employed to determine the
    a. tuberculocidal effectiveness of phenol
    b. effectiveness of a test disinfectant with phenol
    c. safety of phenol on living tissue
    d. market value of phenol with the public

    ans. b.   diff. 7   obj. 1   p. 41

39. Phenol coefficients are of little value when
    a. the testing conditions are inappropriate
    b. they are not on the package
    c. the testing temperature is greater than 25°C
    d. pure phenol is used
    e.
    ans. a.   diff. 6   obj. 1   p. 41

40. Electromagnetic radiation is not effective in killing microorganisms if _____ are used.
    a. ultraviolet rays
    b. gamma or X rays
    c. visible light rays
    d. short wavelengths

    ans. c.   diff. 6   obj. 12   p. 42

41. Gamma rays cause biological damage in living systems by
    a. adding inert molecules to enzyme systems
    b. altering the growth pattern of microbes
    c. producing hyperreactive ions
    d. introducing toxins

    ans. c.   diff. 7   obj. 12   p. 42

42. Gamma irradiation is used in the food industry to
    a. kill vegetative pathogens
    b. sterilize food
    c. destroy spores
    d. improve flavor and texture

    ans. a.   diff. 6   obj. 12   p. 43

43. Gamma radiation has several advantages over ethylene oxide in sterilizing delicate materials. One of the greatest advantages is
    a. no change in texture or flavor
    b. immediate use of sterilized item
    c. vitamin C and/or alcohol increased effectiveness
    d. spore destruction

    ans. b.   diff. 7   obj. 12   p. 43

44. The most effective antimicrobial ultraviolet wavelength is
    a. 0–1,000 nanometers
    b. 75–175 nanometers
    c. 400–475 nanometers
    d. 200–310 nanometers

    ans. d.   diff. 4   obj. 12   p. 45

45. Which of the following microbial life types are least susceptible to ultraviolet light?
    a. actively growing cells
    b. spores
    c. stabilized cultures
    d. cellulose inclusion bodies

    ans. b.   diff. 5   obj. 12   p. 46

True/False

1. At sea level, water temperature rises to 121°C (boiling), which is sufficient to kill endospores.
   false (at sea level water boils at 100°C)

2. The effectiveness of the autoclave is related to the increased ability of moist heat to coagulate proteins more readily than dry heat.
   true

3. Pasteurization is an effective sterilization technique for perishable beverages.
   false (pasteurization is not sterilization)

4. Properly chlorinated drinking water requires 0.3 ppm of free chlorine.
   false (0.5 parts/million)

5. Iodophors are not reliable for disinfecting instruments contaminated by *M. tuberculosis*.
   true

6. Medical instruments can be safely sterilized by immersion in mercury compounds for 30 minutes.
   false (soaking will not sterilize)

7. *Streptococcus pyogenes* is considered the challenge organism for destruction by a germicide.
   false (*Mycobacterium tuberculosis*)

8. Even if an object is not thoroughly cleansed, it can still be successfully disinfected.
   false (must be clean first)

9. Hospital disinfectants labelled "tuberculocidal" can be used successfully to destroy viruses and fungal spores.
   false (not dependable on viruses or spores)

10. Phenol coefficients of less than 1 indicate that the test disinfecting agent is more effective than phenol.
    true

11. Gamma irradiation is effective in sterilizing food products for safe consumption.
    false (Gamma radiation can be used for killing pathogens in food without altering the quality of that food but not sterilizing it.)

    Single use items

12. Microbial cultures in the actively growing stages are most susceptible to UV light.
    true

Chapter 10

Multiple Choice

1. In the 1870s, Cohn published several papers on bacterial classification; his methods grouped bacteria according to their
   a. biochemistry
   b. shape
   c. Gram stain
   d. nucleic acids

   ans. b.   diff. 5   obj. 1   p. 1

2. In 1908, Orla-Jensen proposed a scheme of bacterial classification based on
   a. evolutionary relationships
   b. physiology
   c. nucleic acids
   d. Gram stains

   ans. b.   diff. 5   obj. 1   p. 2

3. In the 1930s, Kluyver and van Neil proposed a classification system based on
   a. presumed evolutionary relationships
   b. biochemical relationships
   c. Gram reactions
   d. DNA sequences

   ans. a.   diff. 5   obj. 1   p. 2

4. In 1970, Stanier proposed that classification could be based on
   a. physiology and fermentation
   b. RNA similarities and DNA differences
   c. DNA molecules and their products
   d. evolution

   ans. c.   diff. 5   obj. 1   p. 2

5. Classification is the arranging of organisms into
   a. separate categories
   b. diverse groups
   c. related groups
   d. evolutionary schemes

   ans. c.   diff. 4   obj. 1   p. 4

6. The fossil record suggests that bacteria inhabited the earth as early as
   a. 100,000 years ago
   b. 2.4×10^9 years ago
   c. 1.2×10^5 years ago
   d. 3.5×10^9 years ago

   ans. d.   diff. 5   obj. 2   p. 4

7. The basic taxonomic unit in the classification schemes of all living things is
   a. class
   b. kingdom
   c. species
   d. genus

   ans. c.   diff. 5   obj. 2   p. 4

8. In higher organisms, successful mating can occur between members of the same
   a. species
   b. genus
   c. class
   d. order

   ans. a.   diff. 6   obj. 2   p. 5

9. The strain designation of a species of bacteria is necessary to classify organisms based on their
   a. specific biochemical differences
   b. individual morphological characteristics
   c. Gram stain variabilities
   d. conjugation activities

   ans. a.   diff. 6   obj. 3   p. 6

10. The kingdom in which all bacteria are officially classified is
    a. Eukaryotae
    b. Monera
    c. Plantae
    d. Prokaryotae

    ans. d.   diff. 4   obj. 3   p. 6

11. Bacteria having the greatest number of characteristics in common with each other belong to the same
    a. kingdom
    b. genus
    c. family
    d. class

    ans. a.   diff. 4   obj. 3   p. 6

12. Bacteria having the fewest characteristics in common would belong to the same
    a. order
    b. family
    c. class
    d. species

    ans. d.   diff. 4   obj. 3   p. 6

13. The reference text for taxonomic descriptions of bacteria is
    a. *Bergey's Manual of Systematic Bacteriology*
    b. *Gray's Anatomy*
    c. your textbook
    d. *Webster's Dictionary of Bacteriology*

    ans. a.   diff. 4   obj. 3   p. 7

14. The *Bergey's Manual* groups bacteria according to their
    a. evolutionary relationships
    b. kingdom
    c. shared properties
    d. staining differences

    ans. c.   diff. 6   obj. 3   p. 7

15. The relatedness between organisms by determining how many characteristics they have in common is the basis of
    a. DNA bases
    b. evolutionary taxonomy
    c. amino acid sequences
    d. numerical taxonomy

    ans. d.   diff. 5   obj. 3   p. 9

16. The total number of characteristics that are common to two strains of bacteria are expressed in terms of their
    a. amino acid descriptions
    b. numerical similarity
    c. base synchronization
    d. similarity coefficient

    ans. d.   diff. 6   obj. 3   p. 9

17. When using the numerical taxonomy system of classification, if strains have a greater than ____ percent similarity they are termed a single species.
    a. 75
    b. 80
    c. 70
    c. 90

    ans. d.   diff. 6   obj. 4   p. 10

18. The common ancestry of two organisms can be determined by their
    a. RNA transcription
    b. DNA sequences
    c. biochemical metabolism
    d. lactose fermentation

    ans. b.   diff. 7   obj. 4   p. 10

19. Modern approaches to evolutionary taxonomy often involve
    a. comparison of DNA and RNA
    b. biochemical differences
    c. protein similarities
    d. mutation frequencies

    ans. a.   diff. 6   obj. 4   p. 10

20. The GC (guanine/cytosine) content varies widely among organisms, with numbers ranging from ____ % to ____ %.
    a. 22, 78
    b. 10, 80
    c. 0, 100
    d. 50, 60

    ans. a.   diff. 6   obj. 4   p. 11

21. The most widely used and accurate way to determine the relatedness of organisms is to
    a. use numerical taxonomy as a reference
    b. measure the length of the DNA chain
    c. count the G-C and A-V bondings
    d. compare the base sequence in DNA

    ans. d.   diff. 6   obj. 4   p. 12

22. The more closely two organisms are related, the greater the
    a. degree of nucleic acid hybridization
    b. mutation rate
    c. possibility of conjugation
    d. temperature at which denaturing will occur

    ans. a.   diff. 6   obj. 4   p. 14

23. Dr. Carl Woese and his colleagues have divided organisms into three groups, which evidence suggests arose from a common ancestor. These three groups are the:
    a. viruses, fungi, and protozoa
    b. Prokaryota, Eukaryota, and viruses
    c. archaebacteria, eubacteria, and eukaryotes
    d. Monera, Protista, and Archaebacteria

    ans. c.   diff. 7   obj. 4   p. 17

24. The three groups into which Dr. Woese divided all organisms is based on related properties and
    a. sugar fermentation
    b. amino acid denaturization
    c. signature sequences
    d. DNA transcriptase

    ans. c.   diff. 6   obj. 4   p. 17

25. Relatedness between organisms that seem quite different based on their phenotypic characteristics has been made by using
    a. numerical taxonomy
    b. signature sequences
    c. similarity coefficient
    d. subspecies specialization

    ans. b.   diff. 7   obj. 4   p. 18

26. The two families Rickettsiaceae and Rhizobiaceae have been shown to be related when their sequences of _____ were compared.
    a. DNA
    b. amino acids
    c. 16S ribosomal RNA
    d. protein synthesis

    ans. c.   diff. 7   obj. 4   p. 19

27. Detection of an organism that grows very slowly *in vitro* may be very difficult; its presence may be more easily detected by
    a. measuring its metabolic by-products
    b. its pigment appearance
    c. its antibiotic sensitivity
    d. using discriminating media

    ans. a.   diff. 5   obj. 4   p. 27

28. Which of the following organisms can be detected by labelling and then measuring their radiotagged by-products rather than their appearance in pure culture?
    1) *M. tuberculosis*   2) *S. aureus*   3) *Clostridium difficile*
    4) *S. marcescens*   5) *B. tetani*
    a. 2, 3, and 4
    b. 2 and 4
    c. 1, 3, and 5
    d. 1 and 3

    ans. d.   diff. 5   obj. 4   p. 27

29. DNA sequence similarity of two organisms can be determined by comparing the patterns of fragments produced when the
    a. numerical taxonomy is similar
    b. host RNA and DNA is released
    c. growth medium is incubated at 40°C
    d. same restrictive enzyme is used

    ans. d.   diff. 6   obj. 4   p. 29

30. Nucleic acid techniques have great power as diagnostic tools because they are
    a. very specific
    b. easy to obtain
    c. dissimilar in different species
    d. subject to degradation

    ans. a.   diff. 6   obj. 4   p. 30

31. DNA probes have been very useful in
    a. identifying organisms in pure culture
    b. detecting toxins difficult to isolate by standard methods
    c. coding the DNA in organisms grown *in vitro*
    d. diagnosing diseases of protozoa

    ans. b.   diff. 6   obj. 4   p. 31

32. The approach of _____ _____ is being used to identify agents causing infectious diseases that cannot be cultured.
    a. DNA probing
    b. signature sequences
    c. 16S RNA determination
    d. Gram staining

    ans. b.   diff. 5   obj. 4   p. 33

33. _____ allows an investigator to determine whether HIV is present in people who cannot be identified as being HIV-positive by conventional means.
    a. DNA probe
    b. PCR technology
    c. Signature sequences
    d. Numerical taxonomy

    ans. b.   diff. 6   obj. 4   p. 33

34. The basic taxonomic unit in bacteria is
    a. order
    b. genera
    c. family
    d. species

    ans. d.   diff. 4   obj. 3   p. 36

35. Which of the following is not a method of identifying bacteria?
    a. detection of microbial by-products
    b. culture techniques
    c. pathogenicity
    d. molecular biological techniques

    ans. c.   diff. 6   obj. 3   p. 37

True/False

1. The only scientifically correct way to classify organisms is by their relatedness through Gram reaction.
   false (relatedness through evolution)

2. Taxonomy and classification schemes overlap and frequently are identical.
   true

Matching

Match the scientist with the method he used to classify bacteria.

1. Ferdinand Cohn
2. Sigurd Orla-Jensen
3. Albert Kluyver and Jan van Neil
4. Roger Stanier
5. Carl Woese

a. evolution (relatedness)
b. physiology (fermentation)
c. RNA sequences
d. shape
e. genetics

1 - d; 2 - b; 3 - a; 4 - e; 5 - c

3. If two organisms differ only in minor phenotypic properties, then one organism may be considered as a separate genus.
   false (subspecies or strain)

4. The reference text for descriptions of bacteria is contained in *Bergey's Manual.*
   true

5. The term *similarity coefficient* is used to describe the results of classification by the system of DNA recombination.
   false (numerical taxonomy)

6. Organisms that are related have DNA base compositions that must be identical.
   false (similar or identical)

7. If the GC content of two organisms differs by more than a few percent, they cannot be related.
   true

8. Rickettsiaceae and Rhizobiaceae have one obvious characteristic in common—they are both extracellular pathogens.
   false (both have intracellular associations with eukaryotes)

9. Woese has proven that the mitochondria of eukaryotic cells have arisen from bacterial parasites.
   false (from endosymbionts)

10. Clinical microbiology labs routinely test for products of metabolism rather than relying on conventional cultural techniques.
    false (tests are relatively new and not widely used)

Chapter 11

Multiple Choice

1. Organisms that are classified as "Gram-negative" when using the Gram stain are colored
   a. purple
   b. blue
   c. red
   d. green

   ans. c.   diff. 4   obj. 1   p. 1

2. A key test that has been used in the initial identification of bacterial species since the 1880s is the
   a. identification in *Bergey's Manual*
   b. use of selective media
   c. simple stain
   d. Gram stain

   ans. d.   diff. 3   obj. 1   p. 2

3. The most complete classification of bacteria can be found in
   a. *Difco Manual*
   b. your text book
   c. *World Book Encyclopedia*
   d. *Bergey's Manual of Systemic Bacteriology*

   ans. d.   diff. 4   obj. 2   p. 3

4. The organisms that are described as spirals with a flexible cell wall and a unique motility mechanism are the
   a. Spirochetes
   b. rods
   c. spirals
   d. cocci

   ans. a.   diff. 5   obj. 3   p. 4

5. Movement of spirochetes occurs by means of structures called
   a. cilia
   b. flagella
   c. axial filaments
   d. spirals

   ans. c.   diff. 4   obj. 3   p. 4

6. Each axial filament is made up of fibrils identical in structure to
   a. pili
   b. flagella
   c. cilia
   d. pseudopods

   ans. b.   diff. 5   obj. 3   p. 4

7. Since many spirochetes are difficult to cultivate, their classification is based on their morphology and
   a. ability to cause disease
   b. Gram reaction
   c. sugar utilization
   d. number and types of flagella

   ans. a.   diff. 5   obj. 3   p. 5

8. Which of the following is mismatched?
   a. *Borrelia recurrentis*—relapsing fever
   b. *Borrelia burgdorferi*—Lyme disease
   c. *Leptospira interrogans*—leptospirosis
   d. *Treponema pallidum*—yaws

   ans. d.   diff. 6   obj. 3   p. 6

9. A major difference between the spirals and the spirochetes is the presence of a(n)
   a. cell membrane
   b. flexible cell wall
   c. flagellum
   d. nonflexible cell wall

   ans. d.   diff. 5   obj. 4   p. 6

10. A microaerophilic helical-shaped pathogen is
    a. *Campylobacter jejuni*
    b. *Spirillum minus*
    c. *Bordetella pertussis*
    d. *Bdellovibrio comma*

    ans. a.   diff. 6   obj. 4   p. 7

208

11. The flagellated spiral that can transmit rat bite fever by bites of infected animals is
    a. *Campylobacter jejuni*
    b. *Spirillum minus*
    c. *Bordetella pertussis*
    d. *Bdellovibrio comma*

    ans. b.　　diff. 7　　obj. 4　　p. 7

12. The tiny, Gram-negative curved rod that is unique as a bacterial parasite is
    a. *Neisseriae*
    b. *Rickettsiae*
    c. *Bdellovibrio*
    d. *Borelliae*

    ans. c.　　diff. 8　　obj. 4　　p. 8

13. The important genera of nitrogen-fixing, Gram-negative aerobic rods is
    a. *Rhizobium*
    b. *Pseudomonas*
    c. *Francisella*
    d. *Legionella*

    ans. a.　　diff. 7　　obj. 4　　p. 9

14. The genera of organisms that are Gram-negative rods that are motile by polar flagella and often produce pigments are the
    a. *Azotobacter*
    b. *Pseudomonas*
    c. *Yersinia*
    d. *Vibrio*
    e. *Proteus*

    ans. b.　　diff. 7　　obj. 4　　p. 11

15. The unique genus of bacteria able to fix atmospheric nitrogen and form cysts
    a. *Pseudomonas*
    b. *Azotobacter*
    c. *Vibrio*
    d. *Proteus*

    ans. b.　　diff. 7　　obj. 4　　p.12

16. Of the 25 recognized genera of enterobacteria, how many include medically important species?
    a. 20–25
    b. 10–15
    c. all
    d. none

    ans. b.　　diff. 6　　obj. 4　　p. 16

17. The bacterial enzyme luciferase has been identified in bacteria that
    a. infect insect larvae
    b. can cause disease
    c. are found deep in the ocean
    d. produce light in the presence of oxygen

    ans. d.　　diff. 6　　obj. 4　　p. 17

18. Organisms that are killed by exposure to oxygen are classified as
    a. aerobes
    b. facultative anaerobes
    c. anaerobes
    d. halophiles

    ans. c.　　diff. 5　　obj. 4　　p. 17

19. Which of the following groups of organisms are obligate intracellular parasites?
    a. rickettsias and chlamydias
    b. viruses and bacteroides
    c. cocci and spirals
    d. viroids and spirochetes

    ans. a.　　diff. 6　　obj. 4　　p. 19

20. Which arthropod is generally responsible for the transmission of rickettsias?
    a. fleas
    b. ticks
    c. mites
    d. flies

    ans. b.　　diff. 5　　obj. 4　　p. 21

21. Which of the following rickettsias is transmissible from animal to animal without necessarily involving a blood-sucking parasite?
    a. *Coxiella burnetii*
    b. *Rickettsia prowazeki*
    c. *Rickettsia rickettsii*
    d. *Chlamydia psittaci*

    ans. a.   diff. 6   obj. 4   p. 22

22. Which of the following eubacteria has the general appearance of a Gram-negative cell wall but lacks the peptidoglycan layer?
    a. *Rickettsia*
    b. *Chlamydia*
    c. *Enterobacter*
    d. *Leptotrichia*

    ans. b.   diff. 7   obj. 7   p. 23

23. Microorganisms that have established a stable intracellular relationship with a eukaryotic host cell are referred to as
    a. chloroplasts
    b. parasites
    c. pathogens
    d. endosymbionts

    ans. d.   diff. 5   obj. 6   p. 24

24. The endosymbiont that grows in the root cells of some plants and fixes atmospheric nitrogen is
    a. *Lyticum flagellatum*
    b. X-bacteria
    c. bipolar bodies
    d. *Rhizobium*

    ans. d.   diff. 5   obj. 6   p. 25

25. The species of gliding bacteria that are able to break down complex polysaccharides such as cellulose are important as
    a. endosymbionts of fish
    b. pathogens of plants
    c. degraders of organic material
    d. precursors of the gliding organisms such as snakes

    ans. c.   diff. 7   obj. 8   p. 27

26. _____ are non-photosynthetic, aerobic, Gram-negative rods that employ gliding motility but are unrelated to the gliding, non-fruiting bacteria.
    a. Eubacteria
    b. Myxobacteria
    c. *Pseudomonas*
    d. *Agrobacterium*

    ans. b.   diff. 7   obj. 8   p. 29

27. The microcysts of Myxobacteria are resistant to heat, drying, and radiation, but not nearly as well as the
    a. endospores of bacteria
    b. capsules of bacteria
    c. endosymbionts of eukaryotes
    d. parasites of insects

    ans. a.   diff. 8   obj. 8   p. 29

28. Phototrophic bacteria obtain their energy from sunlight in contrast to the chemotrophic bacteria that obtain their energy from
    a. oxygen
    b. ultraviolet light
    c. organic or inorganic material
    d. lipoproteins

    ans. c.   diff. 7   obj. 9   p. 31

29. Which of the following organisms probably had ancestors that predominated in the early history of the earth?
    a. oxygenic phototrophic bacteria
    b. sheathed bacteria
    c. anoxygenic phototrophic bacteria
    d. aerobic chemolithotrophic bacteria

    ans. c.   diff. 6   obj. 9   p. 32

30. The Gram-negative prokaryotes that liberate oxygen during photosynthesis are the
    a. Cyanobacteria
    b. *Rhodospirillum*
    c. *Chromatium*
    d. *Pelodictyon*

    ans. a.   diff. 5   obj. 9   p. 34

210

31. Cyanobacteria are found widespread growing on ponds polluted with
    a. nitrogenous compounds
    b. fastidious bacteria
    c. myxobacteria
    d. heat-producing bacteria

    ans. a.   diff. 6   obj. 9   p. 34

32. The chemolithotrophic bacteria grow in the dark using an inorganic substance as an energy source and
    a. light as an alternate energy source
    b. H_2S as a sulfur source
    c. hydrogen as an electron source
    d. CO_2 as a carbon source

    ans. d.   diff. 7   obj. 9   p. 37

33. Bacteria that are important in the nitrogen cycle by oxidizing nitrites to nitrates are the
    a. *Nitrosospira*
    b. *Cyanobacteria*
    c. *Leptospira*
    d. *Nitrobacter*

    ans. d.   diff. 8   obj. 9   p. 39

34. The bacteria that were isolated from hot springs in Japan, and are aerobic and an obligate hydrogen oxidizers belongs to the genera
    a. *Poracoccus*
    b. *Hydrogenobacter*
    c. *Nocardia*
    d. *Bacillus*

    ans. b.   diff. 6   obj. 9   p. 40

35. Species that divide in parallel planes and produce chains of bacteria belong to the genus
    a. *Enterobacter*
    b. *Staphylococcus*
    c. *Streptococcus*
    d. *Pneumococcus*

    ans. c.   diff. 5   obj. 10   p. 43

36. If the red blood cells are partly destroyed around a bacterial colony, the effect is called
    a. ∝-(alpha) hemolysis
    b. β-(beta) hemolysis
    c. γ-(gamma) hemolysis
    d. viridans

    ans. a.   diff. 4   obj. 10   p. 43

37. β-(beta) hemolysis is characterized by _____ of the red blood cells around the bacterial colony.
    a. incomplete hemolysis
    b. complete hemolysis
    c. no hemolysis
    d. variable hemolysis

    ans. b.   diff. 4   obj. 10   p. 44

38. Streptococci that produce no red blood cell destruction are referred to as _____ streptococci.
    a. viridans
    b. ∝-(alpha)
    c. β-(beta)
    d. γ-(gamma)

    ans. d.   diff. 4   obj. 10   p. 44

39. Repeated divisions in different planes that produce clusters of organisms are characteristic of the genus
    a. *Staphylococcus*
    b. *Pseudomonas*
    c. *Streptococcus*
    d. *Diplococcus*

    ans. a.   diff. 3   obj. 10   p. 44

40. Two genera of Gram-positive, rod-shaped bacteria that produce spores are
    a. *Escherichia* and *Salmonella*
    b. *Staphylococcus* and *Streptococcus*
    c. *Bacillus* and *Enterobacter*
    d. *Bacillus* and *Clostridium*

    ans. d.   diff. 4   obj. 10   p. 44

41. The most resistant forms of life exist as
    a. vegetative cells
    b. cysts
    c. spores
    d. capsules

    ans. c.   diff. 3   obj. 10   p. 45

42. Which of the following organisms is not an anaerobe?
    a. Clostridium cereus
    b. Clostridium botulinum
    c. Clostridium tetani
    d. Bacillus anthracis

    ans. d.   diff. 3   obj. 10   p. 45

43. Which of the following Gram-positive rods is among those that colonize the human vagina during childbearing years?
    a. Lactobacillus
    b. Clostridium
    c. Streptococcus
    d. Enterobacter

    ans. a.   diff. 5   obj. 11   p. 47

44. The lactobacilli, in their role as normal flora of the vagina, help the vagina resist infection by contributing to the
    a. food for other normal flora found there
    b. neutrality of the vagina mucus
    c. acidity of the vagina
    d. fertility of the host

    ans. c.   diff. 5   obj. 1   p. 47

45. Which of the following Gram-positive rods is capable of both human disease and growth at refrigerator temperature?
    a. Erysipelothrix rhusiopathiae
    b. Listeria monocytogenes
    c. Bacillus subtiles
    d. Staphylococcus aureus

    ans. b.   diff. 6   obj. 1   p. 48

46. The human pathogen that shows irregular staining, misshapen cells, and even a fission process resulting in a palisade arrangement of cells is
    a. Bacillus anthracis
    b. Listeria monocytogenes
    c. Corynebacterium
    d. Actinomyces israelii

    ans. c.   diff. 7   obj. 1   p. 49

47. The species of organism that are pleomorphic, aerotolerant, and form from fragmentation of branching filaments are the
    a. Corynebacteria
    b. Actinomyces
    c. Arthrobacter
    d. Lactobacilli

    ans. b.   diff. 7   obj. 11   p. 50

48. The species of bacteria largely responsible for the degradation of insecticides and herbicides in the soil and in fresh media, and show characteristic coryneform shape are the
    a. Micrococcus
    b. Arthrobacter
    c. Lactobacilli
    d. corynebacteria

    ans. b.   diff. 7   obj. 1   p. 50

49. The Gram-positive rod that is also acid fast and is a human pathogen is
    a. Mycobacterium tuberculosis
    b. Corynebacterium diphtheria
    c. Streptococcus lactis
    d. Listeria monocytogenes

    ans. a.   diff. 7   obj. 1   p. 51

50. Which of the following organisms is the causative agent of leprosy?
    a. Mycobacterium tuberculosis
    b. Mycobacterium leprae
    c. Corynebacterium diphtheriae
    d. Mycobacterium smegmatis

    ans. b.   diff. 7   obj. 1   p. 51

51. Filamentous organisms that usually form aerial mycelia and then fragment into bacillary forms are the
    a. Nocardioforms
    b. Actinomycetes
    c. Conidia
    d. Frankia

    ans. a.   diff. 6   obj. 11   p. 53

52. The prokaryotes that lack cell walls are the
    a. *Chlamydia*
    b. *Actinomyces*
    c. *Mycoplasma*
    d. *Mycobacteria*

    ans. c.   diff. 6   obj. 12   p. 59

53. Organisms that typically produce colonies with a "fried egg" appearance are the
    a. *Mycoplasma*
    b. *Actinomycetes*
    c. *Chlamydia*
    d. *Mycobacteria*

    ans. a.   diff. 5   obj. 12   p. 60

54. The group of prokaryotes that lack peptidoglycan in their cell walls are the
    a. Archaebacteria
    b. *Eubacteria*
    c. *Mycoplasma*
    d. *Chlamydia*

    ans. a.   diff. 6   obj. 12   p. 62

55. Which of the following is not a member of the Archaebacteria?
    a. thermoacidophiles
    b. halobacteria
    c. eubacteria
    d. methanogens

    ans. c.   diff. 7   obj. 12   p. 66

True/False

1. The microorganisms with a flexible cell wall and axial filaments are the spirochetes.
   true

2. The Gram stain was developed in the laboratory directed by Christian Gram in the 1880s.
   false   (Carl Friedlander)

3. Endosymbionts are microorganisms that have established a stable intracellular relationship with a prokaryotic host cell.
   false   (eukaryotic)

4. Members of the genera *Beggiatoa* and *Thiothrix* obtain energy by oxidizing CO_2 to granules of carbon that are deposited intracellularly.
   false   (H_2S to granules of sulfur)

5. *Sphaerotilus* species are sheathed bacteria that produce brownish scum beneath the surface of polluted streams and interfere with sewage treatment by plugging pipes.
   true

6. Anoxygenic phototrophs grow photosynthetically only under aerobic conditions.
   false   (anaerobic)

7. Bacteria with a flexible cell wall and seen best using dark-field illumination are the spirals.
   false   (spirochetes)

8. Cyanobacteria produce colored scum on the surface of water polluted with nitrogenous and phosphate compounds.
   true

9. Heterocysts of cyanobacteria are specialized structures that contain chlorophyll.
   false   (function in nitrogen fixation)

10. *Thiobacillus* species are Gram-negative, rod-shaped aerobic organisms that obtain energy by oxidizing sulfur compounds.
    true

11. Streptococci are generally catalase positive, in contrast to the staphylococci, which are catalase negative.
    false   (staphylococci are catalase positive, streptococci are catalase negative)

12. Both of the genera *Bacillus* and *Clostridium* are Gram-positive, aerobic spore-forming rods.
    false   (*Clostridium* are anaerobic)

13. *Nocardia orientalis* is best known for its production of the antibiotic tetracycline.
    false   (vancomycin)

Matching

Match the organism with the disease caused:

1.

1.	*Pseudomonas aeruginosa*		a.	tularemia
2.	*Francisella tularensis*	b.		pertussis
3.	*Legionella pneumophila*	c.		gonorrhea
4.	*Bordetella pertussis*	d.		Legionnaires' disease
5.	*Neisseria gonorrhoeae*	e.		burns

1 - e; 2 - b; 3 - d; 4 - b; 5 - c

1.	*Escherichia coli*		a.	diarrhea
2.	*Klebsiella pneumoniae*	b.		bubonic plague
3.	*Yersinia pestis*	c.		pneumonia
4.	*Salmonella typhi*	d.		gastroenteritis
		e.		typhoid fever

1 - a; 2 - c; 3 - b; 4 - e

1.	*Rickettsia rickettsii*		a.	mild pneumonia
2.	*Chlamydia pneumoniae*	b.		Q fever
3.	*Coxiella burnetii*	c.		Rocky Mountain spotted fever
4.	*Chlamydia psittaci*	d.		gastroenteritis
		e.		parrot fever

1 - c; 2 - a; 3 - b; 4 - e

1.	*Bacillus anthracis*		a.	tetanus
2.	*Clostridium tetani*	b.		gas gangrene
3.	*Clostridium botulinum*	c.		anthrax
		d.		botulism

1 - c; 2 - a; 3 - d

Chapter 12

Multiple Choice

1. Research into the cause of Texas cattle fever led Dr. Theobald Smith to discover the paramount importance of _____ in the spread of disease.
   a. crowded populations
   b. environment changes
   c. insect vectors
   d. southern hospitality

   ans. c   diff. 4   obj. 1   p. 1

2. Many differences separate the prokaryotic organisms from the eukaryotic organisms. Among the most important is the presence of:
   a. DNA
   b. RNA
   c. a cell membrane
   d. a nucleus

   ans. d   diff. 4   obj. 2   p. 4

3. Microscopic free-living organisms possessing chlorophyll are classified with the
   a. algae
   b. fungi
   c. bacteria
   d. worms

   ans. a   diff. 3   obj. 3   p. 5

4. Algae are important economically as
   a. infectious agents
   b. local contaminants
   c. saprophytes
   d. primary producers

   ans. d   diff. 4   obj. 2   p. 5

5. Algae are important environmentally as
   a. local contaminants
   b. local flora
   c. major producers of oxygen
   d. major producers of carbon dioxide

   ans. c   diff. 5   obj. 3   p. 7

6. A word used to describe an intimate relationship between members of different species is
   a. symbiosis
   b. synergism
   c. parasitism
   d. mutualism

   ans. a   diff. 3   obj. -   p. 7

7. Free floating, photosynthetic organisms found in marine environments are
   a. blue-green algae
   b. phytoplankton
   c. *Brucella*
   d. krill

   ans. b   diff. 4   obj. -   p. 7

8. The microscopic crustaceans that feed the benthic whales are
   a. phytoplankton
   b. bacteria
   c. krill
   d. diatoms

   ans. c   diff. 3   obj. -   p. 7

9. Algae often grow where other forms of life cannot thrive. They prepare the surface for invasion of more complex members of the biological community and, therefore, are known as _____ species.
   a. pioneer
   b. synergistic
   c. pathogenic
   d. saprophytic

   ans. a   diff. 4   obj. 3   p. 8

10. Factors such as tides, water depth, salinity, and sunlight are considerations in determining the _____ of algae.
    a. growth cycle
    b. prominent pigment
    c. pattern of distribution
    d. evolutionary place

    ans. c   diff. 4   obj. 5   p. 8

11. Microscopic algae can be single-celled organisms floating free or they can be _____, growing in long chains or filaments.
    a. colonial
    b. pathogenic
    c. multi-colored
    d. multi-cellular

    ans. a   diff. 5   obj. 3   p. 8

12. Macroscopic algae possess a special structure that acts as an anchor. It is *commonly* referred to as a(n)
    a. anchor
    b. root system
    c. holdfast
    d. thallus

    ans. c   diff. 5   obj. 5   p. 10

13. All the following statements about algae are true except
    a. they use light for photosynthesis
    b. all use CO_2 as a carbon source
    c. they produce O_2 as a carbon source
    d. all are unicellular

    ans. d   diff. 4   obj. 6   p. 11

14. The typical cell wall of an algae cell is rigid due to the presence of
    a. cellulose
    b. pectin
    c. chitin
    d. carotene

    ans. a   diff. 5   obj. 5   p. 11

15. Shells of diatoms are useful economically as
    a. fertilizers
    b. filters
    c. stabilizers
    d. have no known function

    ans. b   diff. 6   obj. 5   p. 12

16. The site in a photosynthetic cell where photosynthesis takes place is the
    a. carotenoid
    b. chloroplast
    c. nucleus
    d. stipe

    ans. b   diff. 5   obj. 5   p. 13

17. A type of asexual reproduction in multicellular, filamentous algae is
    a. fission
    b. budding
    c. meiosis
    d. fragmentation

    ans. d   diff. 5   obj. 5   p. 14

18. The asexual reproductive cycle of eukaryotes characterized by very distinctive stages is commonly known as
    a. fission
    b. mitosis
    c. budding
    d. meiosis

    ans. b   diff. 6   obj. 5   p. 14

19. Meiosis is also known as
    a. reduction division
    b. fragmentation
    c. mitosis
    d. budding

    ans. a   diff. 5   obj. 6   p. 15

20. The diploid number of chromosomes may be referred to as
    a. 1N
    b. 2N
    c. 3N
    d. 4N

    ans. b   diff. 4   obj. 6   p. 14

216

21. The 1N chromosome number is also referred to as
   a. diploid
   b. polyploid
   c. haploid
   d. triploid

   ans. c   diff. 4   obj. 6   p. 15

22. The fusion of gametes results in a cell called the
   a. zygote
   b. conceptus
   c. cell
   d. fragment

   ans. a   diff. 4   obj. 6   p. 15

23. Organisms that exist as haploid in one generation and diploid in the next exhibit a characteristic described as
   a. confusion
   b. meiotic description
   c. alternation of generation
   d. mitotic exactness

   ans. c   diff. 5   obj. 6   p. 15

24. The dinoflagellates that cause "red tide" belong to all of these except
   a. *Pyrrhophyta*
   b. *Gymnodinium*
   c. *Gonyaulax*
   d. *Dinoflagellata*

   ans. d   diff. 6   obj. 7   p. 16

25. Paralytic shellfish poisoning occurs when humans eat shellfish that
   a. metabolize a neurotoxin
   b. are harvested from approved water
   c. have fed on *Gonyaulax* sp.
   d. have fed on *Gymnodium* sp.

   ans. c   diff. 6   obj. 7   p. 16

26. The chemical nature of neurotoxins produced by *Gonyaulax* is
   a. protein
   b. polypeptide
   c. glycopeptide
   d. nonprotein

   ans. d   diff. 6   obj. 7   p. 16

27. The group of organisms that are characterized as heterotrophic, eukaryotic saprophytes are
   a. algae
   b. fungi
   c. thallophytes
   d. bacteria

   ans. b   diff. 5   obj. 8   p. 18

28. Most fungal diseases affect
   a. plants
   b. animals
   c. bacteria
   d. plants and humans

   ans. a   diff. 5   obj. 8   p. 19

29. The portion of the earth in which life exists is the
   a. stratosphere
   b. atmosphere
   c. aquasphere
   d. biosphere

   ans. d   diff. 6   obj.   p. 19

30. The principal decomposers of organic material in the biosphere are
   a. bacteria and viruses
   b. fungi and bacteria
   c. viruses
   d. fungi and viruses

   ans. a   diff. 6   obj. 8   p. 19

31. The return of carbon dioxide to the atmosphere and nitrogen to the soil is the principal responsibility of the
   a. bacteria and viruses
   b. fungi and bacteria
   c. viruses and plants
   d. fungi and viruses

   ans. a   diff. 6   obj. 8   p. 19

32. The habitats of fungi are most frequently
   a. aquatic
   b. limited to arid areas
   c. versatile
   d. synthetic

   ans. c   diff. 5   obj. 9   p. 19

33. Fungal spores are a major cause of
    a. food spoilage
    b. fermentation
    c. asthma
    d. anaerobic disease

    ans. c   diff. 6   obj. 9   p. 20

34. The pH at which fungi thrive is at or below
    a. 3.0
    b. 8.0
    c. 7.0
    d. 5.0

    ans. d   diff. 5   obj. 9   p. 21

35. Most fungi are
    a. aerobes
    b. anaerobes
    c. autotrophs
    d. facultative anaerobes

    ans. a   diff. 5   obj. 9   p. 21

36. The terms *yeast, mold,* and *mushroom* indicate their
    _____ form
    a. vegetative
    b. classification
    c. morphological
    d. pathogenic

    ans. c   diff. 5   obj. 9   p. 21

37. A tangle of fungal hyphae is generally known as
    a. germ tubes
    b. spores
    c. mycelium
    d. buds

    ans. c   diff. 5   obj. 10   p. 22

38. Filamentous fungi with large fruiting bodies that are
    frequently edible are examples of
    a. mycelium
    b. hyphae
    c. rhizoids
    d. mushrooms

    ans. d   diff. 5   obj. 10   p. 23

39. Fungi capable of dimorphism grow either as _____ or
    as _____.
    a. yeastlike, mycelium
    b. rhizoids, hyphae
    c. germ tubes, buds
    d. spores, mushrooms

    ans. a   diff. 6   obj. 10   p. 24

40. Organisms that have characteristics of both protozoa and
    fungi are the
    a. terrestrial mushrooms
    b. slime molds
    c. yeasts
    d. molds

    ans. b   diff. 6   obj. 10   p. 24

41. The subgroup of terrestrial fungi that has no known sexual
    spore is the
    a. Zygomycetes
    b. Ascomycetes
    c. Mastigomycota
    d. Deuteromycetes

    ans. d   diff. 7   obj. 10   p. 26

42. The human diseases athlete's foot and ringworm are caused
    by the fungal subgroup
    a. Zygomycetes
    b. Ascomycetes
    c. Mastigomycota
    d. Deuteromycetes

    ans. d   diff. 7   obj. 11   p. 26

43. _____ molds are important in the food chain. They
    ingest both bacteria and algae and, in turn, serve as food for
    larger organisms
    a. Water
    b. Terrestrial
    c. Flagellated
    d. Slime

    ans. d   diff. 5   obj. 10   p. 27

44. *Pneumocystis carinii* is classified as a(n)
   a. alga
   b. fungus
   c. protozoan
   d. bacterium

   ans. b

45. Hallucinogenic properties of certain fungi have been known for centuries. One such hallucinogen is
   a. roseola
   b. ergot
   c. aflatoxin
   d. flacula

   ans. b   diff. 6   obj. 11   p. 30

46. Species of *Aspergillus* that produce carcinogenic toxins are called
   a. ergot
   b. aflatoxins
   c. hepatoma
   d. roseola

   ans. b   diff. 6   obj. 11   p. 30

47. Fungal diseases are generally referred to as
   a. infestations
   b. systemics
   c. infections
   d. mycoses

   ans. d   diff. 5   obj. 12   p. 32

48. Among the most common of the fungal infections are those caused by the organism
   a. *Histoplasma capsulatum*
   b. *Coccidioides immitis*
   c. *Cryptococcus neoformans*
   d. *Candida albicans*

   ans. d   diff. 6   obj. 11   p. 32

49. The yeast used in the production of bread is a member of the genus
   a. *Saccharomyces*
   b. *Candida*
   c. *Cryptococcus*
   d. *Phytophthora*

   ans. a   diff. 5   obj. 11   p. 34

50. Human insulin has been produced by genetic engineering using an organism of the _____ group.
   a. yeast
   b. bacteria
   c. worm
   d. virus

   ans. a

51. The most necessary habitat requirement of protozoa is
   a. moisture
   b. light
   c. temperature
   d. UV light

   ans. a   diff. 5   obj. 12   p. 38

52. Protozoan classification is based on their means of
   a. reproduction
   b. obtaining nutrients
   c. locomotion
   d. growth

   ans. c   diff. 5   obj. 11   p. 39

53. *Sarcodina* move by means of
   a. cilia
   b. flagella
   c. pseudopodia
   d. tentacles

   ans. c   diff. 5   obj. 12   p. 40

54. Which class of protozoa exhibit characteristics resembling phagocytosis?
   a. Mastigophora
   b. Sarcodina
   c. Sporozoa
   d. Ciliata

   ans. b   diff. 7   obj. 12   p. 41

55. The ability to exist as either a trophozoite or a cyst is characteristic of the
   a. bacteria
   b. fungi
   c. viruses
   d. protozoa

   ans. d   diff. 7   obj. 13   p. 42

56. Binary fission as a means of asexual cell division occurs longitudinally in the
    a. ciliates
    b. flagellates
    c. phagocytes
    d. amoeboids

    ans. a   diff. 7   obj. 12   p. 44

57. The microscopic group most important in decreasing total sewage solids are the
    a. bacteria
    b. fungi
    c. protozoa
    d. worms

    ans. c   diff. 7   obj. 13   p. 45

58. One of the greatest protozoan sources of human deaths through the ages has been due to
    a. *Toxoplasma*
    b. *Histoplasma*
    c. *Trypanosoma*
    d. *Giordia*

    ans. c   diff. 7   obj. 14   p. 46

True/False

1. Algae possess all the typical eukaryotic structures except flagella.
   false (and are diploid)

2. Algae are unicellular and hermaphroditic.
   false (divide by mitosis)

3. Diatoms have shells of silicon dioxide, which is an important ingredient in abrasives in polishes.
   true

4. Algae are classified by their reproductive structures.
   false (Fungi)

5. Algae are typical saprophytic eukaryotic organisms.
   false   (autotrophic for algae or change to Fungi are saprophyts)

6. Microscopic algae can be either single cells or multicellular.
   false (macroscopic algae)

7. Some dinoflagellates cause paralytic shellfish poisoning.
   true

8. Algae are principally aquatic and fungi terrestine.
   true

9. Principal decomposers of plant material on the planet are the animals.
   false (bacteria)

10. Fungal spores are found in all areas of the biosphere.
    true

11. Yeasts generally reproduce by fragmentation.
    false (budding)

12. Protozoa are classified by the means of locomotion.
    true

13. Genetic engineering is responsible for the development of human insulin by a bacterium.
    true

14. Malaria has been a leading cause of death worldwide for only a few years.
    false (many, many years)

Matching

1. Match the following specialized structures of macroscopic algae with the best function.

1.	holdfast	a.	leaflike attachment
2.	stipe	b.	anchor
3.	blade	c.	photosynthesis
4.	chloroplast	d.	stalk
		e.	absorb water

1 - b; 2 - d; 3 - a; 4 - c

2. Match the division of the fungi with their member(s).

1.	Amastigomycota	a.	slime molds
2.	Mastigomycota	b.	terrestrial fungi
3.	Gymnomycota	c.	dimorphic fungi
4.	*Coccidioides* sp.	d.	flagellated lower fungi

1 - b; 2 - d; 3 - a; 4 - c

3. Match the class of protozoa with its mode of locomotion.

1.	Mastigophora	a.	pseudopodia
2.	Sarcodina	b.	cilia
3.	Sporozoa	c.	flagella
4.	Ciliata	d.	nonmotile

1 - c; 2 - a; 3 - d; 4 - b

4. Match the protozoan with the disease it causes.

1.	*Toxoplasma gondii*	a.	vaginitis
2.	*Plasmodium* sp.	b.	African sleeping sickness
3.	*Trypanosoma* sp.	c.	malaria
4.	*Trichomonas vaginalis*	d.	toxoplasmosis

1 - d; 2 - c; 3 - b; 4 - a

Chapter 13

Multiple Choice

1. Viruses were named by _____ as "filterable viruses."
   a. Beijerinck
   b. Iwanowsky
   c. Twort
   d. d'Herelle

   ans. a.   diff. 4   obj. 1   p. 2

2. The earliest demonstration that infectious agents smaller than viruses existed were done by _____ and _____ early in the twentieth century.
   a. Watson and Crick
   b. Iwanowsky and Beijerinck
   c. Twort and d'Herelle
   d. Griffith and Jacob

   ans. c.   diff. 5   obj. 1   p. 1

3. Crystallization of the tobacco mosaic virus was accomplished by _____ in _____.
   a. Iwanowsky, 1903
   b. Stanley, 1935
   c. Berkley, 1926
   d. Twort, 1928

   ans. b.   diff. 5   obj. 1   p. 2

4. Viruses that infect bacteria are referred to as
   a. bacterial pathogens
   b. viralcidens
   c. bacteriocidins
   d. bacteriophages

   ans. d.   diff. 3   obj. 1   p. 2

5. A virion is a(n)
   a. individual virus particle
   b. pathogenic virus
   c. subviral organelle
   d. enveloped viral particle

   ans. a.   diff. 6   obj. 2   p. 3

6. Which of the following choices lists the essential viral particles?
   a. capsid, nucleic acid, capsule
   b. nucleic acid, capsid, envelope
   c. genome, virion, envelope
   d. genome, capsid

   ans. d.   diff. 5   obj. 2   p. 3

7. Viruses are smaller than cells. How much smaller?
   a. 100–1,000 times smaller
   b. 1–10 times smaller
   c. They are the same size as cells.
   d. $1,000–1 \times 10^5$ times smaller

   ans. a.   diff. 5   obj. 2   p. 3

8. The genome content of most viruses is about
   a. 150–200 genes
   b. 4–5 genes
   c. 4–200 genes
   d. 3,000 genes

   ans. c.   diff. 5   obj. 2   p. 4

9. The term *segmented* refers to viruses that have
   a. budding viruses still attached to the original molecule
   b. a capsid that is geometric in shape
   c. several different RNA molecules in their capsids
   d. an envelope

   ans. c.   diff. 5   obj. 2   p. 4

10. Based on the principles of symbiosis, viruses are categorized as
   a. obligate extracellular parasites
   b. facultative intercellular saprophytes
   c. obligate intracellular parasites
   d. facultative extracellular parasites

   ans. c.   diff. 5   obj. 2   p. 4

11. The shape of the virus is determined by its
    a. tail
    b. nucleic acid
    c. envelope
    d. capsid

    ans. d.   diff. 5   obj. 2   p. 6

12. If the complete virus is called a nucleocapsid, it is lacking its
    a. tail
    b. envelope
    c. nucleic acid
    d. pathogenicity

    ans. b.   diff. 6   obj. 2   p. 6

13. The viral envelope closely resembles which of the cellular organelles listed?
    a. cytoplasmic membrane
    b. cell wall
    c. ribosome
    d. endoplasmic reticulum

    ans. a.   diff. 6   obj. 3   p. 7

14. The protein projections on the surface of viruses that function in attachment to the host cell are called
    a. spikes
    b. suckers
    c. pili
    d. hooks

    ans. a.   diff. 6   obj. 3   p. 7

15. The infectious agent that is devoid of nucleic acid but not protein is
    a. viroid
    b. prion
    c. virus
    d. capsid

    ans. b.   diff. 6   obj. 2   p. 8

16. If the virus leaves its host cell during the release of its cytoplasm, it is described as being
    a. virulent
    b. latent
    c. enveloped
    d. spiked

    ans. a.   diff. 6   obj. 3   p. 12

17. What part of the *E. coli* bacteriophage attaches to the host cell receptors?
    a. pili of the envelope
    b. capsid fragments around the nucleic acid
    c. protein fibers at the end of the phage tail
    d. spikes of the envelope

    ans. c.   diff. 7   obj. 5   p. 13

18. What part of the attached bacteriophage enters through the host cell wall?
    a. the entire virus
    b. the nucleic acid and the capsid
    c. the nucleic acid (DNA)
    d. only the enzymes necessary for replication

    ans. c.   diff. 6   obj. 5   p. 13

19. A key feature of all viral infections is the
    a. removal of the host receptor site from the viral program
    b. union of the nucleic acid with the host cell membrane
    c. separation of the host nucleic acid from its cytoplasm
    d. separation of the nucleic acid from its protein coat

    ans. d.   diff. 7   obj. 5   p. 13

20. Which of the following enzymes is essential for the synthesis of phage DNA and phage coat protein?
    a. glucose degradation enzyme
    b. phage-induced enzymes
    c. amino acid synthesis enzymes
    d. energy release enzymes

    ans. b.   diff. 8   obj. 5   p. 14

21. After invasion of the host cell, where does synthesis of the phage proteins take place?
    a. on bacterial ribosomes
    b. within viral capsid
    c. in host cell nucleus
    d. in host cytoplasm

    ans. a.   diff. 7   obj. 5   p. 14

223

22. The process of separate replication of viral protein and its nucleic acid components is termed its
    a. maturation process
    b. vegetative reproduction
    c. lytic cycle
    d. productive response

    ans. b.   diff. 6   obj. 5   p. 15

23. The assembly of phage proteins with phage DNA to form infective virions is referred to as the
    a. matrix assembly
    b. vegetative reproductive process
    c. lytic cycle
    d. maturation process

    ans. d.   diff. 6   obj. 5   p. 15

24. In the case of the T-4 phages, the burst size, or the number of phages released per cell, is about
    a. 200
    b. 100
    c. 500
    d. 1,000

    ans. a.   diff. 4   obj. 5   p. 16

25. The replication of the bacteriophage from time of attachment to the bacterium to the release of the virions can be conveniently measured by using the
    a. steps of the bacterial growth curve
    b. burst size adjustment
    c. productive response evaluation
    d. one-step growth curve

    ans. d.   diff. 7   obj. 6   p. 16

26. The site at which a phage infected a bacterial cell and subsequently released its viral progeny by lysing and clearing the area of bacterial growth is referred to as
    a. segmentation
    b. spiking
    c. plaque
    d. bursting

    ans. c.   diff. 6   obj. 6   p. 17

27. The phage growth stage in which the viral protein coat and its genome are separated from one another is termed the _____ period.
    a. eclipse
    b. plaque
    c. bursting
    d. latent

    ans. a.   diff. 5   obj. 6   p. 18

28. The length of time between the entry of DNA into the cell and the phage release is termed the _____ period.
    a. segmented
    b. eclipse
    c. latent
    d. capsid

    ans. c.   diff. 4   obj. 6   p. 20

29. A phage that can only replicate inside the host cell and then lyse its host during its release is termed
    a. lysogenic
    b. segmented
    c. temperate
    d. virulent

    ans. d.   diff. 5   obj. 6   p. 20

30. Phages that can either replicate and cause cell lysis or can integrate their DNA into the host cell chromosome are known as _____ phages.
    a. lysogenic
    b. temperate
    c. virulent
    d. segmented

    ans. b.   diff. 6   obj. 6   p. 20

31. The most thoroughly studied temperate bacteriophage is the phage _____, which infects _____.
    a. delta, *S. aureus*
    b. alpha, *S. aureus*
    c. lambda, *E. coli*
    d. beta, *E. coli*

    ans. c.   diff. 7   obj. 6   p. 21

224

32. The integration of phage DNA into the bacterial chromosome occurs because of
   a. the similarity of enzyme productivity in metabolism
   b. the phage's ability to synthesize enzymes to enter the bacterium
   c. similar DNA nucleotides in both
   d. identical DNA sequences in both

   ans. d.   diff. 6   obj. 6   p. 22

33. Once integrated, phage DNA can remain in the prophage state if the
   a. phage genes are inhibited
   b. bacterium is transferred frequently to new culture media
   c. phage genes are excised
   d. bacterial repressor genes are activated

   ans. a.   diff. 7   obj. 6   p. 2

34. The activation of the SOS system in a bacterium infected with a prophage results in the
   a. destruction of the viral genes
   b. complete lysis of the bacterial culture
   c. mutation of both sets of genetic characteristics
   d. excising of the viral genome from the bacterium

   ans. b.   diff. 8   obj. 6   p. 23

35. The phenomenon responsible for the ability of *Corynebacterium diphtheriae* to produce the virulent toxin responsible for the devastating effects of diphtheria is called
   a. lysogenic conversion
   b. self-assembly
   c. matrix conversion
   d. prion protein

   ans. a.   diff. 7   obj. 6   p. 24

36. An analogous situation to lysogenic conversion in the phage-bacterium system can be found in the
   a. tumor virus-animal system
   b. plant virus-fruit system
   c. enzyme sequences produced by the repressor gene
   d. conversion of viral enzyme spikes to increase virulence

   ans. a.   diff. 6   obj. 5   p. 25

37. Which of the following organisms does not have its virulence increased by lysogenic conversions?
   a. *Vibrio cholerae*
   b. *Streptococcus*-B-hemolytic
   c. *Mycobacterium tuberculosis*
   d. *Clostridium botulinum*

   ans. c.   diff. 6   obj. 5   p. 25

38. Which of the following processes involves virus exit from the host bacterial cell without lysing the host cell?
   a. exercision
   b. lysis
   c. excision
   d. extrusion

   ans. d.   diff. 5   obj. 5   p. 25

39. The bacterial viruses that are released by a process termed extrusion are referred to as
   a. lambda viruses
   b. lysogenic viruses
   c. temperate phages
   d. filamentous phages

   ans. d.   diff. 6   obj. 5   p. 25

40. The filamentous phages are all unique in that they all contain
   a. double-stranded DNA
   b. single-stranded DNA
   c. single-stranded RNA
   d. double-stranded RNA

   ans. b.   diff. 7   obj. 5   p. 25

41. Bacteria infected with filamentous phages are termed
   a. temperate cells
   b. carrier cells
   c. plaque producers
   d. virulent strains

   ans. b.   diff. 7   obj. 5   p. 26

42. Examples of viruses that escape their host cells by budding are the
   a. mumps and influenza viruses
   b. measles and polio viruses
   c. Herpes and mumps viruses
   d. polio and yellow fever viruses

   ans. a.   diff. 7   obj. 5   p. 26

43. An exit method used by viruses that does not destroy the host bacterium is
   a. lysis
   b. budding
   c. inversion
   d. excising

   ans. b.   diff. 7   obj. 5   p. 26

44. Transducing virulent phages do not lyse the cells they invade because
   a. their virulence is dependent on bacteria and phage replicating together
   b. bacterial DNA has replaced the viral DNA in the phage so phage enzyme production is lacking
   c. the lytic enzyme genes are unable to enter the bacterial cell but remain in the environment
   d. transformation is taking place in the phage, which is then transferred to the bacterium

   ans. b.   diff. 5   obj. 5   p. 27

45. If the infecting phage lacks all the DNA necessary to form complete virions and lyses the cell, it is termed
   a. vegetative
   b. incomplete
   c. mutated
   d. defective

   ans. d.   diff. 5   obj. 5   p. 27

46. Genetic elements of bacteriophages that can replicate as part of the bacterial chromosome or independently of it are examples of
   a. virulent phages
   b. episomes
   c. mutants
   d. defectives

   ans. b.   diff. 6   obj. 7   p. 28

47. The groups of organisms that a virus can infect are its
   a. target circle
   b. host range
   c. susceptible group
   d. receptor type

   ans. b.   diff. 5   obj. 7   p. 29

48. Which of the following cell types is a virus unable to infect?
   a. virions
   b. plants
   c. bacteria
   d. protozoa

   ans. a.   diff. 4   obj. 7   p. 29

49. The limiting factor for viral infection of a host cell is the
   a. presence of other viruses in the environment
   b. internal metabolic temperature of the host cell
   c. nutrition of the host cell
   d. presence of specific receptor sites on the host cell

   ans. d.   diff. 5   obj. 7   p. 29

50. Which of the following is not a suitable media for the growth of viruses?
   a. tumor cells
   b. chicken embryo
   c. chocolate agar
   d. animal cells

   ans. c.   diff. 4   obj. 8   p. 31

51. Normal tissues taken from animals and prepared immediately as media for viral growth is termed a(n)
   a. primary culture
   b. advantageous group
   c. monolayer
   d. *in vivo* culture

   ans. a.   diff. 4   obj. 8   p. 32

52. After growth in tissue culture, the infected cells become lysed and the cell debris is removed. The virions can be harvested from the liquid portion termed the
   a. remainder
   b. lysate
   c. virade
   d. plaque

   ans. b.   diff. 5   obj. 8   p. 3

53. The changes that occur in virally infected cells are characteristic for a particular virus and are referred to as its
   a. genotypic expression
   b. phenotypic expression
   c. cytopathic effect
   d. cytology

   ans. c.   diff. 5   obj. 8   p. 34

54. If pure samples of virions are collected and then properly prepared, their concentration can be determined by counting the viral particles while using the
   a. electron microscope
   b. photo colorimeter
   c. gas chromatography
   d. electrophoretic equipment

   ans. a.   diff. 6   obj. 8   p. 34

55. The approximate viral concentration in a sample is referred to as a(n)
   a. quantal assay
   b. endpoint
   c. titer
   d. lysate

   ans. a.   diff. 5   obj. 8   p. 35

56. The concentration of virus that results in fifty percent of the inoculated host cells being infected or killed is referred to as the
   a. ID_{70}
   b. LD_{100}
   c. LD_{50}
   d. ID_{150}

   ans. c.   diff. 5   obj. 8   p. 35

57. One group of animal viruses that is able to agglutinate red blood cells is the
   a. picornaviruses
   b. coronoviruses
   c. reoviruses
   d. myxoviruses

   ans. d.   diff. 5   obj. 8   p. 36

True/False

1. Viruses that infect bacteria are virophages.
   false (bacteriophages)

2. The first virus named was the Rous cancer virus.
   false (tobacco mosaic virus)

3. Stanley showed by the crystallization of viruses that they shared many physical and chemical properties with cells.
   false (physical and chemical properties greatly differed)

4. A synonym for virus is virion.
   true

5. Viruses can have as few as 4 genes and as many as 200 genes.
   true

6. The protein subunits of the capsid, the capsomeres, are a lipid and protein combination external to the envelope.
   false (capsomere is protein only and is under the envelope)

7. Only enveloped viruses have spikes.
   false (naked viruses have spikes also)

8. The plant viruses enter the host cell intact and then the nucleic acid separates from the capsid.
   false (animal viruses)

9. A key feature of all viral infections is its capsid separation from its tail.
   false (nucleic acid separation from protein coat)

10. The phage DNA that is integrated into the host cell DNA is called a prophage.
   true

11. The most thoroughly studied bacteriophage is the phage lambda, which infects *E. coli.*
   true

12. The filamentous phages completely take over their host's metabolism and prevent bacterial replication.
   false (do not completely take over bacterial metabolism)

13. Temperate bacteriophages are examples of latent state viruses.
   false (episomes)

14. Normal vertebrate tissue cells can live indefinitely if they are regularly transferred to fresh media.
   false (divide 50 to 100 times, then die)

15. Viral particles can be counted successfully in a specimen prepared for study with an electron microscope.
   true

16. The approximate viral content of a sample is referred to as a quantal assay.
   true

17. In generalized transduction, virulent phages can transfer any part of the genome to a new host.
   true

227

Chapter 14

Multiple Choice

1. The first suggestion that tumors might be caused by invisible agents was reported in the
   a. 1700s
   b. 1800s
   c. 1900s
   d. 1950s

   ans. a.   diff. 4   obj. --   p. 1

2. That tumors in healthy chickens could be produced by exposing them to a suspension of ground chicken tumors was proved by
   a. Sarah Stewart
   b. Richard Shope
   c. Peyton Rous
   d. Bernice Eddy

   ans. c.   diff. 5   obj. --   p. 2

3. The fact that viruses could exist in some hosts without causing tumors but then, when inoculated into foreign hosts caused tumors was discovered by
   a. Shope
   b. Rous
   c. Stewart
   d. Eddy

   ans. d.   diff. 6   obj. --   p. 2

4. Animal viruses are divided into a number of families whose names end in
   a. -viridae
   b. virus
   c. disease caused
   d. host name

   ans. a.   diff. 4   obj. 1   p. 4

5. The common species name of the virus is based on the
   a. presence or absence of an envelope
   b. type of nucleic acid it contains
   c. disease the virus causes
   d. geographic area

   ans. c.   diff. 5   obj. 1   p. 4

6. The family to which the Rhinovirus belongs is
   a. Enterovirus
   b. Picornaviridae
   c. Enteroviridae
   d. Retrovirus

   ans. b.   diff. 6   obj. 1   p. 4

7. When discussing viral interactions with animal cells, an infection that results in the lysis of all cells in the culture is referred to as a _____ infection.
   a. productive
   b. persistent
   c. latent
   d. permissive

   ans. a.   diff. 5   obj. 1   p. 5

8. Adsorption of animal viruses to the host cell is by means of
   a. tail
   b. spikes
   c. envelope
   d. capsid

   ans. b.   diff. 6   obj. 2   p. 5

9. Resistance of some animals to certain viral diseases is based on
   a. phagocytosis of the virus by the host cell
   b. lack of specific receptors on the host cell
   c. the presence of the viral envelope
   d. lack of spikes for attachment

   ans. b.   diff. 6   obj. 2   p. 6

10. When the enveloped virus adsorbs to the host cell with its protein spikes, the virions are taken into the cell by the process of
    a. production
    b. penetration
    c. endocytosis
    d. fusion

    ans. c.   diff. 6   obj. 2   p. 7

11. In all viral systems, replication of the phage, animal, or plant virus depends to varying degrees on the
   a. enzymes of the host cell
   b. cytoplasmic membrane of the host cell
   c. fusion of the virus with a vacuole
   d. presence of protein spikes

   ans. a.   diff. 6   obj. 3   p. 7

12. The final assembly of the virus with its completed genome and protein capsid is referred to as
   a. synthesis
   b. production
   c. permission
   d. maturation

   ans. d.   diff. 5   obj. 3   p. 9

13. The model used in the maturation studies of plant and animal viruses is the
   a. adenovirus
   b. polyma virus
   c. pox virus
   d. tobacco mosaic virus

   ans. d.   diff. 5   obj. 3   p. 9

14. For which of the following processes are enzymes not required?
   a. self-assembly
   b. maturation
   c. adsorption
   d. synthesis

   ans. a.   diff. 6   obj. 3   p. 10

15. Cells infected with animal viruses lyse as a consequence of the cell's death because
   a. the RNA interacts with protein and kills the cell
   b. the virus releases enzymes that lyse the cell
   c. the release of virions depletes the cell of energy
   d. functions required for cell survival are not carried out

   ans. d.   diff. 7   obj. 4   p. 10

16. Another mechanism, in addition to lysis, for the release of animal viruses is
   a. lysis
   b. extrusion
   c. budding
   d. explosion

   ans. c.   diff. 7   obj. 4   p. 11

17. The virion protein that coats the inside of the host cell in the region of budding is called the
   a. lytic matrix
   b. matrix protein
   c. budding site
   d. extrusion area

   ans. b.   diff. 7   obj. 4   p. 11

18. Viruses obtain their envelopes when they exit the cell by budding. The envelope is part of the
   a. cell membrane
   b. cell matrix
   c. enzyme system of the cell
   d. protein vacuole created by the virus

   ans. a.   diff. 7   obj. 4   p. 11

19. The outcome of viral diseases are related to the
   a. defense mechanisms of the host
   b. presence of blocked receptor sites
   c. kinds of cells invaded
   d. toxins produced in response to viral invasion

   ans. a.   diff. 6   obj. 4   p. 12

20. Consequences of the lack of immunity to particular viruses have resulted in epidemics of diseases such as
   a. the common cold
   b. smallpox
   c. polio
   d. hepatitis

   ans. b.   diff. 5   obj. 4   p. 12

21. The condition in which the virus infects and persists within the host with no obvious harm is termed
   a. parasitism
   b. balanced pathogenicity
   c. commensalistic colonization
   d. mutualism

   ans. b.   diff. 6   obj. 4   p. 12

22. Individuals who harbor viruses as well as antibodies against these viruses with no ill effects are termed
    a. carriers
    b. victims
    c. diseased
    d. balanced

    ans. a.   diff. 4   obj. 4   p. 12

23. Diseases of short duration frequently followed by long-term immunity are referred to as
    a. balanced infections
    b. chronic infections
    c. intermittent infections
    d. acute infections

    ans. d.   diff. 4   obj. 4   p. 13

24. An infection in which the virus is continually present in the body is referred to as a(n) _____ infection.
    a. acute
    b. persistent
    c. balanced
    d. determined

    ans. b.   diff. 5   obj. 4   p. 13

25. Animal viruses that are responsible for persistent infections have as their counterparts
    a. filamentous phages
    b. carrier viruses
    c. acute virions
    d. retroviruses

    ans. a.   diff. 7   obj. 4   p. 14

26. The genome of retroviruses is composed of
    a. tRNA
    b. DNA
    c. RNA
    d. protein

    ans. c.   diff. 5   obj. 4   p. 14

27. The best known chronic human infection involves the _____ virus.
    a. hepatitis A
    b. polio
    c. herpes
    d. hepatitis B

    ans. d.   diff. 5   obj. 4   p. 14

28. Subacute sclerosing panencephalitis (SSPE) is an occasional slow viral disease that develops about five years after having
    a. mumps
    b. measles
    c. polio
    d. shingles

    ans. b.   diff. 6   obj. 4   p. 15

29. Latent viral infections may result from integration of
    a. viral DNA into host cell DNA
    b. host enzymes with the viral enzymes
    c. host RNA with viral genome
    d. viral genome with host enzymes

    ans. a.   diff. 8   obj. 4   p. 15

30. The major difference between persistent and latent infections is that in the latent infections
    a. no mature viral particles are produced for long periods
    b. viral particles are released continuously
    c. viral particles are produced and released sporadically
    d. viral particles are released without symptoms of the disease

    ans. a.   diff. 7   obj. 4   p. 15

31. The best known examples of viruses that cause latent infections are caused by members of the _____ virus family.
    a. polio
    b. herpes
    c. measles
    d. chicken pox

    ans. b.   diff. 5   obj. 4   p. 16

32. After replication and upon stimulation, the mature infectious virions of the Herpes family are carried to the skin or mucous membrane by
    a. chemical attraction
    b. axons of nerve cells
    c. dendrites of nerve cells
    d. collagen fibers

    ans. b.   diff. 6   obj. 4   p. 16

33. The latent disease caused by the varicella-zoster virus is
    a. herpes cold sores
    b. chicken pox
    c. shingles
    d. mumps

    ans. c.   diff. 5   obj. 4   p. 16

34. A swelling that results from an abnormal growth of cells indicates a(n)
    a. cancer
    b. tumor
    c. leukemia
    d. malignancy

    ans. b.   diff. 5   obj. 5   p. 18

35. Cancer cells have several properties that distinguish them from normal cells. Which of the following is not characteristic of cancer cells?
    a. RNA changes
    b. metastasis
    c. abnormal appearance
    d. limited growth

    ans. d.   diff. 4   obj. 6   p. 18

36. Cancer cells growing in tissue culture multiply indefinitely and are termed
    a. normals
    b. limited
    c. genetic duplicates
    d. immortal

    ans. d.   diff. 5   obj. 6   p. 18

37. Which of the following is not a characteristic of cancer cells?
    a. grow as a monolayer
    b. may form tumors
    c. grow indefinitely
    d. attach firmly to one another

    ans. a.   diff. 6   obj. 6   p. 19

38. Which of the following is not a characteristic of normal cells?
    a. do not form tumors
    b. grow as a monolayer
    c. grow indefinitely
    d. stay attached to one another

    ans. c.   diff. 6   obj. 6   p. 19

39. Viruses that are able to convert normal cells in tissue culture into tumor cells in host animals are termed _____ viruses.
    a. budding
    b. DNA
    c. RNA
    d. transformed

    ans. d.   diff. 6   obj. 6   p. 20

40. Transforming viruses capable of stimulating tumor cells to form in susceptible host animals are
    a. retroviruses
    b. DNA viruses
    c. budding viruses
    d. prions

    ans. a.   diff. 6   obj. 6   p. 20

41. Reverse transcriptase is an enzyme that copies the viral RNA into a complementary strand of DNA. This enzyme was demonstrated by
    a. Rous and Shope
    b. Watson and Crick
    c. Temin and Baltimore
    d. Stewart and Eddy

    ans. c.   diff. 7   obj. 6   p. 22

42. The double-stranded DNA molecule integrated into the chromosome of the host cell is called a(n)
    a. provirus
    b. virion
    c. prion
    d. tumor

    ans. a.   diff. 6   obj. 6   p. 22

43. A provirus is analagous to the
    a. prophage lambda
    b. retrovirus
    c. neoplasm
    d. reverse transcriptase

    ans. a.   diff. 5   obj. 6   p. 22

44. Drs. Temin and Baltimore received the Nobel prize in 1975 for their discovery of
    a. neoplasms
    b. reverse transcriptase
    c. proviruses
    d. retroviruses

    ans. b.   diff. 6   obj. 6   p. 22

45. Retroviruses transform cells by a number of different mechanisms but in all cases they
    a. allow the cells to reproduce normally
    b. disrupt control of normal cell growth and tissue development
    c. stimulate key enzymes to function properly
    d. allow communication between cells

    ans. b.   diff. 7   obj. 6   p. 23

46. When the retroviruses are integrated into the host chromosome, they are termed
    a. latent
    b. mutations
    c. neoplasms
    d. oncogenes

    ans. d.   diff. 6   obj. 6   p. 23

47. Pioneering studies on oncogenes and protooncogenes were carried out by _____ and _____ who were jointly awarded the Nobel prize in 1989.
    a. Stewart and Eddy
    b. Temin and Baltimore
    c. Bishop and Varmus
    d. Watson and Crick

    ans. c.   diff. 7   obj. 6   p. 23

48. Which of the following DNA viruses has not been associated with tumors in humans?
    a. papillomavirus
    b. Epstein-Barr virus
    c. hepatitis B virus
    d. parvovirus

    ans. d.   diff. 6   obj. 6   p. 24

49. Human T-cell lymphotrophic virus 1 (HTLV-1) causes a rare cancer in a specific white blood cell called the
    a. T cell
    b. leukocyte
    c. B cell
    d. neutrophil

    ans. a.   diff. 7   obj. 6   p. 25

50. The first specific mention of the disease that is now known as AIDS was in the
    a. *Newsweek* in August of 1981
    b. *Time* magazine in August of 1982
    c. *MMWR* in June of 1981
    d. *World Health Association Newsletter* in June of 1982

    ans. c.   diff. 7   obj. 6   p. 26

51. Plants do not recover from viral diseases due to their inability to develop
    a. immunity
    b. seeds
    c. lateral stems
    d. tumors

    ans. a.   diff. 6   obj. 7   p. 27

52. One plant that has been purposely maintained in a virus-infected state is the _____.
    a. geranium
    b. tulip
    c. rose
    d. iris

    ans. b.   diff. 5   obj. 7   p. 27

53. Plant viruses enter the host plant via
    a. non-specific receptors
    b. specific receptors
    c. wound sites
    d. seeds

    ans. c.   diff. 7   obj. 7   p. 28

54. The plant virus most resistant to inactivation is the
    a. tobacco mosaic virus
    b. tulip activation virus
    c. papillomavirus
    d. corn borer virus

    ans. a.   diff. 7   obj. 7   p. 28

55. The most important plant virus vectors are probably
    a. contaminated seeds
    b. insects
    c. contaminated tubers
    d. worms

    ans. b.   diff. 7   obj. 7   p. 30

56. The wound tumor virus of plants is morphologically similar to the
    a. reoviruses of animals
    b. DNA viruses of animals
    c. bacteriophages
    d. myxoviruses

    ans. a.   diff. 7   obj. 7   p. 30

57. Members of which of the following virus groups are now being used as biological insect control agents?
    a. baculoviruses
    b. East African army worm
    c. bacteriophages
    d. parvoviruses

    ans. a.   diff. 7   obj. 8   p. 31

58. A group of pathogens much smaller than the viruses are the
    a. bacteria
    b. virions
    c. rickettsias
    d. viroids

    ans. d.   diff. 5   obj. 8   p. 31

59. The agent causing potato spindle tuber disease has been associated with a(n)
    a. rickettsia
    b. virion
    c. viroid
    d. bacteria

    ans. c.   diff. 6   obj. 8   p. 32

60. Viroids are characteristically
    a. single-stranded DNA molecules
    b. double-stranded RNA molecules
    c. single-stranded RNA molecules
    d. double-stranded DNA molecules

    ans. c.   diff. 6   obj. 8   p. 32

True/False

1. If a virus exists in a host without causing tumors it is referred to as "silent."
    true

2. The taxonomy of viruses is based on their cellular structure.
    false   (they are not cells)

3. Polyhedral viruses are also referred to as cylindrical.
    false   (spherical)

4. There are twelve families of DNA viruses and six families of RNA viruses.
    false   (6 of DNA, 12 of RNA)

5. There is an inverse relationship between the size of the viral genome and the number of host enzymes needed for its replication.
    true

6. The parvovirus shows complete dependence on its host enzymes to synthesize its DNA.
    true

7. Budding of the virus from the host cell causes cell death by tearing a hole in the cell membrane.
    false   (the cell membranes will repair after budding)

8. If a virus is present in a host with no ill effects to the host, the host is termed a carrier.
    true

9. Only DNA viruses can cause latent infections.
    false   (either DNA or RNA)

10. Chicken pox and measles are different diseases caused by the same virus.
    false   (shingles)

11. A frequently used synonym for tumor is malignancy.
    false   (neoplasm)

12. The conversion of normal cells to cancer cells can frequently be attributed to retroviruses.
    true

13. The original transforming DNA virus can be separated from the altered (cancer) cell and used to transform other cells.
    false   (once incorporated the virus cannot be detected)

14. Experiments have shown that both DNA and RNA viruses cause tumors in humans.
    true

15. The polyoma virus has been shown to definitely cause tumors in humans but not experimental animals.
    false   (tumors in experimental animals, not humans)

16. Plant viral infections are few and are found in plants propagated by seeds.
    false   (many viral infections in plants, especially in those propagated vegetatively)

17. Several of the plant and insect-infecting viruses show similarities to viruses that cause diseases in bacteria.
    false   (humans)

18. Viroids are frequently found as animal pathogens.
    false   (plant)

19. Viroids are characterized as being single-stranded RNA molecules.
    true

Chapter 15

Multiple Choice

1. The term "phagocyte" was coined by _____ after observing the activity of cells within a starfish larva after being pierced with a rose thorn.
   a. Lister
   b. Pasteur
   c. Koch
   d. Metchnikoff

   ans. d.   diff. 5   obj. --   p. 2

2. Metchnikoff proposed that the cells primarily responsible for the body's ability to destroy invading microorganisms were
   a. phagocytes
   b. lymphocytes
   c. red blood cells
   d. lymphocytes

   ans. a.   diff. 5   obj. --   p. 3

3. The study of vertebrate protection against invasion by foreign organisms or toxins is called
   a. embryology
   b. phagocytosis
   c. defense
   d. immunity

   ans. d.   diff. 5   obj. 1   p. 3

4. The body's defense mechanisms, which include the skin and mucus membranes, are usually referred to as
   a. specific
   b. nonspecific
   c. autoimmune
   d. irregular

   ans. b.   diff. 5   obj. 1   p. 4

5. The physiological defense mechanisms of the body do not include
   a. fever
   b. skin and mucus membranes
   c. chemotaxis
   d. inflammation

   ans. b.   diff. 5   obj. 1   p. 5

6. Factors directed against any foreign substance entering the host are described as
   a. immune
   b. specific
   c. irregular
   d. nonspecific

   ans. d.   diff. 5   obj. 1   p. 5

7. The defense system that depends on highly specialized cells and chemicals that respond to specific organisms and foreign materials is the
   a. specific line of defense
   b. idealized defense system
   c. lymphatic system of defense
   d. autoimmune system

   ans. a.   diff. 5   obj. 1   p. 5

8. The interaction of all organisms within a biological community is called a(n)
   a. nature
   b. dialogue
   c. ecosystem
   d. biological relationship

   ans. c.   diff. 4   obj. 1   p. 5

9. The microorganisms that are found on or in the body but do not produce obvious harmful effects are known as
   a. transient flora
   b. abnormal flora
   c. normal flora
   d. variety

   ans. c.   diff. 5   obj. 1   p. 6

10. The microorganisms that are found on or in the body only occasionally are termed
    a. transient flora
    b. resident flora
    c. variety
    d. pathogens

    ans. a.   diff. 5   obj. 1   p. 6

11. Interactions that occur between different organisms that live closely on a more or less permanent basis are referred to as
    a. mutualism
    b. symbiosis
    c. normal flora
    d. parasitism

    ans. b.   diff. 5   obj. 1   p. 6

12. The symbiotic relationship in which one partner benefits but the other remains unaffected is described as
    a. commensalism
    b. parasitism
    c. mutualism
    d. independence

    ans. a.   diff. 5   obj. 1   p. 7

13. A commensalistic relationship between organisms can be described as one in which
    a. one benefits while the other remains unaffected
    b. both partners benefit
    c. one benefits and the other is harmed
    d. organisms are unaffected by the presence of each other

    ans. a.   diff. 5   obj. 1   p. 7

14. The symbiotic relationship in which both partners benefit is referred to as
    a. independence
    b. parasitism
    c. mutualism
    d. commensalism

    ans. c.   diff. 5   obj. 1   p. 7

15. Mutualism is described as a relationship in which
    a. one benefits while the other remains unaffected
    b. both partners benefit
    c. one benefits while the other is harmed
    d. organisms are unaffected by the presence of each other

    ans. b.   diff. 5   obj. 1   p. 7

16. The symbiotic relationship in which one organism benefits at the expense of the other organism is described as
    a. mutualism
    b. commensalism
    c. independence
    d. parasitism

    ans. d.   diff. 5   obj. 1   p. 7

17. Parasitism is described as a relationship in which
    a. members remain unaffected by each other
    b. both members of the relationship are benefited
    c. one member is benefited while the other is harmed
    d. one member is benefited while the other is unaffected

    ans. c.   diff. 5   obj. 1   p. 7

18. Most disease-causing organisms are described as
    a. parasites
    b. saprophytes
    c. commensals
    d. symbionts

    ans. a.   diff. 5   obj. 1   p. 7

19. The resident microbial population of the human fetus is
    a. related to the mother's immunity
    b. nonexistent
    c. small
    d. symbiotic

    ans. b.   diff. 5   obj. 2   p. 8

20. The establishment of a microbial population on or in the host is referred to as
    a. independence
    b. parasitism
    c. colonization
    d. mutualism

    ans. c.   diff. 5   obj. 2   p. 8

21. Which of the following is not a mechanism by which normal flora contribute to the health of the host?
    a. establishing competition for nutrients and vitamins
    b. provide a surface incompatible for attachment of the invader
    c. growth of one organism is prevented or slowed by growth of another organism living close by
    d. production of toxins harmful to the organism

    ans. d.   diff. 5   obj. 2   p. 8

22. Growth of *Corynebacterium diphtheriae* on the host is inhibited by hydrogen peroxide produced by certain groups of _____ in the normal flora.
    a. antibiotics
    b. staphylococci
    c. streptococci
    d. bacilli

    ans. c.   diff. 6   obj. 2   p. 9

23. Which of the following can suppress the growth of normal flora, which then allows pathogens to colonize the host?
    a. toxins
    b. peroxides
    c. antibiotics
    d. fatty acids

    ans. c.   diff. 6   obj. 2   p. 9

24. Proteins produced by the body in response to invasion by a foreign substance that interact specifically with that substance are called
    a. antibodies
    b. antibiotics
    c. peroxides
    d. toxins

    ans. a.   diff. 5   obj. 2   p. 9

25. Microbial populations of the intestine change with diet. Carnivores have a greater population of _____ than do herbivores.
    a. *Bacteroides*
    b. *Salmonella* sp.
    c. staphylococci
    d. streptococci

    ans. a.   diff. 6   obj. 2   p. 10

26. Establishment of microbial growth on or in the host is referred to as
    a. parasitism
    b. diseased
    c. pathogenic
    d. colonization

    ans. d.   diff. 5   obj. 2   p. 12

27. A disease-causing microorganism or virus is referred to as a(n)
    a. avirulent
    b. colony
    c. infection
    d. pathogen

    ans. d.   diff. 4   obj. 2   p. 12

28. Opportunists are referred to as organisms that are
    a. mutualistic, and benefit the host
    b. able to cause disease under special circumstances
    c. common, well-known pathogens
    d. usually saprophytes

    ans. b.   diff. 5   obj. 2   p. 13

29. A normal flora that becomes an opportunist if it invades the urinary bladder is
    a. *Corynebacterium diphtheriae*
    b. *P. vulgaris*
    c. *E. coli*
    d. *Salmonella typhimurium*

    ans. c.   diff. 6   obj. 2   p. 14

30. The proposal of a "germ theory of disease" was introduced in 1546 by
    a. Koch
    b. Fracastoro
    c. Redi
    d. Leeuwenhoek

    ans. b.   diff. 4   obj. 3   p. 14

31. In 1876, Robert Koch offered convincing evidence of the germ theory of disease by showing the cause of anthrax as
    a. *Clostridium anthrax*
    b. *Bacillus cereus*
    c. *Bacillus anthracis*
    d. *M. tuberculosis*

    ans. c.   diff. 4   obj. 3   p. 15

32. The series of steps that scientists use to determine the causative agents for most infectious disease are known as
    a. Koch's postulates
    b. Pasteur's systematic
    c. Leeuwenhoek's animalcules
    d. Lister's aseptics

    ans. a.   diff. 4   obj. 3   p. 15

33. If a microorganism cannot be grown outside the human body, which of Koch's postulates cannot be satisfied?
   a. postulate #4
   b. postulate #1
   c. postulate #3
   d. postulate #2

   ans. d.   diff. 5   obj. 3   p. 16

34. If a disease affects only the human, and no other animal can be infected even experimentally, then postulate ___ cannot be satisfied.
   a. 4
   b. 1
   c. 2
   d. 3

   ans. d.   diff. 5   obj. 3   p. 16

35. Transmission of an infectious disease by touching, handshaking, or kissing is considered _____ contact.
   a. infectious
   b. indirect
   c. direct
   d. communicable

   ans. c.   diff. 6   obj. 4   p. 18

36. Which of the following types of transmission is described as indirect?
   a. insect bites
   b. touching
   c. handshaking
   d. sexual intercourse

   ans. a.   diff. 5   obj. 4   p. 18

37. The number of organisms necessary to insure infection is termed the
   a. infectious dose
   b. fatal number
   c. minimum lethal dose
   d. pathogenic number

   ans. a.   diff. 5   obj. 4   p. 18

38. All surfaces of the body exposed to the outside world are covered or lined with
   a. endothelium
   b. epithelium
   c. skin
   d. peritoneum

   ans. b.   diff. 6   obj. 4   p. 19

39. The first step in the establishment of infection is that the organism must
   a. produce toxins
   b. invade host tissues
   c. evade phagocytes
   d. attach to host cells

   ans. d.   diff. 5   obj. 4   p. 20

40. Projections from the surfaces of infectious agents that aid in attachment to host cells are
   a. attachers
   b. receptors
   c. ligands
   d. toxins

   ans. c.   diff. 7   obj. 4   p. 20

41. On which of the following organisms are ligands not found?
   a. *Neisseria gonorrhoeae*
   b. *E. coli*
   c. *Bacillus anthracis*
   d. *Streptococcus pyogenes*

   ans. c.   diff. 7   obj. 4   p. 20

42. Ligand molecules attach to the host receptor by a mechanism referred to as
   a. base pairing
   b. lock and key
   c. complementarity
   d. mirroring

   ans. b.   diff. 7   obj. 4   p. 20

43. The glycocalyx is a sticky network of fibers produced by streptococci and is responsible for its ability to produce
   a. sore throat
   b. tooth decay
   c. pneumonia
   d. bronchitis

   ans. b.   diff. 8   obj. 4   p. 21

44. It is believed that in some infectious agents the ligands are carried on the
   a. cell membranes
   b. plasmids
   c. flagella
   d. capsule

   ans. b.   diff. 7   obj. 4   p. 21

45. The severe diarrhea occasionally caused by *E. coli* is the result of the production of
   a. enterotoxin
   b. neurotoxin
   c. plasmids
   d. ligands

   ans. a.   diff. 6   obj. 4   p. 22

46. The process by which infectious agents are ingested by host cells is
   a. exocytosis
   b. phagocytosis
   c. pinocytosis
   d. endocytosis

   ans. d.   diff. 6   obj. 4   p. 23

47. After invading the host cells, infectious agents may exit by
   a. exocytosis
   b. phagocytosis
   c. pinocytosis
   d. endocytosis

   ans. a.   diff. 6   obj. 4   p. 23

48. The process of entering and then exiting the host cell by infectious agents is
   a. endocytosis, exocytosis
   b. pinocytosis, phagocytosis
   c. ligand, receptor
   d. endotoxin, exotoxin

   ans. a.   diff. 5   obj. 4   p. 23

49. Which of the following is not considered essential for colonization of infectious agents on host cells?
   a. acidity
   b. temperature
   c. proper moisture
   d. nutrient supply

   ans. b.   diff. 7   obj. 4   p. 25

50. Organisms that grow and reproduce outside of the host cells are
   a. extracellular
   b. facultatively extracellular
   c. intracellular
   d. obligately intracellular

   ans. a.   diff. 5   obj. 4   p. 26

51. Which of the following organisms is not considered an obligate intracellular parasite?
   a. *V. cholerae*
   b. rickettsia
   c. chlamydia
   d. viruses

   ans. a.   diff. 4   obj. 4   p. 26

52. Which of the following leukocytes are considered especially important in phagocytosis?
   a. monocytes and eosinophils
   b. lymphocytes and leukocytes
   c. erythrocytes and platelets
   d. neutrophils and macrophages

   ans. d.   diff. 5   obj. 4   p. 27

53. After a microorganism is engulfed by a phagocyte, it resides in a structure known as a(n)
   a. lysosome
   b. polysome
   c. phagosome
   d. phagolysosome

   ans. c.   diff. 6   obj. 4   p. 28

54. The phagolysosome is the result of fusion of a lysosome with a
   a. hydrosome
   b. vacusome
   c. polysome
   d. phagosome

   ans. d.   diff. 4   obj. 4   p. 28

55. Which of the following pathogens is capable of reproducing inside the phagolysosome despite its hostile environment?
   a. *M. tuberculosis*
   b. *S. aureus*
   c. *Yersinia pestis*
   d. *S. pneumoniae*

   ans. c.   diff. 6   obj. 4   p. 29

56. Which of the following is not a characteristic of exotoxins?
   a. released from cell during metabolism
   b. can be used to make toxins
   c. heat stable
   d. protein

   ans. c.   diff. 7   obj. 5   p. 30

57. Which of the following does not describe an endotoxin?
   a. heat stable
   b. protein
   c. lipopolysaccharide
   d. produced by Gram-negative species

   ans. b.   diff. 7   obj. 5   p. 30

58. Most exotoxins are inactivated by heat. Which of the following organisms produces an exotoxin that is the notable exception?
   a. *S. aureus*
   b. *C. diphtheriae*
   c. *S. pyogenes*
   d. *V. cholerae*

   ans. a.   diff. 5   obj. 5   p. 31

59. Exotoxin production by the organisms causing diphtheria and scarlet fever are produced by
   a. plasmids
   b. temperate phages
   c. transduction
   d. transformation

   ans. b.   diff. 6   obj. 5   p. 32

60. All endotoxins are part of the _____ of Gram-negative bacteria.
   a. pili
   b. capsule
   c. flagella
   d. cell wall

   ans. d.   diff. 5   obj. 5   p. 32

61. Endotoxins consist of a two-part molecule. The two parts are
   a. lipid A and polysaccharide
   b. protein and lipid A
   c. carbohydrate and protein B
   d. enzyme and substrate
   e. Lipid and protein A

   ans. a.   diff. 6   obj. 5   p. 33

62. Which portion of the endotoxin is responsible for its toxicity?
   a. polysaccharide
   b. protein
   c. carbohydrate
   d. lipid A

   ans. d.   diff. 7   obj. 5   p. 33

63. Which of the following tests is used to verify that intravenous fluids are not contaminated with endotoxins?
   a. *Limulus*
   b. litmus
   c. Benedict's
   d. Ames

   ans. a.   diff. 7   obj. 5   p. 34

64. Which of the following bacterial products is not considered an extracellular product that can contribute to virulence?
    a. collagenase
    b. coagulase
    c. toxin
    d. hyaluronidase

    ans. c.   diff. 8   obj. 8   p. 35

65. Which of the following is a fungal toxin that is responsible for liver damage in dogs?
    a. coagulase
    b. aflatoxin
    c. leukocidin
    d. phospholipase

    ans. b.   diff. 6   obj. 5   p. 35

66. The defense that is directed in a specific way against an invader is described as the _____ response.
    a. immune
    b. inflammatory
    c. autoimmune
    d. infection

    ans. a.   diff. 5   obj. 6   p. 37

67. Which of the following cells is not considered originating from a hematopoietic stem cell in the bone marrow?
    a. osteocyte
    b. erythrocyte
    c. platelet
    d. leukocyte

    ans. a.   diff. 5   obj. 6   p. 37

68. The cells primarily responsible for defense of the body against microorganisms are the
    a. erythrocytes
    b. leukocytes
    c. platelets
    d. osteocytes

    ans. b.   diff. 4   obj. 6   p. 37

69. The leukocyte that is involved in the inflammatory response by releasing histamine and serotonin is the
    a. basophil
    b. monocyte
    c. eosinophil
    d. neutrophil

    ans. a.   diff. 4   obj. 6   p. 39

70. The leukocyte present in the greatest number in circulating blood is the
    a. eosinophil
    b. lymphocyte
    c. monocyte
    d. neutrophil

    ans. d.   diff. 4   obj. 6   p. 39

71. Which of the following leukocytes is the most efficient of the phagocytes?
    a. eosinophil
    b. lymphocyte
    c. neutrophil
    d. basophil

    ans. c.   diff. 5   obj. 6   p. 39

72. The life span of tissue macrophages has been judged to be approximately
    a. a few days
    b. several months to several years
    c. several hours
    d. a few weeks

    ans. b.   diff. 5   obj. 6   p. 39

73. Which of the following is not considered a name for macrophages in specific tissues?
    a. monocytes
    b. dust cells
    c. Kupffer cells
    d. histiocytes

    ans. a.   diff. 6   obj. 6   p. 40

74. Which of the following matches of macrophages is incorrect?
    a. microglia—central nervous system
    b. Kupffer cells—liver
    c. dust cells—lung
    d. histiocytes—skin

    ans. d.   diff. 7   obj. 6   p. 40

75. The leukocyte that is described as having a small amount of cytoplasm surrounding a round nucleus is the
    a. neutrophil
    b. monocyte
    c. basophil
    d. lymphocyte

    ans. d.   diff. 6   obj. 6   p. 40

76. The leukocyte responsible for the specific immune response is the
    a. eosinophil
    b. monocyte
    c. lymphocyte
    d. neutrophil

    ans. c.   diff. 5   obj. 6   p. 41

77. The first host response to a nonspecific tissue injury is described as
    a. reaction
    b. inflammation
    c. antibodies
    d. trauma

    ans. b.   diff. 5   obj. 7   p. 41

78. The first kind of leukocyte lured to the site of inflammation is the
    a. macrophage
    b. monocyte
    c. polymorphonuclear neutrophil
    d. basophil

    ans. c.   diff. 6   obj. 7   p. 42

79. The compound that attacks the cytoplasmic membrane of bacterial cells during inflammation is
    a. peptidoglycan
    b. lysozyme
    c. beta-lysin
    d. antibodies

    ans. c.   diff. 7   obj. 7   p. 43

80. The low molecular weight glycoprotein produced by animal cells in response to viral infections is
    a. lysozyme
    b. complement
    c. beta-lysin
    d. interferon

    ans. d.   diff. 6   obj. 7   p. 44

81. Interferons function to make cells
    a. lyse when exposed to viruses
    b. resistant to viral replication
    c. nonmotile when infected by viruses
    d. resistant to phagocytosis

    ans. b.   diff. 6   obj. 7   p. 44

82. Which of the following statements about interferon is incorrect?
    a. do not directly inactivate viruses
    b. makes cells resistant to viral replication
    c. is species specific
    d. is virus specific

    ans. d.   diff. 7   obj. 7   p. 45

83. A group of interacting serum proteins that provide a non-specific defense mechanism is
    a. chemotoxic agents
    b. interferon
    c. glycoproteins
    d. complement

    ans. d.   diff. 7   obj. 7   p. 45

84. The complement pathway that is initiated by the antigen-antibody complex is referred to as the _____ pathway.
    a. properdin
    b. classical
    c. alternate
    d. divergent

    ans. b.   diff. 5   obj. 7   p. 46

85. Which of the host's metabolic responses is not involved in mobilization of the body to fight infection?
    a. redistribution of iron in the body
    b. fever
    c. nutrition
    d. changes in protein metabolism

    ans. c.   diff. 7   obj. 7   p. 47

86. One of the strongest indications of infectious disease is
    a. fever
    b. rash
    c. pustules
    d. vesicles

    ans. a.   diff. 5   obj. 7   p. 47

87. The temperature-regulating center of the body is located in the
    a. hypothalamus
    b. lymph nodes
    c. immune system
    d. skin

    ans. a.   diff. 6   obj. 7   p. 47

88. The messenger that alerts the hypothalamus to an infection is
    a. interferon
    b. interleukin-1
    c. coagulase
    d. fever

    ans. b.   diff. 6   obj. 7   p. 47

89. People with sickle-cell anemia are resistant to malarial infections, but are more susceptible to
    a. fever
    b. *Salmonella* infections
    c. Staphylococcal infections
    d. malnutrition

    ans. b.   diff. 8   obj. 7   p. 51

90. Which of the following is not important in increasing the body's susceptibility to infectious disease?
    a. adequate nutrition
    b. stress
    c. genetic factors
    d. fever

    ans. a.   diff. 6   obj. 7   p. 53

Matching

1.
1. one organism benefits while the other is unaffected
2. both organisms benefit
3. one organism benefits while the other is harmed
4. living together

    a. mutualism
    b. commensalism
    c. parasitism
    d. symbiosis
    e. independence

1 - b; 2 - a; 3 - c; 4 - d

2.
1. establish microbial growth on a body surface
2. able to produce disease in host with impaired defense
3. microbe that lacks ability to cause disease
4. disease causing

    a. avirulent
    b. colonization
    c. opportunist
    d. pathogen
    e. virulence

1 - b; 2 - c; 3 - a; 4 - d

3. Match the disease with the causative agent:
1. dysentery
2. common cold
3. conjunctivitis
4. blood infection

    a. Rhinovirus
    b. *Chlamydia* species
    c. *Shigella* species
    d. *Salmonella* species
    e. *Neisseria gonorrhoeae*

1 - c; 2 - a; 3 - b; 4 - d

4. Match the type of blood cell with its major function:

1. lymphocytes
2. basophils
3. eosinophils
4. polymorphonuclear neutrophils

a. phagocytosis
b. inflammation and immunity to parasites
c. release histamine
d. antibodies
e. carry oxygen

1 - d; 2 - c; 3 - b; 4 - a

True/False

1. The cells primarily responsible for the destruction of invading microorganisms within the body are lymphocytes.
   false (phagocytes)

2. The relationship in which organisms interact with each other is termed symbiosis.
   true

3. Parasitism is the symbiotic relationship in which both organisms are benefited.
   false (mutualism)

4. The relationship in which one organism is benefited and the other is harmed is commensalism.
   false (parasitism)

5. A microorganism that is able to produce disease only in a host with an impaired defense system is avirulent.
   false (opportunist)

6. Tissue invasion by microorganisms can be increased by frequent washing.
   false (by high concentrations)

7. An organism capable of producing a change in phagolysosome surface and surviving destruction is *Yersinia pestis.*
   false (*M. tuberculosis*)

8. *S. pneumoniae* can produce both coagulase and leukocidins, which aid in its virulence.
   false (*S. aureus*)

9. Endotoxins are necessary for cell division in Gram-negative bacteria.
   true

10. Endotoxins generally play a primary role rather than a contributing role in pathogenicity.
    false (contributory rather than primary)

11. A large number of Gram-positive and Gram-negative bacteria produce exotoxins.
    false (small)

12. The leukocyte involved in allergies and in rejection of parasitic worms is the eosinophil.
    true

13. Monocytes that migrate to the tissues are known as phagocytes.
    false (macrophages)

14. The least frequently found leukocyte in circulating blood is the basophil.
    true

15. Interferon is produced by the body in response to viral invasion.
    true

16. Complement is involved in inflammation by stimulating chemotaxis and phagocytosis.
    true

17. Interleukin-1 activates the hypothalamus, "telling" it that immunity is developing.
    false (it warns that an infection is present)

Chapter 16

Multiple Choice

1. The practice of deliberately stimulating the immune system
   to protect individuals against disease is referred to as
   a. acquired
   b. memory
   c. immunization
   d. disease
   e. reckless

   ans. c.   diff. 4   p. 1

2. The procedure described in old Chinese writings that
   deliberately exposed people to smallpox is termed
   a. oriental pox
   b. vaccinia
   c. variolation
   d. humoral immunity

   ans. c.   diff. 5   p. 2

3. Variolation as practiced until the late 1700s was dangerous
   because
   a. the person actually had smallpox and was contagious
   b. infections could result
   c. no permanent immunity developed
   d. allergies were a frequent complication

   ans. a.   diff. 6   p. 2

4. Which of the following terms did Pasteur use to describe
   any type of protective inoculation?
   a. immunity
   b. vaccination
   c. inoculation
   d. variolation

   ans. b.   diff. 6   p. 3

5. The last case of naturally contracted smallpox occurred in
   Africa in
   a. 1980
   b. 1987
   c. 1967
   d. 1977

   ans. d.   diff. 4   p. 3

6. The virus used for vaccination against smallpox is named
   the
   a. smallpox toxoid
   b. cowpox virus
   c. smallpox virus
   d. vaccinia virus

   ans. d.   diff. 5   p. 4

7. Which of the following scientists was responsible in 1796 for
   the development of the smallpox immunization technique?
   a. Pasteur
   b. Jenner
   c. Montagu
   d. Sung

   ans. b.   diff. 4   p. 4

8. The type of immunity that does not involve the production
   of antibodies or cellular immunity is
   a. innate
   b. acquired
   c. active
   d. passive

   ans. a.   diff. 4   obj. 1   p. 6

9. The type of immunity that causes one animal species to be
   resistant to the microorganisms affecting another species is
   a. acquired
   b. innate
   c. active
   d. passive

   ans. b.   diff. 4   obj. 1   p. 6

10. Which of the following is not considered an important
    factor in innate immunity?
    a. general health
    b. genetics
    c. antibodies
    d. nutrition

    ans. c.   diff. 5   obj. 1   p. 6

11. The result of exposure to microorganisms or an infectious agent by an immunologically competent person is described as
   a. acquired immunity
   b. susceptibility
   c. innate protection
   d. genetic immunity

   ans. a.   diff. 6   obj. 2   p. 6

12. The specific proteins manufactured by the immune system in response to processed infectious agents are
   a. antibodies
   b. antigens
   c. lymphocytes
   d. macrophages

   ans. a.   diff. 5   obj. 2   p. 6

13. The type of immunity that is acquired after contracting measles, chicken pox or other infectious diseases is _____ immunity.
   a. artificially acquired passive
   b. artificially acquired active
   c. naturally acquired passive
   d. naturally acquired active

   ans. d.   diff. 5   obj. 2   p. 6

14. Antibodies acquired from maternal blood by the fetus across the placenta is described as _____ immunity.
   a. artificially acquired passive
   b. artificially acquired ~~passive~~ active
   c. naturally acquired active
   d. naturally acquired passive

   ans. d.   diff. 5   obj. 2   p. 7

15. Immunity acquired by the fetus across the placenta is termed
   a. genetic
   b. active
   c. innate
   d. passive

   ans. d.   diff. 5   obj. 2   p. 7

16. The result of deliberately introducing a foreign substance into the body to produce an immune response is described as
   a. artificial
   b. natural
   c. innate
   d. passive

   ans. a.   diff. 5   obj. 2   p. 7

17. Artificially acquired active immunity is produced in the person by exposing him or her to a vaccine containing which of the following?
   a. serum of an immune subject
   b. antitoxin
   c. attenuated organisms
   d. phagocytes

   ans. c.   diff. 5   obj. 2   p. 7

18. Artificially acquired passive immunity involves the
   a. exposure to the disease agent
   b. transfer of antibodies formed in one person to another person
   c. placental transfer of antibodies from mother to fetus
   d. person being a victim of the disease

   ans. b.   diff. 5   obj. 2   p. 7

19. The most frequently used immune serum globulin protects people from
   a. scurvy
   b. hepatitis B
   c. diarrhea
   d. hepatitis A

   ans. d.   diff. 6   obj. 2   p. 8

20. Artificially acquired passive immunity generally does not last longer than
   a. 4 weeks
   b. 10 days
   c. 3 months
   d. 1 year

   ans. c.   diff. 5   obj. 2   p. 8

21. Which of the following defense mechanisms is not part of the nonspecific defenses?
    a. lymphoid system
    b. skin
    c. inflammation
    d. fever

    ans. a.   diff. 6   obj. 3   p. 8

22. Which of the following defense mechanisms is not considered part of the specific defense?
    a. lymphoid system
    b. fever
    c. antibodies
    d. cell-mediated response

    ans. b.   diff. 6   obj. 3   p. 8

23. The second line of defense in vertebrates is referred to as the
    a. allergy
    b. phagocytic response
    c. immunization
    d. specific immune response

    ans. d.   diff. 4   obj. 3   p. 9

24. Which of the following systems is most important in the specific immune response?
    a. lymphatic
    b. endocrine
    c. circulatory
    d. respiratory

    ans. a.   diff. 5   obj. 3   p. 10

25. Which of the following terms is synonymous with antibody generator?
    a. autoimmune
    b. antibody
    c. immunogen
    d. immunoglobulin

    ans. c.   diff. 6   obj. 4   p. 10

26. Which of the following molecules is not considered a good antigen?
    a. lipid
    b. protein
    c. polysaccharide
    d. enzyme

    ans. a.   diff. 5   obj. 4   p. 10

27. Substances that make good antigens should have a molecular weight of no less than _____ daltons.
    a. 100,000
    b. 10,000
    c. 50,000
    d. 5,000

    ans. b.   diff. 6   obj. 4   p. 10

28. The body's adverse response to its own body constituents with the potential of tissue damage is termed
    a. responsiveness
    b. allergy
    c. immunity
    d. autoimmune disease

    ans. d.   diff. 6   obj. 4   p. 10

29. Specific chemical groups on the antigen molecule to which the immune response is directed are called
    a. antigenic determinants
    b. autoimmune response
    c. monomers
    d. allergens

    ans. a.   diff. 6   obj. 4   p. 11

30. Although a bacterial cell wall may have as many as 100 or more antigenic determinants, how many amino acids can act as specific antigenic determinants?
    a. all the sequences
    b. 50 to 60
    c. 70 to 80
    d. 10 to 20

    ans. d.   diff. 7   obj. 4   p. 11

31. Low molecular weight molecules that can combine with larger molecules, such as proteins, to become antigenic are termed
    a. antigens
    b. haptens
    c. polysaccharides
    d. epitopes

    ans. b.   diff. 8   obj. 4   p. 12

32. When a small molecule like penicillin becomes combined with a large protein "carrier" to become antigenic it is known as a(n)
   a. hapten
   b. epitope
   c. polysaccharide
   d. complex

   ans. b.　　diff. 8　　obj. 4　　p. 12

33. In order for antigens to incite the immune response, they must be
   a. processed
   b. reduced
   c. combined with other proteins
   d. directed

   ans. a.　　diff. 6　　obj. 4　　p. 12

34. Antigens may be processed by
   a. erythrocytes
   b. monocytes
   c. phagocytes
   d. nullcells

   ans. c.　　diff. 6　　obj. 4　　p. 12

35. The humoral immune system involves the manufacture and use of
   a. T cells
   b. antibodies
   c. lymphokines
   d. antigens

   ans. b.　　diff. 5　　obj. 4　　p. 14

36. In which of the body fluids will antibodies not be dissolved?
   a. cytoplasm
   b. blood
   c. lymph
   d. mucous

   ans. a.　　diff. 6　　obj. 5　　p. 14

37. Antibody production is directed by lymphocytes known as
   a. program mastic
   b. T cells
   c. phagocytes
   d. B cells

   ans. d.　　diff. 4　　obj. 5　　p. 14

38. The cells that actually secrete the antibodies are
   a. T cells
   b. B cells
   c. phagocytes
   d. plasma cells

   ans. d.　　diff. 6　　obj. 5　　p. 14

39. The T cells are responsible for
   a. humoral immunity
   b. cell-mediated immunity
   c. generalized defense system
   d. allergic responses

   ans. b.　　diff. 5　　obj. 9　　p. 14

40. The immune response that is most effective against intracellular parasites and infections with eukaryotes is
   a. humoral immunity
   b. cell-mediated immunity
   c. generalized immune response
   d. allergies

   ans. b.　　diff. 6　　obj. 9　　p. 14

41. Another term for the helper T cells, which are involved in AIDS, is
   a. $CD4^+$
   b. CD1
   c. $CD2^+$
   d. $CD3^+$

   ans. a.　　diff. 8　　obj. 9　　p. 15

42. Stems cells that mature into B cells mature in the _____ of the fetus.
   a. yellow bone marrow
   b. yolk sac
   c. liver
   d. red bone marrow

   ans. c.　　diff. 6　　obj. 5　　p. 16

43. Where in the adult human are B cells not likely to be found?
   a. skin
   b. Peyer's patches
   c. appendix
   d. tonsils

   ans. a.　　diff. 6　　obj. 5　　p. 16

44. An actively secreting plasma cell can produce approximately how many antibody molecules per second?
   a. 4,000
   b. 2,000
   c. 1,000
   d. as many as needed

   ans. b.   diff. 5   obj. 5   p. 16

45. The technique that allows the body to produce only the antibodies that are needed in any specific disease is termed
   a. B cell specificity
   b. plasma cell production
   c. clonal expansion
   d. antibody control

   ans. c.   diff. 6   obj. 5   p. 17

46. Another name by which antibodies are known is
   a. immunoglobulins
   b. antigens
   c. allergens
   d. plasma cells
   e. immunogens

   ans. a.   diff. 5   obj. 5   p. 18

47. Antibodies are divided into groups based on their chemical structure. These groups are commonly called
   a. groups
   b. types
   c. classes
   d. categories

   ans. c.   diff. 6   obj. 7   p. 18

48. The two chains of an antibody molecule are held together by a _____ bond.
   a. hydrogen
   b. disulfide
   c. oxygen
   d. ionic

   ans. b.   diff. 6   obj. 7   p. 18

49. The classic antibody is the
   a. IgG
   b. IgA
   c. IgD
   d. IgE

   ans. a.   diff. 4   obj. 7   p. 18

50. Each antibody molecule is _____-shaped and has two identical halves. Each half consists of a _____ and a _____ chain of amino acids.
   a. Y, heavy and light
   b. X, medium and short
   c. X, heavy and light
   d. Y, medium and short

   ans. a.   diff. 6   obj. 7   p. 18

51. Antigenic determinants can bond with antibodies at their variable region, which is located
   a. at the junction of the heavy and light chains
   b. on the outer end of each arm of the Y
   c. at different regions on each class of antibody
   d. at the terminal end of the heavy chain

   ans. b.   diff. 7   obj. 7   p. 19

52. Which of the following antibodies is primarily responsible for protection from infectious disease?
   a. IgM
   b. IgG
   c. IgD
   d. IgE

   ans. b.   diff. 5   obj. 7   p. 20

53. Which of the following antibodies is the first one produced in response to infectious disease?
   a. IgM
   b. IgG
   c. IgD
   d. IgE

   ans. a.   diff. 5   obj. 7   p. 19

54. Which of the following antibodies is found primarily in external secretions like milk and saliva?
   a. IgD
   b. IgM
   c. IgA
   d. IgG

   ans. c.   diff. 6   obj. 7   p. 19

55. Which of the following antibodies is primarily involved in hypersensitivity?
   a. IgA
   b. IgG
   c. IgD
   d. IgE

   ans. d.   diff. 6   obj. 7   p. 19

56. For an antigen-antibody reaction to be triggered there must be a close physical and chemical association. This is described as
   a. gene and expression
   b. substrate and enzyme
   c. lock and key
   d. clonal expansion

   ans. c.   diff. 7   obj. 7   p. 22

57. The only antibody that can cross the placenta is
   a. IgM
   b. IgG
   c. IgE
   d. IgD

   ans. b.   diff. 6   obj. 7   p. 23

58. Each class of antibody differs especially in its
   a. ability to cross the placenta
   b. specific disease response
   c. amino acid sequence and carbohydrate content
   d. disulfide bonds

   ans. c.   diff. 8   obj. 7   p. 23

59. The class of immunoglobulin that contains the Rh immunoglobulin, which can cross the placenta and possibly damage the fetus is
   a. IgM
   b. IgG
   c. IgD
   d. IgE

   ans. b.   diff. 5   obj. 7   p. 24

60. The only immunoglobulin produced in response to the capsular polysaccharide antigen is
   a. IgM
   b. IgG
   c. IgE
   d. IgD

   ans. a.   diff. 8   obj. 7   p. 24

61. The polypeptide transport piece of the IgA is synthesized by
   a. epithelial cells
   b. immune system
   c. blood
   d. lymph

   ans. a.   diff. 8   obj. 7   p. 25

62. The function of the secretory component of the IgA is to
   a. protect IgA from being destroyed by proteolytic enzymes
   b. coat the antigen
   c. opsonize it
   d. fix the IgA to the antigen and stimulate lysis

   ans. a.   diff. 8   obj. 7   p. 25

63. The overall functions of IgD are not completely known, but it is quite probable that IgD serves
   a. to protect the fetus against infectious disease
   b. to ready the body in a severe allergic response
   c. as a specific membrane receptor for its intended antigenic determinants
   d. as a model for monoclonal antibodies

   ans. c.   diff. 7   obj. 7   p. 26

64. The immunoglobulin that stimulates the release of the granular content of mast cells and basophils after combination with the antigen is
   a. IgM
   b. IgA
   c. IgE
   d. IgD

   ans. c.   diff. 8   obj. 7   p. 27

65. Hybridomas are the result of the fusion of
   a. enzymes produced by cancer cells and IgEs
   b. antigens and IgDs
   c. tumor cells and the oncogene
   d. antibody-producing B cells and myeloma cells

   ans. d.   diff. 8   obj. 7   p. 27

66. Antibodies that have been effective in neutralizing the T cells that are responsible for kidney rejection are
   a. IgG and IgM
   b. IgD
   c. IgG
   d. monoclonal antibodies

   ans. d.   diff. 7   obj. 8   p. 28

67. In 1975, Kohler and Milstein developed techniques for producing
   a. safe vaccines
   b. genetically engineered antibodies
   c. hybridomas
   d. an antibody that has the characteristics of IgG and IgA

   ans. c.   diff. 8   obj. 8   p. 28

68. For which of the following diseases are monoclonal antibodies not useful in diagnosis?
   a. influenza
   b. syphilis
   c. hepatitis
   d. herpes simplex

   ans. b.   diff. 6   obj. 8   p. 28

69. The cells responsible for the secondary immune response to the antigen are
   a. plasma cells
   b. B cells
   c. memory cells
   d. T cells

   ans. c.   diff. 5   obj. 6   p. 29

70. Antibodies known as opsonins function to inactivate antigens by
   a. combining them with many other protein molecules for agglutination
   b. neutralizing the binding sites to prevent attachment
   c. lysing the antigens so phagocytosis is enhanced
   d. coating them to increase macrophage ingestion of them

   ans. d.   diff. 7   obj. 9   p. 31

71. T cells are manufactured in the
   a. thymus
   b. bone marrow
   c. Peyer's patches
   d. lymphoid tissue

   ans. b.   diff. 6   obj. 9   p. 31

72. T cells have _____ receptors on their cell membranes.
   a. glycoprotein
   b. lipopolysaccharide
   c. nucleic acid
   d. lipoprotein

   ans. a.   diff. 7   obj. 9   p. 31

73. The stimulation of the B cell to divide and mature into an antibody-producing plasma cell is provided by a
   a. suppressor T cell
   b. helper T cell
   c. cytotoxic T cell
   d. macrophage

   ans. b.   diff. 8   obj. 9   p. 32

74. Which of the following cytokines is not produced by T cells?
   a. interleukin 1
   b. interleukin 2
   c. interleukin 3
   d. interleukin 4

   ans. a.   diff. 7   obj. 9   p. 33

75. Cytotoxic T cells destroy infected cells by secreting the protein _____, which induces cell _____.
   a. cytotoxin, division
   b. interleukin 2, death
   c. perforin, lysis
   d. suppresin, toxicity

   ans. c.   diff. 8   obj. 9   p. 34

76. Regulation of the immune response in both B and T cells is done by
    a. helper T cells
    b. cytotoxic T cells
    c. suppressor T cells
    d. killer T cells

    ans. c.　　diff. 7　　obj. 10　　p. 35

77. Lymphocytes without membrane surface markers and that appear to kill certain types of tumors and virus-infected cells are designated
    a. NK cells
    b. macrophages
    c. B cells
    d. T cells

    ans. a.　　diff. 8　　obj. 10　　p. 35

78. Monocytes that migrate out of blood to the tissues are called
    a. mast cells
    b. lymphocytes
    c. Kupffer cells
    d. macrophages

    ans. d.　　diff. 7　　obj. 10　　p. 36

79. Calculations have placed the number of possible recognition sites on B cells at
    a. 10^7
    b. 10^8
    c. 10^6
    d. 10^5

    ans. b.　　diff. 8　　obj. 10　　p. 38

80. During clonal expansion, memory cells are produced by
    a. both B cells and T cells
    b. neither B cells nor T cells
    c. B cells only
    d. T cells only

    ans. a.　　diff. 6　　obj. 10　　p. 38

81. The cells responsible for the anamnestic response are the
    a. memory cells
    b. T cells
    c. B cells
    d. antigens

    ans. a.　　diff. 6　　obj. 1　　p. 39

82. A treatment that can kill the microorganism but leave its antigenicity is frequently employed in the manufacture of
    a. inactive immunological agents
    b. active immunological agents
    c. passive immunological agents
    d. toxins

    ans. a.　　diff. 7　　obj. 1　　p. 40

83. Which of the following vaccines is an example of an active attenuated vaccine?
    a. diphtheria toxoid
    b. Salk polio
    c. Sabin polio
    d. gamma globulin

    ans. c.　　diff. 7　　obj. 11　　p. 40

84. Parasites present problems for vaccine development because they have several developmental stages and each stage
    a. will produce a different immune response
    b. may be toxic
    c. has different antigens
    d. presents several alternatives

    ans. c.　　diff. 8　　obj. 11　　p. 41

85. Recognition of one's own tissues and failure to respond to them immunologically is a form of
    a. immune incompetency
    b. autoimmune disease
    c. vaccination
    d. immunological tolerance

    ans. d.　　diff. 8　　obj. 12　　p. 42

86. An active immune response is not stimulated by
    a. foreign proteins
    b. substances encountered during embryonic development
    c. active vaccines
    d. inactive vaccines

    ans. b.　　diff. 7　　obj. 12　　p. 42

87. The average half-life of most IgG molecules in human circulation is about
    a.   1–5 days
    b.   a lifetime
    c.   19–25 days
    d.   two weeks

    ans. c.   diff. 7   obj. 12   p. 44

88. Almost all antibodies of the newborn are
    a.   the result of infection
    b.   self made
    c.   IgMs
    d.   of maternal origin
    e.   none of the above
    ans. d.   diff. 8   obj. 12   p. 44

89. The cells that inhibit B cells from maturing to plasma cells are
    a.   cytotoxic T cells
    b.   suppressor T cells
    c.   antigens
    d.   tolerant cells

    ans. b.   diff. 8   obj. 12   p. 45

True/False

1. Immunity as the result of genetics is referred to as acquired.
    false   (innate)

2. Fever is part of the nonspecific defense system.
    true

3. Proteins, polysaccharides, and lipids are generally good antigens.
    false   (lipids do not make good antigens)

4. Substances of a molecular weight of less than 10,000 daltons make good antigens.
    false   (more than 10,000 daltons)

5. The specific glycoprotein found on the membrane of all macrophages responsible for attachment of antigenic determinants is known as the MLA (major lymphocyte antigen).
    false   (MHC-major histocompatibility complex)

6. The life span of plasma cells is approximately ten days.
    false   (very short-lived)

7. The major circulating antibody found in blood, lymph, and tissue fluids is IgM.
    false   (IgG)

8. There are five classes of immunoglobulins.
    true

9. In 1986, the FDA approved a monoclonal antibody for use on humans that reduces heart transplant rejection.
    false   (kidney)

10. T cell receptors are identical to antibody molecules.
    false   (similar not identical)

11. When monocytes multiply in the connective tissue, they are called macrophages.
    false   (they do not multiply, they migrate)

12. The anamnestic response is responsible for the effects of vaccination.
    true

13. Inactivated bacterial toxins are the least effective vaccine.
    false   (very effective)

14. Inactivated bacterial toxins used as vaccines are antitoxins.
    false   (toxoids)

Matching

1. Match the immunoglobulin with its most significant function.

1.	IgG	a.	protection against infectious disease
2.	IgM	b.	agglutinates antigens
3.	IgA	c.	protection of mucous membranes
4.	IgD	d.	involved in hypersensitivity
5.	IgE	e.	development and maturation of antibody response

1 - a; 2 - b; 3 - c; 4 - e; 5 - d

2. Match the situation with the type of immunity.

1.	immunity to distemper	a.	naturally acquired passive
2.	antibody present in the newborn	b.	innate
3.	child recovered from chickenpox	c.	naturally acquired active
4.	gamma globulin	d.	artificially acquired passive
5.	immunization with measles vaccine	e.	artificially acquired active

1 - b; 2 - a; 3 - c; 4 - d; 5 - e

3. Match the activity with the immunoglobulin.

1.	found in saliva and mucous	a.	IgA
2.	binds to cell membranes of mast cells	b.	IgG
3.	found in blood and lymph	c.	IgE
4.	antibody of the A, B, and O blood groupings	d.	IgM

1 - a; 2 - c; 3 - b; 4 - d

Chapter 17

Multiple Choice

1. The scientist responsible for the first successful experiments (1901) in blood typing was
   a. Landois
   b. Landsteiner
   c. Lancefield
   d. Weiner

   ans. b.   diff. 6   obj.   p. 2

2. The Nobel prize of 1930 was awarded to Landsteiner for his discovery of the
   a. A, B, O blood groups
   b. pH factor
   c. immunodeficiency complexes
   d. agents of hypersensitivity

   ans. a.   diff. 7   obj.   p. 3

3. If the immune system responds inadequately to antigenic stimulation, the result is
   a. hypersensitivity
   b. immunodeficiency
   c. autoimmunity
   d. Rh incompatibility

   ans. b.   diff. 7   obj.   p. 3

4. An inappropriate response by the immune system is termed
   a. A, B, O incompatibility
   b. Rh incompatibility
   c. disease state
   d. hypersensitivity

   ans. d.   diff. 6   obj.   p. 3

5. Immunodeficiency cannot be directly traced to
   a. acquisition as a result of infection
   b. genetic defects
   c. poor nonspecific defense mechanisms
   d. trauma or starvation

   ans. c.   diff. 6   obj.   p. 4

6. Primary immunodeficiency can also be described as
   a. acquired
   b. secondary
   c. congenital
   d. allergy

   ans. c.   diff. 6   obj. 1   p. 4

7. A developmental defect in the lymphoid system is referred to as a(n) _____ defect.
   a. congenital
   b. acquired
   c. alternate
   d. infection-mediated

   ans. a.   diff. 7   obj. 1   p. 4

8. Primary immunodeficiencies will not result if there is a defect in the
   a. mononuclear phagocytes
   b. B cells
   c. T cells
   d. spleen

   ans. d.   diff. 6   obj. 1   p. 4

9. If the thymus fails to develop, then which of the following cells will be affected?
   a. phagocytes
   b. B cells
   c. T cells
   d. red blood cells

   ans. c.   diff. 6   obj. 1   p. 5

10. If the thymus fails to develop, the resulting disease is
    a. DiGeorge syndrome
    b. AIDS
    c. monoclonal gammopathy
    d. Chediak-Higashi disease

    ans. a.   diff. 6   obj. 1   p. 5

11. If a patient lacks B cells, the resulting disease is
    a. infantile X-linked agammaglobulinemia
    b. AIDS
    c. DiGeorge syndrome
    d. SCID

    ans. a.   diff. 7   obj. 1   p. 5

12. Patients with infantile X-linked agammaglobulinemia are highly susceptible to _____ infections.
    a. fungal
    b. streptococcal
    c. eukaryotic pathogenic
    d. obligate intracellular bacteria

    ans. b.   diff. 8   obj. 1   p. 5

13. Defects in the bone marrow stem cells result in a condition known as
    a. Chediak-Higashi disease
    b. AIDS
    c. SCID
    d. DiGeorge syndrome

    ans. c.   diff. 7   obj. 1   p. 5

14. A condition that results from ineffective phagocytes is
    a. DiGeorge syndrome
    b. AIDS
    c. Chediak-Higashi disease
    d. congenital agammaglobulinemia

    ans. c.   diff. 7   obj. 1   p. 6

15. Secondary immunodeficiency diseases are not the result of
    a. malignancies
    b. genetic defects
    c. advanced age
    d. malnutrition

    ans. b.   diff. 5   obj. 2   p. 6

16. Which of the following diseases has been shown to deplete the body temporarily of lymphoid cells?
    a. malaria
    b. syphilis
    c. leprosy
    d. measles

    ans. d.   diff. 5   obj. 2   p. 6

17. The most serious of the secondary immunodeficiency diseases is
    a. Chediak-Higashi disease
    b. DiGeorge syndrome
    c. SCID
    d. AIDS

    ans. d.   diff. 5   obj. 2   p. 6

18. An RNA virus that infects and destroys helper T cells has been determined to be the cause of
    a. AIDS
    b. SCID
    c. HBV
    d. HAV

    ans. a.   diff. 5   obj. 2   p. 7

19. When the body responds inappropriately to an antigen, the condition is described as
    a. monoclonal gammopathy
    b. AIDS
    c. secondary immunodeficiency disease
    d. hypersensitivity

    ans. d.   diff. 6   obj. 3   p. 7

20. The immunoglobulin involved in Type I hypersensitivity is
    a. IgE
    b. IgB
    c. IgM
    d. IgG

    ans. a.   diff. 4   obj. 3   p. 8

21. In which type of hypersensitivity is IgE involved?
    a. Type III
    b. Type II
    c. Type I
    d. Type IV

    ans. c.   diff. 4   obj. 3   p. 8

22. Exposure to which of the following antigens will not result in Type I hypersensitivity?
    a. pollen
    b. dust
    c. transfusion
    d. animal dander

    ans. c.   diff. 5   obj. 3   p. 9

23. The release of chemicals that are responsible for hives, hay fever, shock, and other allergic reactions are secreted by _____ under the influence of IgEs.
   a. lymphokines
   b. B cells
   c. mast cells
   d. phagocytes

   ans. c.   diff. 6   obj. 3   p. 9

24. Wheal formation occurs when sensitized mast cells release histamine. The skin condition is referred to as
   a. complement
   b. urticaria
   c. necrosis
   d. dermatitis

   ans. b.   diff. 6   obj. 3   p. 10

25. The Type I hypersensitivity described as asthma results from sensitized mast cells releasing chemical mediators such as
   a. leukotrienes
   b. histamine
   c. bronchiospasms
   d. antihistamines

   ans. a.   diff. 6   obj. 3   p. 11

26. One word that can be used to describe generalized anaphylaxis is
   a. shock
   b. wheal
   c. flare
   d. rash

   ans. a.   diff. 6   obj. 3   p. 11

27. Most cases of generalized anaphylaxis are attributable to
   a. injectable penicillin and bee stings
   b. aspirin and oral penicillin
   c. fire ant stings and haptens
   d. pollen and shellfish

   ans. a.   diff. 6   obj. 3   p. 11

28. The hypersensitivity treatment that stimulates an increase of IgG and suppressor T cells, and a decrease of IgE is known as
   a. immunity
   b. immunotherapy
   c. sensitization
   d. exposure

   ans. b.   diff. 6   obj. 3   p. 12

29. The type of hypersensitivity expressed with the lysing of red blood cells is
   a. Type I
   b. Type III
   c. Type IV
   d. Type II

   ans. d.   diff. 6   obj. 4   p. 13

30. In which of the following blood types is there a lack of coding for glycosyltransferase?
   a. B
   b. A
   c. O
   d. AB

   ans. c.   diff. 7   obj. 4   p. 14

31. The natural antibodies present in blood of the anti-A and anti-B types are of the class
   a. IgE
   b. IgG
   c. IgM
   d. IgA

   ans. c.   diff. 6   obj. 4   p. 14

32. Antibodies present in blood plasma without any obvious or deliberate stimulus are described as
   a. natural
   b. acquired
   c. injurious
   d. bacterial

   ans. a.   diff. 7   obj. 4   p. 15

33. A transfusion reaction can occur if a patient receives
_____ differing from his own during a blood
transfusion.
   a. leukocytes
   b. erythrocytes
   c. plasma
   d. platelets

   ans. b.   diff. 5   obj. 4   p. 15

34. Hemolytic disease of the newborn may result if the pregnant
woman forms _____ antibodies.
   a. anti-Rh
   b. anti-A
   c. anti-B
   d. either anti-A or anti-B

   ans. a.   diff. 6   obj. 4   p. 16

35. The cell types responsible for the hypersensitivity known as
transfusion reaction are
   a. complement
   b. T cells
   c. macrophages
   d. B cells

   ans. d.   diff. 6   obj. 4   p. 18

36. If an antibody reacts with a specific soluble antigen and
fixes complement, the resulting response is termed a(n)
   a. dermatitis
   b. transfusion reaction
   c. immune complex
   d. serum sickness

   ans. c.   diff. 6   obj. 5   p. 18

37. An area of the body not frequently complicated by immune
complexes is
   a. skin
   b. kidney
   c. joints
   d. heart

   ans. d.   diff. 7   obj. 5   p. 19

38. Which of the following is not involved in the
hypersensitivity reaction described as immune complex?
   a. IgE
   b. IgG
   c. IgM
   d. B cells

   ans. a.   diff. 6   obj. 5   p. 19

39. When there is localized injury or death of tissue resulting
from repeated injections of an antigen into a previously
immunized animal the condition is referred to as
   a. farmer's lung
   b. Arthus reaction
   c. German measles
   d. malaria

   ans. b.   diff. 7   obj. 5   p. 20

40. Which of the following disease states is not among those
described as immune complexes?
   a. farmer's lung
   b. serum sickness
   c. hay fever
   d. bacterial endocarditis

   ans. c.   diff. 7   obj. 5   p. 21

41. The response mechanism that is responsible for delayed
hypersensitivity involves the
   a. IgG and IgM
   b. B cells
   c. IgEs
   d. T cells

   ans. d.   diff. 6   obj. 6   p. 21

42. Delayed hypersensitivity is also known as
   a. Type IIII
   b. Type IV
   c. Type II
   d. Type I

   ans. b.   diff. 5   obj. 6   p. 21

43. Which of the following conditions is described as contact hypersensitivity?
    a. poison ivy
    b. serum sickness
    c. hay fever
    d. hives

    ans. a.   diff. 6   obj. 6   p. 21

44. A familiar example of delayed hypersensitivity reaction is a(n)
    a. positive tuberculin test
    b. immune complex reaction
    c. serum sickness
    d. hives

    ans. a.   diff. 6   obj. 6   p. 21

45. The redness and induration of delayed skin hypersensitivity reactions are the result of the specific antigen reacting with
    a. IgEs
    b. sensitized T cells
    c. IgG and IgM
    d. complement

    ans. b.   diff. 7   obj. 6   p. 22

46. Patch tests are used to detect
    a. serum sickness
    b. contact hypersensitivity
    c. immune complexes
    d. hives

    ans. b.   diff. 5   obj. 6   p. 22

47. Delayed hypersensitivity is involved in the cell destruction and progressive impairment frequently seen in
    a. rheumatic fever
    b. measles
    c. leprosy
    d. hepatitis A

    ans. c.   diff. 6   obj. 6   p. 23

48. The tissue antigens most commonly involved in graft rejections are of the _____ system.
    a. A, B, D
    b. MHC
    c. Rh
    d. disease

    ans. b.   diff. 7   obj. 7   p. 23

49. Graft versus host disease is an example of
    a. delayed hypersensitivity
    b. the Arthur reaction
    c. Type I hypersensitivity
    d. Type III hypersensitivity

    ans. a.   diff. 6   obj. 7   p. 24

50. Killing of graft cells occurs through direct contact with
    a. sensitized T cells
    b. serum incompatibility
    c. MHC
    d. IgEs

    ans. a.   diff. 7   obj. 7   p. 24

51. The cluster of genes found on chromosome #6 is important in tissue transplantation. It is known as
    a. DNA
    b. MLA
    c. HLA
    d. graft rejection complex

    ans. c.   diff. 8   obj. 7   p. 24

52. The main role of the MHC appears to be important in
    a. recognition of macrophages
    b. preventing anaphylaxis
    c. cell-to-cell communication
    d. regulatory inflammation

    ans. c.   diff. 7   obj. 7   p. 25

53. Which of the following is inconsistent with the properties identified as those of Class 1 MHC?
    a. involved with virus-infected cells
    b. found on all nucleated cells
    c. found principally on macrophages
    d. involved in antigen presentation to cytotoxic T cells

    ans. c.   diff. 8   obj. 7   p. 25

54. Which of the following is inconsistent with the properties identified as those of Class II MHC?
    a. found principally on macrophages
    b. involved in antigen presentation to cytotoxic T cells
    c. involved in cells that process antigens
    d. involved in presenting antigen to helper T cells

    ans. b.   diff. 8   obj. 7   p. 25

55. If both the MHC and the T cells are necessary to induce an immune response, then the antigen is described as
    a. pathogenic
    b. T independent
    c. T dependent
    d. communicable

    ans. c.   diff. 8   obj. 7   p. 26

56. If the immune system of the body responds to the tissues of the body as foreign, the resulting condition is known as
    a. anaphylaxis
    b. rheumatoid arthritis
    c. poison ivy
    d. an autoimmune disease

    ans. d.   diff. 9   obj. 8   p. 28

57. The autoimmune disease that affects the joints, kidney, skin, and heart is termed
    a. Lupus erythematosus
    b. rheumatoid arthritis
    c. hemolytic anemia
    d. Graves' disease

    ans. a.   diff. 8   obj. 8   p. 28

58. Myasthenia gravis is a disease that involves
    a. sensitized T cells
    b. humoral antibodies
    c. cytotoxic T cells
    d. IgDs

    ans. b.   diff. 8   obj. 8   p. 28

59. An example of a disease in which delayed hypersensitivities are involved is
    a. Graves' disease
    b. rheumatoid arthritis
    c. myasthenia gravis
    d. sympathetic ophthalmia

    ans. d.   diff. 8   obj. 8   p. 28

60. The muscle weakness of myasthenia gravis results from antibodies to acetylcholine receptors being produced. The antibodies are of the _____ type.
    a. IgE
    b. IgG
    c. IgA
    d. IgD

    ans. b.   diff. 8   obj. 8   p. 28

61. Which of the following autoimmune diseases is mediated by delayed hypersensitivity?
    a. myasthenia gravis
    b. sympathetic ophthalmia
    c. rheumatoid arthritis
    d. rheumatic heart disease

    ans. b.   diff. 8   obj. 8   p. 30

62. Diagnostic tests that involve interactions between antigens and antibodies are known collectively as
    a. serology
    b. complement fixing
    c. precipitation
    d. agglutination

    ans. a.   diff. 5   obj. 9   p. 31

63. Antibodies that combine with soluble antigens *in vitro* to form visible precipitates are called
    a. Ouchterlony techniques
    b. agglutinins
    c. serology
    d. precipitins

    ans. d.   diff. 6   obj. 9   p. 32

64. Which of the following serology tests can be used to detect antibody against more than one antigen at the same time?
    a. immunoelectrophoresis
    b. precipitation
    c. Ouchterlony technique
    d. agglutination

    ans. c.   diff. 7   obj. 9   p. 32

65. The serology test used to detect bacterial meningitis is
    a. indirect agglutination
    b. precipitation
    c. Ouchterlony technique
    d. countercurrent immunoelectrophoresis

    ans. d.   diff. 8   obj. 9   p. 33

66. The direct agglutination test that can be used to conduct a number of antigen-antibody tests at the same time is called a(n)
    a. Widal test
    b. serial dilution test
    c. microtiter plate
    d. gel diffusion

    ans. c.   diff. 7   obj. 9   p. 34

67. Hemagglutination tests can be used to test for diseases caused by
    a. viruses
    b. protozoans
    c. worms
    d. fungi

    ans. a.   diff. 8   obj. 9   p. 35

68. The indicator system that employs sheep red blood cells and their specific antibodies is used in the serology test referred to as
    a. complement fixation
    b. hemagglutination
    c. fluorescent antibody
    d. immunoassay

    ans. a.   diff. 7   obj. 9   p. 35

69. A serological test frequently used to test for virus infections is
    a. complement fixation
    b. neutralization
    c. microtiter plate
    d. Widal test

    ans. b.   diff. 6   obj. 9   p. 37

70. The serology test that shows the antigen-antibody complex as yellow green under the microscope while using an ultraviolet light is known as the
    a. radioimmunoassay
    b. fluorescent antibody test
    c. ELISA test
    d. AIDS test

    ans. b.   diff. 6   obj. 9   p. 38

71. Blood for transfusion is frequently tested for AIDS by using the
    a. complement fixation test
    b. radioimmunoassay test
    c. ELISA method
    d. RIA test

    ans. c.   diff. 8   obj. 9   p. 38

True/False

1. Hypersensitivity and allergy are synonymous terms.
   true

2. Immunodeficiency disease would develop if the spleen did not develop correctly.
   false   (thymus)

3. DiGeorge syndrome develops in patients with defective phagocytes.
   false   (who lack a thymus)

4. If T cells fail to develop, the patient cannot produce antibodies.
   false   (B cells)

5. Localized anaphylaxis may be fatal within minutes.
   false   (general anaphylaxis)

6. The antibody responsible for most cases of immediate hypersensitivity is IgE.
   true

7. The chemical mediators released by mast cells sensitized by IgEs are responsible for the skin reactions in Type II hypersensitivity.
   false   (Type I)

8. The most serious expression of Type I hypersensitivity is asthma.
   false   (generalized anaphylaxis)

9. Transfusion reactions of the ABO blood groups are examples of Type I hypersensitivity.
   false (Type II)

10. Complications of pregnancy occur when an Rh-negative mother is carrying an Rh-positive fetus.
    true

11. Immune complex formation and the Arthus reaction are examples of Type III hypersensitivity.
    true

12. Type IV hypersensitivity involves sensitized T cells.
    true

13. The tuberculin skin test is an example of Type III hypersensitivity.
    false (Type IV)

14. Tissue graft rejection can be attributed to a poor match of the ABO blood groups between donor and recipient.
    false (MHC antigens)

15. The autoimmune disease myasthenia gravis is due to delayed hypersensitivity reactions.
    false (due to the production of antibodies)

16. The Ouchterlony technique is a specific example of a precipitation reaction.
    true

Chapter 18

Multiple Choice

1. Another name for childbed fever is
    a. *S. aureus* intoxication
    b. puerperal fever
    c. young folks syndrome
    d. Klein's disease

    ans. b.   diff. 5   obj. 1   p. 2

2. The doctor who was responsible for introducing hand washing with chloride of lime before attending patients was
    a. Robert Koch
    b. Louis Pasteur
    c. Professor Klein
    d. Ignaz Semmelweis

    ans. d.   diff. 5   obj. 1   p. 3

3. A dramatic increase in the incidence of a specific disease in a given population is referred to as a(n)
    a. epidemic
    b. pandemic
    c. endemic disease
    d. mortality
    e. infectious

    ans. a.   diff. 5   obj. 2   p. 4

4. A disease that rapidly spreads to include most of the world population is called a(n)
    a. endemic disease
    b. epidemic
    c. pandemic
    d. controllable disease

    ans. c.   diff. 5   obj. 2   p. 4

5. Sporadic cases of an infectious disease in a population over a long period is referred to as a(n)
    a. epidemic
    b. contagion
    c. endemic disease
    d. pandemic

    ans. c.   diff. 5   obj. 2   p. 4

6. The study of the spread of a disease through a population is referred to as
    a. epidemiology
    b. contagiology
    c. morbidity
    d. mortality

    ans. a.   diff. 5   obj. 2   p. 4

7. Defining the characteristics of the affected population who are the victims of an epidemic is a first step in determining the
    a. direction of the flow of new cases
    b. rate of spread of the disease
    c. cause or origin of the epidemic
    d. mortality rate expected

    ans. c.   diff. 5   obj. 2   p. 5

8. A technique used to differentiate among strains of a particular organism is referred to as
    a. taxonomy
    b. species typing
    c. differentiation   Serotyping
    d. bacteriophage typing

    ans. d.   diff. 6   obj. 3   p. 6

9. Bacteriocin typing is a technique used to distinguish among strains of
    a. *Pseudomonas*
    b. *Streptococci*
    c. *Escherichia*
    d. *Clostridium*

    ans. a.   diff. 6   obj. 3   p. 7

10. The process of identifying strains of an organism by patterns of biochemical activity against a variety of substrates is called
    a. bacteriocins
    b. antibiograms
    c. electrophoresis
    d. biotyping

    ans. d.   diff. 6   obj. 3   p. 9

11. Viral strains can often be distinguished from one another by using
    a. antisera
    b. antibiograms
    c. biotypes
    d. restriction enzymes

    ans. a.   diff. 6   obj. 3   p. 9

12. The study of the epidemiology of genital herpes has been aided by the study band patterns produced in the technique of
    a. biotyping
    b. gel electrophoresis
    c. serology
    d. restriction enzymes

    ans. b.   diff. 6   obj. 3   p. 10

13. The possible sources of an infectious disease agent are referred to as its
    a. fragmentation sequence
    b. vector
    c. transmission route
    d. reservoir
    e. vehicle

    ans. d.   diff. 7   obj. 4   p. 10

14. Infected humans are the ___reservoir___ for diphtheria, syphilis, and typhoid fever,
    a. fomites
    b. vectors
    c. foci
    d. reservoirs

    ans. d.   diff. 5   obj. 4   p. 10

15. The spread of a disease from its reservoirs may involve insects or other animals. These insects or animals are known as
    a. vehicles
    b. vectors
    c. sentinels
    d. indices

    ans. b.   diff. 5   obj. 4   p. 11

16. Food or other nonliving materials that are involved in the spread of disease are referred to as
    a. vectors
    b. vehicles
    c. sentinels — none of the above
    d. infectors
    e. fomites

    ans. b.   diff. 6   obj. 4   p. 11

17. Mosquito-borne epidemic diseases such as viral encephalitis are frequently studied by use of
    a. sentinels
    b. indices
    c. spread
    d. vehicles

    ans. a.   diff. 7   obj. 4   p. 11

18. The extent and speed of the spread of an epidemic are not influenced by the
    a. incubation period
    b. mode of transmission
    c. dose
    d. isolation of the organism
    e. all of the above are involved

    ans. d.   diff. 7   obj. 4   p. 12

19. Which of the following means is not involved in the spread of infectious agents by air?
    a. coughing
    b. sneezing
    c. feces
    d. singing

    ans. c.   diff. 6   obj. 4   p. 12

20. Important sources of contaminations in crowded locations are
   a. droplets of saliva or mucus
   b. restaurants
   c. rest rooms
   d. blood

   ans. a.   diff. 5   obj. 4   p. 12

21. The first identified case of an infectious disease is called the
   a. traceable case
   b. starter case
   c. index case
   d. nucleus

   ans. c.   diff. 7   obj. 4   p. 13

22. The spread of the highly infectious agents of chicken pox, measles, and smallpox can frequently be traced to
   a. indices
   b. droplet nuclei
   c. air currents
   d. animal pets

   ans. b.   diff. 7   obj. 4   p. 14

23. Intestinal pathogens frequently enter the air by the passage of
   a. saliva
   b. food
   c. drinks
   d. flatus

   ans. d.   diff. 6   obj. 4   p. 14

24. In order to prevent the flow of air and its contaminants from the hospital microbiology laboratory into other parts of the hospital, the air in the laboratory is
   a. equipped with exhaust fans
   b. subject to positive air flow
   c. under a slight vacuum
   d. connected to the patient rooms not the operating room

   ans. c.   diff. 8   obj. 4   p. 15

25. Satisfactory control of airborne infection is usually achieved by
   a. sweeping and mopping
   b. good ventilation and dust control
   c. UV light
   d. air filtration

   ans. b.   diff. 6   obj. 4   p. 16

26. In the study of Legionnaires' disease, the epidemiologists determined the source of the infectious agents to be a(n)
   a. air conditioning unit in a local hotel
   b. mist machine in a local grocery
   c. member of the Legionnaires' convention
   d. food served at a delegate dinner

   ans. b.   diff. 8   obj. 4   p. 17

27. The most common source of water pollution is
   a. sewage
   b. food
   c. water animals
   d. fish

   ans. a.   diff. 8   obj. 4   p. 18

28. The number of cases of a specific disease per hundred people exposed is commonly termed the
   a. attack rate
   b. index case
   c. droplet nucleus
   d. vehicle

   ans. a.   diff. 8   obj. 4   p. 18

29. An important step in preventing the spread of infectious agents from one person to another is
   a. hand washing
   b. pasteurization
   c. home cooking
   d. proper food preparation temperature

   ans. a.   diff. 7   obj. 4   p. 20

30. The number of infectious agents sufficient to establish an infection is the _____ dose.
   a. minimum lethal
   b. lethal
   c. infectious
   d. transit

   ans. c.   diff. 7   obj. 4   p. 20

31. Inanimate objects capable of transferring infectious disease agents are
    a. reservoirs
    b. vehicles
    c. fomites
    d. vectors

    ans. c.   diff. 6   obj. 4   p. 20

32. The surgical scrub performed by doctors and nurses always employ a(n) _____ while washing for _____ minutes.
    a. hand soap, twenty
    b. disinfectant, fifteen
    c. antibiotic, five to ten
    d. antimicrobial agent, ten

    ans. d.   diff. 6   obj. 4   p. 21

33. An apparently healthy individual who harbors a pathogen and is capable of transferring it to other humans is termed a(n)
    a. transmitter
    b. vector
    c. carrier
    d. vehicle

    ans. c.   diff. 6   obj. 4   p. 21

34. Carriers do not include those who
    a. are in the incubation period of the disease
    b. are acutely ill with the disease
    c. have recovered from the disease
    d. continue to excrete the pathogen

    ans. b.   diff. 6   obj. 4   p. 21

35. Carriers of an infectious disease who are contaminated with the agent for only a brief time are referred to as _____ carriers.
    a. transient
    b. chronic
    c. active
    d. indirect

    ans. a.   diff. 5   obj. 4   p. 21

36. Chronic carriers of a disease may continue to excrete the infectious agent
    a. during antibiotic treatment
    b. while actively fighting the disease
    c. for long periods
    d. to their offspring at conception

    ans. c.   diff. 6   obj. 4   p. 22

37. The human population characteristics that are not significant to the nature of epidemics are
    a. prior exposure to the same or related organisms
    b. genetic background
    c. nutritional status
    d. climate conditions
    e. over-crowding

    ans. d.   diff. 7   obj. 5   p. 23

38. Which members of a population provide the best chance of the spread of an infectious disease?
    a. older people
    b. newborns
    c. mature adults
    d. young children

    ans. d.   diff. 7   obj. 5   p. 23

39. Protection of the newborn to most infectious disease is provided by
    a. being an only child
    b. maternal antibodies
    c. good nutrition
    d. being enrolled in licensed day care centers

    ans. b.   diff. 8   obj. 5   p. 23

40. To which of the following infectious diseases is the newborn not protected by maternal antibodies?
    a. diphtheria
    b. whooping cough
    c. tetanus
    d. rubella

    ans. b.   diff. 8   obj. 5   p. 23

41. The genetic factor involved in the immunity of black Africans to malaria is the lack of a(n)
    a. specific red blood cell receptor
    b. susceptibility to the bite of the *Anopheles* mosquito
    c. plasma environment conducive to the growth of the protozoan
    d. cultural susceptibility

    ans. a.   diff. 7   obj. 5   p. 25

42. Which of the following factors appears to have the least influence in the character of epidemics?
    a. malnutrition
    b. climate
    c. poor hygiene
    d. over crowding

    ans. b.   diff. 6   obj. 5   p. 25

43. Epidemics of infectious disease depend on both the virulence of the infecting strain and the
    a. number of plasmids in the agent
    b. exposure of the population
    c. quantity of the agent
    d. toxoid available to the agent

    ans. c.   diff. 7   obj. 5   p. 26

44. The differences in the rate and intensity of an infection are most likely due to differences in the _____ of the strain.
    a. virulence
    b. serotype
    c. frequency
    d. type

    ans. a.   diff. 7   obj. 5   p. 26

45. Which of the following situations may produce serious disease in an individual who has immunity to ordinary doses of the pathogen?
    a. a laboratory accident
    b. previous immunization
    c. pregnancy
    d. stress

    ans. a.   diff. 7   obj. 5   p. 27

46. Which of the following is not a significant factor in the length of the incubation period of an infectious disease?
    a. dose
    b. defense mechanisms
    c. virulence
    d. immunity
    e. none of the above

    ans. d.   diff. 8   obj. 5   p. 27

47. Which of the following is not an important consideration in increasing resistance to infectious disease?
    a. adequate immunization
    b. good health
    c. high stress
    d. proper nutrition

    ans. c.   diff. 6   obj. 5   p. 28

48. Lack of adequate immunization showed a rise in reported measles cases from about 1,500 in the early 1980s to over _____ in 1990.
    a. 10,000
    b. 20,000
    c. 30,000
    d. 40,000

    ans. b.   diff. 7   obj. 5   p. 28

49. The current national goal is to immunize at least sixty percent of the elderly against
    a. hepatitis
    b. measles
    c. influenza
    d. tuberculosis

    ans. c.   diff. 7   obj. 5   p. 28

50. Which of the following methods would be used to increase host resistance to an epidemic of hepatitis A if one is suspected?
    a. active immunization
    b. passive immunization
    c. adequate nutrition
    d. quarantine

    ans. b.   diff. 6   obj. 5   p. 30

51. Which of the following is not an effective method of reducing an infectious disease reservoir?
    a. quarantine of the susceptible population
    b. eliminate the reservoir
    c. treating carriers
    d. medical treatment of patients

    ans. a.   diff. 7   obj. 5   p. 31

52. Cross-contamination of hospitalized patients with communicable diseases was common in the late 1800s. This was most likely due to
    a. poor nutrition
    b. lack of asepsis
    c. contaminated food
    d. open windows

    ans. b.   diff. 7   obj. 6   p. 31

53. The spread of infection in a hospital setting is not inhibited if
    a. isolation techniques are employed
    b. private rooms are used
    c. asepsis is practiced
    d. hand washing techniques are not stressed

    ans. d.   diff. 7   obj. 6   p. 32

54. Which of the following diseases could be acquired by health care workers by coming in contact with the blood or body fluids of a patient having a blood-borne disease?
    a. hepatitis B
    b. measles
    c. candidiasis
    d. cancer

    ans. a.   diff. 6   obj. 6   p. 33

55. To protect health care workers from the danger of blood-borne diseases, the CDC has recommended
    a. private rooms for contagious disease patients
    b. using scrubs instead of uniforms while caring for these patients
    c. the use of disposable eating utensils for the patient's food
    d. universal blood and body-fluid precautions

    ans. d.   diff. 7   obj. 6   p. 33

56. Which of the following activities is not included in the recommended universal precautions as stated by the CDC?
    a. controlling spurting blood
    b. starting I.V. fluids
    c. assisting in childbirth
    d. taking the patient's temperature

    ans. d.   diff. 7   obj. 6   p. 34

57. The newer approach taken by hospital personnel to the universal precautions as stated by the CDC is referred to as
    a. active immunization of personnel
    b. complete isolation of the patient
    c. private room principle
    d. body substance isolation

    ans. d.   diff. 8   obj. 6   p. 34

58. An infection acquired during hospitalization is referred to as
    a. medication stimulated
    b. antibiotic induced
    c. therapeutic
    d. nosocomial

    ans. d.   diff. 8   obj. 6   p. 35

59. Which of the following would not be involved in acquiring a nosocomial infection?
    a. hospital environment
    b. diagnostic procedures
    c. asepsis
    d. medical personnel

    ans. c.   diff. 7   obj. 6   p. 25

60. What percent of hospitalized patients acquired a nosocomial infection?
    a. 10 to 25
    b. 5 to 15
    c. 2 to 10
    d. 50

    ans. c.   diff. 7   obj. 6   p. 35

61. Hospital patients taking antibiotics are susceptible to nosocomial infections by resident pathogens because
   a. of their hospital stay
   b. their normal flora are suppressed
   c. of the nature of their illness
   d. of isolation procedures

   ans. b.　　diff. 6　　obj. 6　　p. 36

62. The member of the hospital team responsible for identifying and controlling nosocomial infections is designated as the
   a. infection control practitioner
   b. head nurse
   c. staff committee
   d. hospital administrator

   ans. a.　　diff. 7　　obj. 7　　p. 36

63. Which of the following is not included among the responsibilities of the hospital epidemiologist?
   a. follow the precautions of the public relations office of the hospital
   b. investigate outbreaks of infections in the hospital
   c. prepare and present information to the infection control committee
   d. report infections to the regional health department

   ans. a.　　diff. 8　　obj. 7　　p. 38

64. The National Centers for Disease Control and Prevention is part of the
   a. Food and Drug Administration
   b. Department of Health and Human Services
   c. Department of Human Resources
   d. Federal University System

   ans. b.　　diff. 8　　obj. 7　　p. 41

65. The publication of the CDC that reports the new cases of infectious disease to its subscribers is titled the
   a. *Federal Report of Infectious Disease*
   b. *State Health News Letter*
   c. *Morbidity and Mortality Weekly Report*
   d. *Laboratory Data File*

   ans. c.　　diff. 7　　obj. 7　　p. 41

True/False

1. The reservoir for botulism and coccidioidomycosis is infected humans.
   false (soil)

2. The most frequent reservoirs of disease are infected humans.
   false (animals)

3. Examples of fomites capable of transmitting infectious disease are any inanimate objects.
   true

4. If surgical gloves are worn, then hand washing can be very brief.
   false (ten minutes scrubbing is necessary)

5. The genetic basis for resistance to infectious disease is unknown.
   false
   　　(in many diseases it is related to cell receptor sites)

6. The worldwide incidence of an infectious disease is termed an epidemic.
   false (a pandemic)

7. The reservoir of an infectious disease is the sum of its active cases.
   false (potential sources)

8. Water-borne infections tend to be of lower incidence than food-borne infections due to a lower concentration of the infectious agents.
   true

9. The most frequent transfer mechanism for person-to-person infectious disease agents is hands.
   true

10. The number of infectious agents sufficient to establish an infection is termed the infectious dose.
   true

11. Fomites are described as animals that are capable of transmitting infectious disease agents.
   false (inanimate objects)

12. People who may briefly harbor a potential pathogen and transfer it to susceptible individuals are described as intermittent carriers.
   false (transient carriers)

13. Herd immunity is unlikely to protect a population from the spread of a recurrent infectious disease.
   false (herd immunity is good protection)

14. Malnutrition, overcrowding, and fatigue increase
    susceptibility to infectious diseases, and their spread.
    true

15. The *Morbidity and Mortality Weekly Report* is the
    reporting tool of the World Health Organization.
    false   (Centers for Disease Control and Prevention)

Chapter 19

Multiple Choice

1. In 1906, Howard Ricketts described the causative agent of Rocky Mountain spotted fever as a(n)
   a. obligate extracellular bacterium
   b. rickettsia
   c. obligate intracellular virus
   d. obligate intracellular coccobacillus

   ans. d.   diff. 5   obj.   p. 1

2. The rickettsial disease that caused the death of Howard Ricketts was
   a. Rocky Mountain spotted fever
   b. typhus
   c. typhoid
   d. endemic typhus

   ans. b.   diff. 5   obj.   p. 2

3. Which of the following is considered a very important function of the skin?
   a. hold muscle to bone
   b. control of body temperature
   c. manufacture blood cells
   d. produce antibodies

   ans. b.   diff. 5   obj. 1   p. 2

4. Which of the following is not an accepted function of the skin?
   a. contribute to the synthesis of vitamin D
   b. regulate body temperature
   c. produce T cells
   d. prevent loss of fluid from body tissues

   ans. c.   diff. 6   obj. 1   p. 2

5. The surface layer of the skin is the _____. The deeper layer of skin cells is the _____.
   a. epidermis, dermis
   b. cutaneous, keratin
   c. keratin, dermis
   d. dermis, cutaneous

   ans. a.   diff. 6   obj. 1   p. 3

6. The skin protein that helps to make it tough and waterproof is
   a. keratin
   b. carotene
   c. melanin
   d. vitamin D

   ans. a.   diff. 7   obj. 1   p. 3

7. The oily secretion that lubricates the hair follicles of the skin is
   a. acne
   b. eczema
   c. suder
   d. sebum

   ans. d.   diff. 6   obj. 1   p. 3

8. The disorder in which the sebaceous glands become inflamed and plugged with excessive sebum is termed
   a. carbuncles
   b. eczema
   c. acne
   d. boils

   ans. c.   diff. 7   obj. 1   p. 4

9. Although skin secretions provide nutrients for microbial growth, the _____ break down into _____, which inhibit the growth of many potential pathogens.
   a. carbohydrates, monosaccharides
   b. proteins, amino acids
   c. lipids, fatty acids
   d. lipids, glycerol

   ans. c.   diff. 8   obj. 1   p. 4

10. The antimicrobial action of the skin's fatty acids is neutralized by oozing of the plasma in an allergy termed
    a. poison ivy
    b. acne
    c. eczema
    d. psoriasis

    ans. c.   diff. 8   obj. 1   p. 4

11. Which part of the body has the highest density of normal flora?
    a. back
    b. scalp
    c. arms
    d. thigh

    ans. b.   diff. 7   obj. 2   p. 5

12. Which of the following organisms is not among the principal members of normal skin flora?
    a. staphylococci
    b. *Malassezia* species
    c. *Candida* species
    d. diphtheroids

    ans. c.   diff. 6   obj. 2   p. 5

13. Which of the following normal flora is a small yeast?
    a. diphtheroids
    b. staphylococci
    c. *Malassezia* species
    d. *Candida* species

    ans. c.   diff. 6   obj. 2   p. 5

14. The Gram-positive, nonmotile, pleomorphic, non-spore-forming rods found in abundance on the skin are the
    a. pneumococci
    b. bacilli
    c. diphtheroids
    d. pseudomonads

    ans. c.   diff. 7   obj. 2   p. 6

15. A common normal flora, *Propionibacterium acnes*, is present within hair follicles. Its production of fatty acids is somewhat protective to the skin, but in some individuals has been associated with _____.
    a. eczema
    b. acne
    c. carbuncles
    d. boils

    ans. b.   diff. 7   obj. 2   p. 6

16. The principal species of staphylococci found on the skin is
    a. *pyogenes*
    b. *aureus*
    c. *acnes*
    d. *epidermidis*

    c. Saprophyticus

    ans. d.   diff. 6   obj. 2   p. 7

17. The member of the normal flora that seems to be important in maintaining the balance among the members of this group is
    a. staphylococci
    b. diphtheroids
    c. yeasts
    d. fungi

    ans. a.   diff. 7   obj. 2   p. 7

18. The member of normal flora responsible for tinea versicolor in some people is
    a. *Staphylococcus* sp.
    b. *Malassezia* sp.
    c. *diphtheroids*
    d. *Tinea* sp.

    ans. b.   diff. 7   obj. 2   p. 8

19. Of the following, which is deemed the most serious staphylococcal skin infection?
    a. boil
    b. carbuncle
    c. folliculitis
    d. furuncle

    ans. b.   diff. 7   obj. 3   p. 10

20. Which of the following can be added to ordinary laboratory media to convert it into one selective for *S. aureus*?
    a. 0.5% HCl
    b. 7.5% NaCl
    c. 1.07% glucose
    d. 6.0% mannitol

    ans. b.   diff. 6   obj. 3   p. 11

21. An enzyme produced by virtually all strains of *S. aureus* and important in identifying it is
    a. leukocidin
    b. coagulase
    c. mannose
    d. streptokinase

    ans. b.   diff. 7   obj. 3   p. 11

22. The *S. aureus* protein that functions to interfere with phagocytosis is
    a. protein A
    b. capsular protein
    c. protein M
    d. collagen

    ans. a.   diff. 7   obj. 3   p. 13

23. Which of the following components of *S. aureus* is mismatched?
    a. proteases - degrade collagen
    b. leukocidins - kill white blood cells
    c. hyaluronic acid - prevent capsule formation
    d. lipase - degrade lipids

    ans. c.   diff. 8   obj. 3   p. 13

24. The *S. aureus* product that separates layers of epidermis, causing scalded skin syndrome is
    a. protein A
    b. epidermolytic toxin
    c. proteases
    d. coagulase

    ans. b.   diff. 7   obj. 3   p. 14

25. Although most moist skin areas of the body can be inhabited by *S. aureus,* the preferred area seems to be the
    a. nasal chambers
    b. throat
    c. vagina
    d. urethra

    ans. a.   diff. 7   obj. 3   p. 15

26. Which of the following is not usually a method used to characterize strains of *S. aureus?*
    a. bacteriophage typing
    b. antibiotic susceptibility
    c. coagulase production
    d. plasmid identification

    ans. c.   diff. 7   obj. 3   p. 15

27. A potentially fatal toxin-mediated *S. aureus* disease occurring primarily in infants is
    a. scalded skin syndrome
    b. septicemia
    c. pyoderma
    d. toxic shock syndrome

    ans. a.   diff. 6   obj. 3   p. 16

28. The affecting toxin of Ritter's disease is an epidermolytic toxin that
    a. is responsible for lysis of epidermal cells
    b. causes necrosis of the epidermis
    c. destroys the mitotic layers of epidermal cells
    d. destroys the binding between the layers of skin cells

    ans. d.   diff. 8   obj. 3   p. 16

29. The disease characterized by a tender red rash with sandpaper texture that covers 20 to 100% of the body, and is frequently accompanied by malaise, irritability, fever, and skin peeling is
    a. pyoderma
    b. impetigo
    c. burns infected with *S. aureus*
    d. Ritter's disease

    ans. d.   diff. 7   obj. 3   p. 18

30. A frequent complication of SSSS is a secondary infection caused by
    a. *Pseudomonas* sp.
    b. Streptococci
    c. diphtheroids
    d. rickettsia

    ans. a.   diff. 7   obj. 3   p. 20

273

31. The most common type of pyoderma is described as
    a. carbuncle
    b. impetigo
    c. furuncle
    d. Ritter's disease

    ans. b.   diff. 6   obj. 4   p. 20

32. Most cases of impetigo are caused by
    a. *Bacillus subtilis*
    b. *Staphylococcus aureus*
    c. *Streptococcus pyogenes*
    d. *Staphylococcus epidermidis*

    ans. c.   diff. 6   obj. 4   p. 20

33. The disease most frequently caused by β-hemolytic group A streptococci is
    a. scalded skin syndrome
    b. impetigo
    c. Ritter's disease
    d. toxic shock syndrome

    ans. b.   diff. 7   obj. 4   p. 20

34. Which of the following is not a virulence-related extra-cellular product of *S. pyogenes?*
    a. M-protein
    b. proteases
    c. nucleases
    d. hyaluronidase

    ans. a.   diff. 7   obj. 4   p. 22

35. *S. pyogenes* has surface components that tend to protect it from destruction by phagocytes. Which of the following cell surface components fills this role?
    a. nuclease
    b. protease
    c. impetigo
    d. M-protein

    ans. d.   diff. 6   obj. 4   p. 22

36. A serious complication of pyoderma (caused by *S. pyogenes*) is
    a. severe scarring of the skin
    b. pneumonia
    c. acute glomerulo nephritis
    d. toxic shock syndrome

    ans. c.   diff. 7   obj. 4   p. 22

37. _____ is a serious, worldwide rickettsial disease transmitted by ticks.
    a. typhus
    b. typhoid fever
    c. Rocky Mountain spotted fever
    d. Q fever

    ans. c.   diff. 5   obj. 5   p. 23

38. In which of the following diseases is there a characteristic rash that begins on the palms and soles, and progresses toward the trunk of the body?
    a. typhoid
    b. epidemic typhus
    c. Rocky Mountain spotted fever
    d. Brill-Zinsser disease

    ans. c.   diff. 6   obj. 5   p. 24

39. The obligate intracellular parasite *Rickettsia rickettsii* is the causative agent of _____.
    a. rickettsial pox
    b. epidemic typhus
    c. murine typhus
    d. Rocky Mountain spotted fever

    ans. d.   diff. 5   obj. 5   p. 25

40. Which of the following tests is used for an early diagnosis of Rocky Mountain spotted fever?
    a. enzyme-linked immunoassay using antigen from killed rickettsia
    b. demonstrate *R. rickettsii* in biopsies of skin lesions
    c. detection of antibodies to *R. rickettsii* in blood
    d. DNA tests for *R. rickettsii*

    ans. b.   diff. 7   obj. 5   p. 26

41. Rocky Mountain spotted fever is caused by *Rickettsia rickettsii* and is transmitted to the human by the bite of an infected _____, which has been feeding on that person's capillary blood for _____ hours.
    a. tick, twenty-four
    b. mite, twenty-four
    c. tick, four to ten
    d. mite, four to six

    ans. c.   diff. 8   obj. 5   p. 26

42. Rocky Mountain spotted fever is an example of a(n)
    a. zoonosis
    b. necrotic disease
    c. accidental intoxication
    d. food-borne illness

    ans. a.   diff. 6   obj. 5   p. 27

43. The major vector of Rocky Mountain spotted fever in the western United States is
    a. *Dermacentor variabilis*
    b. *Dermacentor andersoni*
    c. *Rickettsia rickettsii*
    d. *Rickettsia prowazeki*

    ans. b.   diff. 6   obj. 5   p. 27

44. The overall mortality of Rocky Mountain spotted fever if not treated is about ____%; with early diagnosis and treatment, however, mortality is less than ____%.
    a. 40, 10
    b. 20, 5
    c. 50, 10
    d. 70, 25

    ans. b.   diff. 7   obj. 5   p. 28

45. _____ is classified as a chronic, tick-borne rickettsial disease.
    a. Q fever
    b. Rocky Mountain spotted fever
    c. Lyme disease
    d. epidemic typhus

    ans. c.   diff. 5   obj. 5   p. 30

46. The causative agent of Lyme disease is
    a. *Borrelia burgdorferi*
    b. *Rickettsia prowazeki*
    c. *Treponema lymei*
    d. the Lyme virus

    ans. a.   diff. 6   obj. 5   p. 30

47. The hallmark of Lyme disease is
    a. the inflammatory reactions in the skin
    b. the progression through the 3 stages, each with its own symptoms
    c. "the great imitator"
    d. erythema chronicum migrans

    ans. d.   diff. 7   obj. 5   p. 32

48. The variable strains of *Borrelia burgdorferi* can be distinguished from one another by
    a. growing them on special media under aerobic conditions
    b. using fluorescent dyes to mark antibodies
    c. growth differences in embryonated eggs
    d. using monoclonal antibodies

    ans. d.   diff. 8   obj. 5   p. 33

49. The spirochete of Lyme disease is capable of entering the blood stream and disseminating to all parts of the body in the _____ stage of the disease.
    a. third
    b. second
    c. first
    d. fourth

    ans. c.   diff. 5   obj. 5   p. 33

50. The _____ stage of Lyme disease that is characterized by arthritis.
    a. second
    b. third
    c. first
    d. fourth

    ans. b.   diff. 5   obj. 5   p. 33

51. The most common vector-borne disease in the United States is _____.
    a. malaria
    b. Rocky Mountain spotted fever
    c. rabies
    d. Lyme disease
    e. plague

    ans. d.   diff. 6   obj. 5   p. 34

52. The most important vector of Lyme disease in the eastern United States is
    a. *Ixodes dammini*
    b. *Dermacentor variabilis*
    c. *Dermacentor andersoni*
    d. *Borrelia burgdorferi*

    ans. a   diff. 5   obj. 5   p. 34

53. The growth stage of the tick that is mainly responsible for transmitting Lyme disease is the
   a. moulter
   b. egg stage
   c. nymph stage
   d. adult

   ans. c.   diff. 8   obj. 5   p. 34

54. The preferred host of *Ixodes dammini* is the
   a. white-footed mouse
   b. white-tailed deer
   c. human
   d. wood rat

   ans. a.   diff. 8   obj. 5   p. 35

55. Many childhood diseases caused by viruses can usually be diagnosed by
   a. inspection of the rash
   b. type of cough
   c. type of fever
   d. the incubation period

   ans. a.   diff. 7   obj. 6   p. 38

56. The most widespread of the childhood rashes in the United States is
   a. rubeola
   b. variola
   c. rubella
   d. varicella

   ans. d.   diff. 8   obj. 6   p. 39

57. The varicella virus is a member of the _____ virus family.
   a. Toga
   b. Paramyxo
   c. Herpes
   d. Papilloma

   ans. c.   diff. 7   obj. 6   p. 39

58. The diagnostic rash of _____ begins as red macules, papules, and vesicles surrounded by a narrow zone of redness. These first appear on the back of the head.
   a. fifth disease
   b. measles
   c. German measles
   d. chicken pox

   ans. d.   diff. 6   obj. 6   p. 40

59. Although mild in young children, about 20% of infected adults develop pneumonia during the progress of _____.
   a. rubeola
   b. variola
   c. rubella
   d. varicella

   ans. d.   diff. 8   obj. 6   p. 40

60. Reactivation of varicella is called herpes zoster, or
   a. shingles
   b. cold sores
   c. exanthems
   d. pneumonia

   ans. a.   diff. 7   obj. 6   p. 42

61. The reactivation of varicella correlates with a(n)
   a. decline in cell-mediated immunity
   b. re-exposure to chicken pox
   c. epidemic of measles
   d. inflammation of sensory neurons

   ans. a.   diff. 8   obj. 6   p. 43

62. The reservoir of varicella zoster virus is the
   a. household pet, usually a dog
   b. human
   c. deer
   d. white-tailed rat

   ans. b.   diff. 7   obj. 6   p. 43

63. When the varicella virus is introduced into a community, it can persist indefinitely and cause recurrent epidemics of chicken pox. The source of the infectious virus in the community is
   a. pediatricians
   b. newborns
   c. cold sore infections
   d. active cases of shingles

   ans. d.   diff. 7   obj. 6   p. 44

64. The incubation period for chicken pox is
   a. 10–21 days
   b. 7–13 days
   c. 24 hours
   d. 15–35 days

   ans. a.   diff. 6   obj. 6   p. 46

65. The childhood disease that begins with fever, runny nose, cough, and swollen, weepy, red eyes and progresses to a fine red rash that appears on the forehead and spreads outward over the body is _____.
    a. variola
    b. rubella
    c. rubeola
    d. varicella

    ans. c.   diff. 7   obj. 6   p. 47

66. The childhood disease that damages the body defenses and is frequently complicated by secondary infections caused by Gram-positive cocci is
    a. German measles
    b. mumps
    c. chicken pox
    d. measles

    ans. d.   diff. 7   obj. 6   p. 47

67. A serious complication of measles, which occurs in about 1 out of a thousand cases, is
    a. epilepsy
    b. pneumonia
    c. earache
    d. encephalitis

    ans. d.   diff. 8   obj. 6   p. 47

68. Subacute sclerosing panencephalitis is a rare complication of _____ that can develop two to ten years after the disease subsides.
    a. measles
    b. chicken pox
    c. shingles
    d. German measles

    ans. a.   diff. 8   obj. 6   p. 47

69. Measles is caused by an RNA virus that is a member of the _____ family.
    a. paramyxovirus
    b. herpes zoster
    c. toga virus
    d. papilloma

    ans. a.   diff. 9   obj. 6   p. 48

70. An important diagnostic sign of measles is
    a. fluffy lymphocytes
    b. giant cells
    c. Koplik's spots
    d. fever

    ans. c.   diff. 5   obj. 6   p. 49

71. Death from measles in the United States occurs in about 1 to 2 out of every thousand cases, mainly from
    a. leukemia and pernicious anemia
    b. pneumonia and encephalitis
    c. cancer
    d. shingles

    ans. b.   diff. 7   obj. 6   p. 50

72. Vaccination against measles became widespread in the United States in the
    a. 1950s
    b. 1970s
    c. 1960s
    d. 1980s

    ans. c.   diff. 5   obj. 6   p. 50

73. The MMR vaccine contains antigens against
    a. mononucleosis, mumps, and rubella
    b. mononucleosis, mange, and roseola
    c. measles, mange, and rubeola
    d. measles, mumps, and rubella

    ans. d.   diff. 5   obj. 6   p. 51

74. The most serious consequence of rubella is
    a. deafness
    b. encephalitis
    c. meningitis
    d. birth defects

    ans. d.   diff. 6   obj. 6   p. 54

75. Rubella is caused by a member of the _____ virus family.
    a. parvo
    b. toga
    c. coccaxi
    d. papilloma

    ans. b.   diff. 7   obj. 6   p. 54

76. The rash and joint symptoms of German measles are probably due to the humoral and cell-mediated immunity that develop against the virus and result in
    a. lysis of cells due to cytotoxic activity
    b. antigen-antibody reaction
    c. complement interaction
    d. muscle weakness

    ans. b.   diff. 8   obj. 6   p. 56

77. Fetal infection with the rubella virus occurs during the period of viremia. The virus readily crosses the placenta and infects the fetus during _____ of pregnancy.
    a. the third trimester
    b. the second trimester
    c. the first trimester
    d. any of the forty weeks

    ans. c.   diff. 6   obj. 6   p. 56

78. Prevention of rubella depends on a subcutaneous injection of a(n)
    a. live attenuated virus
    b. inactive virus
    c. toxoid of the virus
    d. antitoxin to the virus

    ans. a.   diff. 5   obj. 6   p. 57

79. Acquisition of rubella is by the _____ route.
    a. fecal-oral
    b. respiratory
    c. genitourinary
    d. gastrointestinal

    ans. b.   diff. 6   obj. 6   p. 59

80. The disease that occurs in children and young adults, and is caused by parvovirus B-19 is
    a. erythema infectiosum
    b. exanthem subitum
    c. roseola infantum
    d. rubella

    ans. a.   diff. 7   obj. 6   p. 60

81. Another name for exanthem subitum is
    a. fifth disease
    b. erythema infectiosum
    c. roseola infantum
    d. German measles

    ans. c.   diff. 7   obj. 6   p. 60

82. The disease that occurs chiefly in infants six months to three years old and is characterized by the abrupt onset of a fever that may reach 105° F and cause convulsions is
    a. fifth disease
    b. erythema infectiosum
    c. exanthem subitum
    d. red measles

    ans. c.   diff. 7   obj. 6   p. 61

83. Warts are caused by
    a. enveloped RNA viruses
    b. parvoviruses
    c. herpes viruses
    d. papillomaviruses

    ans. d.   diff. 7   obj. 6   p. 61

84. The skin-invading molds belong mainly to the genera *Trichophyton*, *Microsporum*, and *Epidermophyton* are collectively called
    a. dermatophytes
    b. jock itch virus
    c. candida
    d. ringworm

    ans. a.   diff. 6   obj. 7   p. 63

85. Generally, the dermatophytes are unable to grow in living tissue but multiply in structures that contain
    a. collagen
    b. carapace
    c. keratin
    d. sensitized T cells

    ans. c.   diff. 7   obj. 7   p. 63

True/False

1. Complete regeneration of the skin occurs once a month.
   true

2. The protein that makes skin tough and waterproof is keratin.

   true

3. Diphtheroids are among the major flora of the skin but occasionally produce abscesses of the skin or gut.

   false   (bone, liver, or heart)

4. The large group of Gram-positive cocci found on the skin have low virulence and are members of the group *Staphylococci aureus.*

   false   (*S. epidermidis*)

5. The staphylococci of the normal flora have been shown to produce an antimicrobial substance helpful in preventing acne from developing on the skin.

   true

6. Infectious agents can invade the skin via injuries, blood, or hair follicles.

   true

7. Squeezing facial lesions caused by *S. aureus* is an effective and safe way of evacuating pus and bacteria from the site.

   false   (dangerous; should be surgically lanced to prevent spread of infection)

8. Another name for scalded skin syndrome is systemic pyoderma.

   false   (Ritter's disease)

9. The mortality rate of SSSS can range up to 40 percent, depending on age and immunocompetency.

   true

10. The cell walls of β-hemolytic group A streptococci contain a protein called group A amino acid.

    false   (polysaccharide called group A carbohydrate)

11. Lyme disease is caused by the bite of a tick infected with the rickettsia *B. burgdorferi.*

    false   (spirochete)

12. The nymph form of the tick transmits Lyme disease.

    true

13. The chicken pox virus usually causes a fine red rash on the forehead that spreads outward over the rest of the body.

    false   (measles)

14. The natural host for the rubeola virus is the rat.

    false   (the human)

15. To prevent a rubella infection, the active attenuated vaccine is administered to females during the first trimester of pregnancy.

    false   (administered to children at 15 months of age) (never to pregnant women for fear of its crossing the placenta)

16. Another name for erythema infectiosum is roseola infantum.

    false   (fifth disease)

17. Warts are small tumors that appear on the skin and are commonly called papillomas.

    true

18. The dermatophytes are fungi that invade superficial tissues.

    true

19. Staphylococcal scalded skin syndrome results from endotoxins produced by certain strains of *Staphylococcus aureus.*

    false   (exotoxins)

20. The most effective antibiotic for *Streptococcus pyogenes* infections is tetracycline.

    false   (penicillin or erythromycin)

Matching

1. Match the *S. Aureus* product with its contribution to virulence.

   1. protease
   2. protein A
   3. leukocidin
   4. epidermolytic toxin

   a. kills white blood cells
   b. separates layers of epidermis
   c. inhibits phagocytosis
   d. degrades collagen

   1 - d; 2 - c; 3 - a; 4 - b

2. Match the organism with the disease caused by it.

   1. *Staphylococcus aureus*
   2. *Rickettsia rickettsii*
   3. *Streptococcus pyogenes*
   4. *Borrelia burgdorferi*

   a. scalded skin syndrome
   b. acute glomerulonephritis
   c. Rocky Mountain spotted fever
   d. Lyme disease

   1 - a; 2 - c; 3 - b; 4 - d

3. Match the disease with the type of causative agent.

   1. Lyme disease
   2. rubeola
   3. ringworm
   4. impetigo

   a. virus
   b. spirochete
   c. fungi
   d. bacteria

   1 - b; 2 - a; 3 - c; 4 - d

Chapter 20

Multiple Choice

1. Diphtheria was first described by Arataeus in the _____ century A.D.
   a. second
   b. first
   c. third
   d. fourth

   ans. a.   diff. 4   obj. 1   p. 2

2. In recognition of their research into this disease, the organism that causes _____ became known as the Klebs-Loeffler bacillus.
   a. botulism
   b. tuberculosis
   c. diphtheria
   d. whooping cough

   ans. c.   diff. 5   obj. 1   p. 2

3. Emile Roux and Alexandre Yersin published findings of their studies of diphtheria, which proved that the systemic aspects of the disease were due to a(n)
   a. circulatory toxin
   b. organism circulating in the blood
   c. secondary infection with a virus
   d. weakening of the immune system

   ans. a.   diff. 6   obj. 1   p. 2

4. A toxoid for diphtheria was developed in 1890 by von Behring, Ehrlich, and Kitasato by treating the diphtheria toxin with
   a. a neutralizing agent
   b. alum
   c. heat
   d. dilute formalin

   ans. d.   diff. 6   obj. 1   p. 2

5. Which of the following scientists is not mentioned in connection with diphtheria?
   a. Loeffler
   b. Pasteur
   c. Brettonneau
   d. Ehrlich

   ans. b.   diff. 6   obj. 1   p. 2

6. Which of the following structures is not considered part of the upper respiratory tract?
   a. tonsils
   b. trachea
   c. middle ear
   d. nasolacrimal ducts

   ans. b.   diff. 7   obj. 2   p. 4

7. Enlargement or engorgement of which of the following structures can contribute to ear infections by interfering with eustachian tube drainage?
   a. epiglottis
   b. nasal conchae
   c. nasal mucosa
   d. adenoids

   ans. d.   diff. 7   obj. 2   p. 4

8. Which of the following respiratory structures does not use cilia as a means of removing debris and preventing it from settling in the deep respiratory structures?
   a. epiglottis
   b. tear ducts
   c. sinuses
   d. middle ear

   ans. a.   diff. 7   obj. 2   p. 5

9. Which of the following functions is not directly associated with the upper respiratory system?
   a. production of bacteriocidal product
   b. warming of inhalations
   c. humidifying of inhaled air
   d. regulation of body temperature

   ans. a.   diff. 6   obj. 2   p. 6

10. Which of the following is not among the principal types of bacteria inhabiting the upper respiratory system?
    a. *Corynebacterium*
    b. *Staphylococcus*
    c. *Clostridium*
    d. *Bacteroides*

    ans. c.   diff. 6   obj. 3   p. 7

11. The eyes are protected from large populations of microorganisms settling on the conjunctiva by the
    a. eyelids, eyelashes, and eyebrows providing excellent physical barriers
    b. eyelid blink reflex and lysozyme-rich tears
    c. use of protective lenses (glasses or contacts)
    d. constant seepage of tears from the lacrimal duct up over the eye

    ans. b.   diff. 7   obj. 3   p. 8

12. Which of the following pathogens is not among the common inhabitants of the nasopharynx?
    a. *Neisseria meningitidis*
    b. *Heamophilus influenzae*
    c. *Mycoplasma pneumoniae*
    d. *Bacteroides* sp.

    ans. c.   diff. 7   obj. 3   p. 9

13. Rheumatic fever is a serious complication of
    a. streptococcal pharyngitis
    b. glomerulonephritis
    c. β-hemolytic streptococci of group C
    d. the common cold

    ans. a.   diff. 7   obj. 4   p. 10

14. The cause of strep throat is
    a. *Streptococcus pneumoniae*
    b. *Staphylococcus aureus*
    c. *Streptococcus pyogenes*—α hemolytic, group B
    d. *Streptococcus pyogenes*—β-hemolytic, group A

    ans. d.   diff. 6   obj. 4   p. 10

15. Streptolysin O and streptolysin S (enzymes released by *S. pyogenes*) are responsible for
    a. enlarged lymph nodes in the neck
    b. α-hemolysis
    c. pus patches on the throat
    d. β-hemolysis

    ans. d.   diff. 7   obj. 4   p. 11

16. Immunity to *S. pyogenes* depends on the hosts' production of antibodies to the
    a. cell wall carbohydrates
    b. cell wall proteins
    c. streptolysin S
    d. M protein

    ans. d.   diff. 8   obj. 4   p. 12

17. The process used for identifying the different carbohydrates of streptococci is called
    a. Lancefield grouping
    b. CHO typing
    c. streptococci specifics
    d. fermentation testing

    ans. a.   diff. 6   obj. 4   p. 13

18. Rebecca Lancefield showed that almost all the strains of β-hemolytic streptococci from human infections
    a. had variable cell wall carbohydrates
    b. could cause rheumatic fever
    c. had the same cell wall carbohydrate "A"
    d. responded to penicillin

    ans. c.   diff. 7   obj. 4   p. 14

19. The scientist who developed a system of identifying the variety of strains of *S. pyogenes* was
    a. O.T. Avery
    b. Hans Zinner
    c. Donald Sutherland
    d. Rebecca Lancefield

    ans. d.   diff. 6   obj. 4   p. 14

20. A common means of spreading streptococcal throat infections is through the
    a. ingestion of contaminated food
    b. use of contaminated needles
    c. urogenital system
    d. fecal-oral route

    ans. a.   diff. 7   obj. 4   p. 15

21. Which of the following is significant in considering the incidence of sore throat caused by *S. pyogenes?*
    a. geographic location
    b. age of the person exposed
    c. time of the year of infection
    d. exposure to an infected person

    ans. d.   diff. 6   obj. 4   p. 15

22. If group A streptococcal infection is confirmed, treatment to prevent further dissemination and to stimulate elimination of the organism is with
    a. streptolysin
    b. penicillin
    c. bacitracin
    d. Keflex

    ans. b.   diff. 7   obj. 4   p. 16

23. _____ is diagnosed in part by the appearance of a toxin-mediated erythrogenic rash and a whitish coating on the tongue.
    a. Scarlet fever
    b. Measles
    c. Strep throat
    d. Rheumatic fever

    ans. a.   diff. 6   obj. 4   p. 17

24. The erythrogenic toxin seen in scarlet fever is produced by a specific bacteriophage of *S. pyogenes*. This is an example of
    a. allergic response
    b. lysogenic conversion
    c. febrile reaction
    d. secondary infection syndrome

    ans. b.   diff. 6   obj. 4   p. 17

25. The kidney disease that results from antibodies reacting with streptococcal products is
    a. nephronecrosis
    b. urethritis
    c. acute glomerulonephritis
    d. chronic nephroglomeruloneurosis

    ans. c.   diff. 7   obj. 4   p. 17

26. The most serious effects of rheumatic fever are on the
    a. heart
    b. brain
    c. kidney
    d. joints

    ans. a.   diff. 5   obj. 4   p. 19

27. Rheumatic fever may leave the heart valves weakened and the heart susceptible to infection by members of the normal bacterial flora. The resulting disease is called
    a. chronic rheumatic heart disease
    b. subacute bacterial endocarditis
    c. infectious myocarditis
    d. sequelae of an acute infection

    ans. b.   diff. 6   obj. 4   p. 19

28. It is believed that the tissue damage in rheumatic fever is due to a(n)
    a. toxin of the capsule of *S. pyogenes*
    b. autoimmune reaction
    c. reaction with the penicillin used to treat the disease
    d. natural reaction in diseases of this type

    ans. b.   diff. 6   obj. 4   p. 20

29. To prevent a reinfection with *S. pyogenes,* people recovering from rheumatic fever are advised to
    a. avoid contact with crowds during strep epidemics
    b. lead low-stress lives for ten years
    c. take penicillin continuously for five years or longer
    d. paint the throat with gentian violet if tender

    ans. c.   diff. 6   obj. 4   p. 20

30. Generally, the incubation period of "strep" throat is
    a. 2–5 days
    b. 7–12 days
    c. 2–3 weeks
    d. 24 hours

    ans. a.   diff. 7   obj. 4   p. 22

283

31. _____ is a club-shaped, nonmotile, Gram-positive rod that is commonly arranged in a "Chinese letter" or "palisade" pattern and contains metachromatic granules.
    a. *Clostridium botulinum*
    b. *Mycobacterium tuberculosis*
    c. *Streptococcus pyogenes*
    d. *Corynebacterium diphtheriae*

    ans. d.   diff. 6   obj. 5   p. 24

32. Loeffler's medium contains potassium tellurite, which inhibits the growth of many bacterial species while allowing colonies of _____ to take on a distinctive grey or black color.
    a. *Corynebacterium diphtheriae*
    b. *Mycobacterium tuberculosis*
    c. *Mycobacterium leprae*
    d. *Streptococcus pyogenes*

    ans. a.   diff. 7   obj. 5   p. 25

33. The seriousness of diphtheria infections is due to a powerful, diffusing _____.
    a. enterotoxin
    b. endotoxin
    c. virus
    d. exotoxin

    ans. d.   diff. 6   obj. 5   p. 25

34. _____ is responsible for the bacteriophage of *Corynebacterium diphtheriae* producing the powerful exotoxin associated with the disease
    a. A specific chemical
    b. Selective media
    c. Lysogenic conversion
    d. Plasmid formation

    ans. c.   diff. 6   obj. 5   p. 26

35. The primary reservoir for *C. diphtheriae* is (are)
    a. domestic pets
    b. humans
    c. wild birds
    d. infected food

    ans. b.   diff. 5   obj. 5   p. 26

36. Although *C. diphtheriae* has little invasive ability, the powerful exotoxin released is
    a. only damaging to the tissues in which the organism is found
    b. absorbed by the blood stream
    c. harmful when the organism is killed by antibiotics
    d. neutralized by IgAs attached to the mucus membranes

    ans. b.   diff. 7   obj. 5   p. 26

37. Absorption of the exotoxin of diphtheria by body cells results in
    a. damage to the heart, nerves, and kidneys
    b. the formation of a pseudomembrane in the throat
    c. invasion of these cells by the bacterium
    d. secondary infection caused by the bacteriophage

    ans. a.   diff. 7   obj. 5   p. 27

38. The diphtheria toxin is a small protein compound of two parts—A and B. The B portion attaches to receptors on the cell membrane and is taken into the cell by
    a. endocytosis
    b. diffusion
    c. active transport
    d. osmosis

    ans. a.   diff. 6   obj. 5   p. 27

39. The general pattern of the diphtheria toxin's action against cells is similar to the toxins of other bacteria, such as
    a. *Staphylococcus aureus*
    b. *S. pneumoniae*
    c. *S. pyogenes*
    d. *Vibrio cholerae*

    ans. d.   diff. 6   obj. 5   p. 27

40. Effective treatment of diphtheria depends on
    a. prompt treatment with antibiotics
    b. administration of antiserum after cultural proof of disease
    c. giving antiserum against diphtheria toxin as soon as possible
    d. immunizing the susceptible population

    ans. c.   diff. 6   obj. 5   p. 28

41. With antibiotic treatment, the mortality rate for diphtheria is
    a. 1 out of 10
    b. 5 out of 100
    c. 10 out of 1,000
    d. about 25%

    ans. a.   diff. 7   obj. 5   p. 28

42. Immunization against diphtheria is accomplished by inoculation with diphtheria toxoid treated with _____ to reduce toxicity.
    a. heat
    b. formalin
    c. alum
    d. subculturing in nonsusceptible animals

    ans. b.   diff. 6   obj. 5   p. 28

43. The DPT shot given as protection against diphtheria, tetanus, and whooping cough consists of
    a. diphtheria toxoid, and tetanus and pertussis vaccine
    b. diphtheria and tetanus toxoids, and pertussis vaccine
    c. toxoids of diphtheria, tetanus, and pertussis
    d. a mixture of antitoxins to the three diseases

    ans. b.   diff. 6   obj. 5   p. 29

44. The incubation period for diphtheria is
    a. 7–21 days
    b. 1–2 weeks
    c. 2–6 days
    d. 3 weeks

    ans. c.   diff. 6   obj. 5   p. 30

45. Otitis media and sinusitis typically arise from _____ infections.
    a. nasopharyngeal
    b. oropharyngeal
    c. middle ear
    d. streptococcal

    ans. a.   diff. 6   obj. 6   p. 31

46. Otitis media develops from an infection of the nose and throat in which the infection spreads
    a. along the sensory neurons of the middle ear
    b. from the inner ear outward to the middle ear
    c. upward through the eustachian tube
    d. across the tympanic membrane, which may burst

    ans. c.   diff. 7   obj. 6   p. 31

47. The most important bacterial pathogens in otitis media are
    a. *Moraxella catarrhalis* and *Streptococcus pyogenes*
    b. *Staphylococcus aureus* and *Mycoplasma pneumoniae*
    c. *Streptococcus pneumoniae* and *Haemophilus influenzae*
    d. several varieties of Gram-negative cocci

    ans. c.   diff. 7   obj. 6   p. 32

48. A complication of colds caused by pathogenic bacteria or nasopharyngeal opportunists is
    a. endocarditis
    b. glomerulonephritis
    c. meningitis
    d. myocarditis

    ans. c.   diff. 6   obj. 6   p. 33

49. The eyes connect directly to the upper respiratory system via the
    a. nasolacrimal ducts
    b. frontal sinuses
    c. middle ear
    d. tonsils

    ans. a.   diff. 5   obj. 7   p. 34

50. Most of the common eye infections are caused by
    a. respiratory pathogens
    b. viruses
    c. fungi
    d. throat opportunists

    ans. a.   diff. 6   obj. 7   p. 34

51. Conjunctivitis caused by rhinoviruses or adenoviruses can be determined by length of infection and
    a. extreme severity of the pain
    b. copious mucus secretions from the infected eye
    c. pain and lack of light sensitivity
    d. lack of yellow exudate from the infected eye

    ans. d.   diff. 6   obj. 7   p. 34

52. Infectious agents of conjunctivitis are frequently transmitted by contact with
    a. infected fomites
    b. secretions of infected persons
    c. fecal material of an infected person
    d. hands of infected persons

    ans. b.   diff. 7   obj. 7   p. 35

53. Another name for the common cold is
    a. otitis media
    b. sinusitis
    c. acute afebrile infectious coryza
    d. horny nose

    ans. c.   diff. 5   obj. 8   p. 36

54. Most colds are probably caused by
    a. orthomyxoviruses
    b. bacteria
    c. rhinoviruses and members of the picorna group
    d. normal flora

    ans. c.   diff. 6   obj. 8   p. 36

55. The most common reservoir of the common cold are
    a. family pets
    b. humans
    c. infected fomites
    d. discarded used tissues

    ans. b.   diff. 6   obj. 8   p. 36

56. The highest concentrations of virus are found in the nasal secretions and possibly on the hands of infected people during
    a. the first two or three days of a cold
    b. the recovery period of the cold
    c. the incubation period of the disease
    d. communication with a nonimmune population

    ans. a.   diff. 6   obj. 8   p. 36

57. The most effective preventive method for avoiding the common cold is
    a. less stress
    b. avoiding a chill
    c. isolation from children
    d. frequent hand washing

    ans. d.   diff. 5   obj. 8   p. 39

58. The disease that closely resembles strep throat but is not of bacterial origin is
    a. otitis media
    b. the common cold
    c. adenoviral pharyngitis
    d. tonsillitis

    ans. c.   diff. 6   obj. 8   p. 41

59. Pharyngoconjunctival fever is caused by
    a. *Staphylococcus aureus*
    b. rhinoviruses
    c. picorna viruses
    d. adenoviruses

    ans. d.   diff. 6   obj. 8   p. 43

60. A vaccine for the common cold is not feasible because
    a. more than eighty-nine types of rhinovirus may be the cause
    b. the surface antigens of the agent mutate frequently
    c. the cause is unknown
    d. the vaccine would be dangerous to some members of the population

    ans. a.   diff. 7   obj. 8   p. 43

True/False

1. The name of the disease that is derived from the Greek word for leather is impetigo.
   false (diphtheria)

2. The use of horses to produce large amounts of antitoxin in the treatment of diphtheria was introduced in 1892 by Louis Pasteur.
   false (Emile Roux)

3. Among the most common reasons for people in the U.S. to seek medical care is a sore throat.
   true

4. Patients with streptococcal sore throats recover rapidly if treated with antistreptolysin.
   false (recover spontaneously after about a week)

5. Streptolysin O and streptolysin S (enzymes released by *S. pyogenes* cells) are responsible for pus-filled mucus membranes in the throat.
   false (β-hemolysis)

6. Rebecca Lancefield spent the majority of her scientific years in the study of the classification of the various strains of *Staphylococcus aureus*.
false (streptococci)

7. A potentially fatal outcome of streptococcal infection is caused by inflammation of the nephron capillaries.
true

8. It is common to find *S. pyogenes* in blood samples of patients suffering from acute glomerulonephritis.
false (rare, damage from toxin)

9. Antibiotic treatment of diphtheria is effective in reducing the mortality rate in victims of the disease.
false (only effective to stop spread of the disease)

10. Immunization against diphtheria is accomplished by administration of the attenuated toxoid.
true

11. The tissue damage in diphtheria is due to diffusion of the toxoid in the blood stream.
true

12. The most frequent cause of death in diphtheria is due to blood poisoning.
false (asphyxiation)

13. When infection spreads from the nasopharynx to the ear through the eustachian tube, otitis media results.
true

14. The more severe forms of conjunctivitis are of viral origin.
false (bacterial)

Chapter 21

Multiple Choice

1. Among the evidences of tuberculosis as an ancient disease is the
   a. tubercular damage of Stone Age skeletons
   b. records of Hippocrates
   c. descriptions of tubercular disease by Laënnec in 1804
   d. journal of Robert Koch

   ans. a.   diff. 5   obj. 1   p. 1

2. Which of the following scientists is responsible for the report that tuberculosis was a single disease not separate diseases as previously considered?
   a. Ehrlich
   b. Villemin
   c. Laënnec
   d. Koch

   ans. c.   diff. 6   obj. 1   p. 2

3. Experimentation showed that tuberculosis was caused by a bacterium. This was proved by
   a. Albert Calmette
   b. Renè Laënnec
   c. Paul Ehrlich
   d. Robert Koch

   ans. d.   diff. 6   obj. 1   p. 2

4. In 1905, Robert Koch received the Nobel prize for his work on the
   a. etiology of tuberculosis
   b. Gram stain methodology
   c. acid-fast stain technique
   d. anthrax bacillus

   ans. a.   diff. 5   obj. 1   p. 2

5. The most successful vaccine against tuberculosis was developed by
   a. Robert Koch
   b. Calmette and Guérin
   c. Renè Laënnec
   d. Ehrlich and Koch

   ans. b.   diff. 6   obj. 1   p. 2

6. Which of the following are considered diseases of the lower respiratory tract?
   a. influenza and diphtheria
   b. tuberculosis and pneumonia
   c. diphtheria and pneumonia
   d. influenza and tuberculosis

   ans. b.   diff. 5   obj. 2   p. 3

7. The thin-walled air sacs that make up the lungs are called
   a. bronchioles
   b. membranes
   c. mucosacs
   d. alveoli

   ans. d.   diff. 5   obj. 2   p. 4

8. The film of mucus in the lower respiratory tract that sweeps foreign material and microorganisms toward the throat is termed the
   a. pseudostratified epithelium
   b. ciliated cells of the trachea
   c. mucociliary escalator
   d. serous fluid

   ans. c.   diff. 6   obj. 2   p. 4

9. The cough reflex, which is helpful in expelling foreign material, is depressed by products such as
   a. tobacco smoke
   b. alcohol
   c. allergies
   d. pepper

   ans. b.   diff. 5   obj. 2   p. 4

10. Microorganisms of the lower respiratory tract are
    a. viruses and fungi
    b. bacilli and pneumococci
    c. normally absent
    d. streptococci and staphylococci

    ans. c.   diff. 5   obj. 2   p. 5

11. The most common cause of bacterial pneumonia is
    a. *Streptococcus pyogenes*
    b. *Streptococcus pneumoniae*
    c. *Staphylococcus aureus*
    d. *Klebsiella pneumoniae*

    ans. b.   diff. 5   obj. 3   p. 6

12. Over 60% of the bacterial pneumonias of adults that require hospitalization are caused by
    a. *Streptococcus pneumoniae*
    b. *Streptococcus pyogenes*
    c. *Staphylococcus aureus*
    d. *Klebsiella pneumoniae*

    ans. a.   diff. 6   obj. 3   p. 6

13. The most striking characteristic of *S. pneumoniae* is its
    a. capsular antigen
    b. Gram reaction
    c. colony formation
    d. thick capsule

    ans. d.   diff. 5   obj. 3   p. 6

14. *S. pneumoniae* when growing aerobically on blood agar produces
    a. α^1-hemolysis
    b. β-hemolysis
    c. γ-hemolysis
    d. α-hemolysis

    ans. d.   diff. 5   obj. 3   p. 7

15. Which of the following characteristics is not significant to the identification of *S. pneumoniae?*
    a. α-hemolysis
    b. growth inhibited by optochin
    c. lysed by bile
    d. identification of capsular proteins

    ans. a.   diff. 7   obj. 3   p. 7

16. The lung damage in pneumococcal pneumonia is due to the rapid multiplication of the bacteria, which causes a(n)
    a. accumulation of sputum
    b. tubercle lesion
    c. inflammatory response
    d. attraction of phagocytes

    ans. c.   diff. 6   obj. 3   p. 8

17. Which of the following is mismatched?
    a. bacteria in the blood stream—septicemia
    b. inflammation of the alveoli—pleurisy
    c. infection of the heart valves—endocarditis
    d. infection of the membranes covering the brain and spinal cord—meningitis

    ans. b.   diff. 7   obj. 3   p. 8

18. Which of the following is not necessarily a pre-existing condition of hospitalized pneumococcal pneumonia patients with positive blood cultures that can lead to death?
    a. penicillin sensitivity
    b. alcoholism
    c. cardiac condition
    d. chronic lung disease

    ans. a.   diff. 8   obj. 3   p. 9

19. The commercially available vaccine to protect against pneumococcal pneumonia is directed against the _____ antigens.
    a. cell membrane
    b. capsular
    c. cell wall
    d. pili

    ans. b.   diff. 8   obj. 3   p. 9

20. The causative agent of the pneumonia that causes permanent lung damage and a mortality of 50% due to complications is
    a. *Mycoplasma pneumoniae*
    b. *Streptococcus pneumoniae*
    c. *Klebsiella pneumoniae*
    d. *Haemophilus influenzae*

    ans. c.   diff. 5   obj. 3   p. 11

21. The virulence of *Klebsiella* is due partly to the
    a. antiphagocytic property of their capsules
    b. exotoxin produced
    c. engorgement of blood vessels
    d. motility of the organism

    ans. a.   diff. 6   obj. 3   p. 11

22. Alcoholism complicates *Klebsiella* pneumonia. Which of the following is not a complication due to alcoholism?
    a. exotoxin production
    b. impairment of ciliary action
    c. inhibition of the cough reflex
    d. reduction of phagocytosis

    ans. a.   diff. 7   obj. 3   p. 12

23. Laboratory diagnosis of *K. pneumoniae* cannot be based on colonies of *Klebsiella* in a sputum culture because
    a. it ferments glucose and lactose
    b. it may be an avirulent strain
    c. alcoholism may not be a factor
    d. *K. pneumoniae* can be present in the sputum of healthy people

    ans. d.   diff. 7   obj. 3   p. 12

24. In hospitals, the organisms that are now the principal source of R factors transferable to other pathogens are of the genus
    a. *Klebsiella*
    b. *Streptococcus*
    c. *Haemophilus*
    d. *Staphylococcus*

    ans. a.   diff. 6   obj. 3   p. 13

25. The spread of *Klebsiella pneumoniae* is due to
    a. transmission by an insect vector
    b. the fecal-oral route
    c. inhalation of infected droplets
    d. direct contact with an infected fomite

    ans. c.   diff. 6   obj. 3   p. 14

26. The common cause of "walking pneumonia" is
    a. *Klebsiella pneumoniae*
    b. *Mycoplasma pneumoniae*
    c. *Haemophilus influenzae*
    d. *Streptococcus pneumoniae*

    ans. b.   diff. 5   obj. 3   p. 14

27. The typical appearance of *Mycoplasma* colonies growing on solid media is
    a. round and raised with fluted edges
    b. that of a fried egg
    c. flat, regular, and metallic looking
    d. pale pink with raised centers

    ans. b.   diff. 6   obj. 3   p. 15

28. Transmission of *Mycoplasma pneumoniae* is
    a. via the fecal-oral route
    b. from person to person
    c. generally from discarded tissues holding respiratory secretions
    d. from ingestion of contaminated food

    ans. b.   diff. 6   obj. 3   p. 16

29. What percent of the bacterial pneumonias are caused by *Mycoplasma pneumoniae*?
    a. 25%
    b. 30%
    c. 10%
    d. 20%

    ans. d.   diff. 7   obj. 3   p. 16

30. Diagnosis of *Mycoplasma pneumoniae* can generally be made by identification of the
    a. complement-fixing antibodies in the patient's blood
    b. specific symptoms
    c. distinctive rash
    d. inflammatory response during testing of the skin rash

    ans. a.   diff. 7   obj. 3   p. 16

31. Why are the penicillins and cephalosporins ineffective in treatment of "walking pneumonia"?
    a. cause of disease is viral
    b. do not neutralize the toxins
    c. lack of bacterial cell wall in causative agent
    d. organism is penicillin resistant

    ans. c.   diff. 6   obj. 3   p. 17

32. The incubation period of *Mycoplasma pneumoniae* is
    a. 2–3 weeks
    b. 7–14 days
    c. 1–4 days
    d. 1 month

    ans. a.   diff. 6   obj. 3   p. 18

33. The causative agent of whooping cough is
    a. whooping cough virus
    b. *Mycoplasma pneumoniae*
    c. *Bordetella pertussis*
    d. *Staphylococcus aureus*

    ans. c.   diff. 6   obj. 3   p. 19

34. The respiratory secretions of whooping cough contain large numbers of the organism early in the illness. These can be detected by
    a. the complement fixation test
    b. using the specific fluorescent antibody
    c. the agglutination test
    d. precipitating the antibody

    ans. b.   diff. 7   obj. 3   p. 19

35. Fatal complications in whooping cough usually result from superinfection of the damaged lung by
    a. *Streptococcus pneumoniae*
    b. *Mycoplasma pneumoniae*
    c. *Haemophilus influenzae*
    d. *Bordetella pertussis*

    ans. a.   diff. 6   obj. 3   p. 20

36. The areas of the respiratory system affected by whooping cough are the trachea and bronchi, which compose the
    a. facial sinuses
    b. nasopharynx
    c. oropharynx
    d. tracheobronchial system

    ans. d.   diff. 5   obj. 3   p. 20

37. The sudden, violent, uncontrollable cough of pertussis is described as
    a. contagious
    b. productive
    c. paroxysmal
    d. infective

    ans. c.   diff. 5   obj. 3   p. 21

38. The chronic bacterial infection of the lungs is
    a. tuberculosis
    b. whooping cough
    c. bronchitis
    d. pneumonia

    ans. a.   diff. 4   obj. 5   p. 24

39. The cause of tuberculosis is *Mycobacterium tuberculosis,* a(n)
    a. typical Gram-positive rod with spores
    b. anaerobic rod, Gram-positive with pili
    c. fast growing, Gram-negative, anaerobic rod
    d. slow growing, Gram-positive, slender, strictly aerobic rod

    ans. d.   diff. 6   obj. 4   p. 25

40. One of the diagnostic tools in the identification of the causative agent of tuberculosis is the
    a. Gram stain
    b. acid-fast stain
    c. hemolysis on blood agar
    d. complement fixation test

    ans. b.   diff. 5   obj. 5   p. 25

41. Although it is unusually resistant to drying, the tubercle bacillus is easily killed by
    a. pasteurization
    b. disinfectants
    c. strong acids
    d. strong alkalis

    ans. a.   diff. 6   obj. 5   p. 25

42. The virulence of the tubercle bacillus is due to its
    a. toxin
    b. survival within macrophages
    c. lysogenic conversion
    d. resistance to antibiotics

    ans. b.   diff. 6   obj. 5   p. 25

43. *M. tuberculosis* grows very slowly on special laboratory media. The growth rate can be speeded up considerably by using a liquid medium containing
    a. radioactive palmitic acid
    b. sheep's blood
    c. a 1% solution of acetic acid
    d. physiological saline

    ans. a.   diff. 7   obj. 5   p. 27

44. The destructive nature of tuberculosis can be characterized as a(n)
   a. delayed hypersensitivity reaction
   b. immune complex reaction
   c. inflammatory response
   d. endotoxin pyrogenic response

   ans. a.   diff. 6   obj. 5   p. 28

45. The characteristic response of the body to the collection of walled tubercle bacilli is referred to as a(n)
   a. myeloma
   b. granuloma
   c. tuberoma
   d. cyst

   ans. b.   diff. 6   obj. 5   p. 28

46. When the granulomas of tuberculosis become reactive and there is renewed growth of the organisms, the resulting disease is referred to as
   a. caseous necrosis
   b. reactivation tuberculosis
   c. tubercle reaction
   d. tuberculosis systemia

   ans. b.   diff. 7   obj. 5   p. 29

47. The test for hypersensitivity to the tubercle bacilli is accomplished by injecting PPD into the skin. This test is called the
   a. Ritz-Carlton test
   b. hypersensitivity test
   c. Mantoux test
   d. Schick test

   ans. c.   diff. 7   obj. 5   p. 30

48. A positive reaction to the tuberculin test generally indicates that
   a. humoral antibodies to tuberculosis are present
   b. the person is immune to tuberculosis
   c. the PPD fluid caused a local rash
   d. active tuberculosis in that person is possible

   ans. d.   diff. 6   obj. 5   p. 30

49. Which of the following is probably not true of tuberculosis?
   a. about one-third of the world's population is infected each year
   b. the acute infection is followed by permanent immunity
   c. three million people die from tuberculosis each year
   d. the tuberculin test is positive

   ans. b.   diff. 5   obj. 5   p. 30

50. Infection rates of tuberculosis are highest among
   a. whites and very young Hispanics
   b. non-whites and elderly poor people
   c. infants and the elderly
   d. American Indians and white infants

   ans. b.   diff. 6   obj. 5   p. 30

51. The drugs _____, commonly used to treat tuberculosis are effective against both actively growing organisms in cavity tissue and relatively inactive intracellular organisms.
   a. rifampin and INH
   b. pyrazinamide and rifampin
   c. INH and isoniazid
   d. isoniazid and pyrazinamide

   ans. a.   diff. 6   obj. 5   p. 30

52. An effective vaccine developed against tuberculosis is used in Europe but not in the U.S. It is derived from *M. bovis* and is known as
   a. TBAG
   b. anti-TB
   c. BCG
   d. gamma globulin

   ans. c.   diff. 7   obj. 5   p. 33

53. The incubation period for tuberculosis is
   a. 1 year
   b. 1–2 weeks
   c. 3–5 months
   d. 2–10 weeks

   ans. d.   diff. 6   obj. 5   p. 34

54. The bacterial lung infection acquired by inhalation of aerosols of contaminated water is known as
   a. pneumonia
   b. Legionnaires' disease
   c. influenza
   d. whooping cough

   ans. b.   diff. 5   obj. 4   p. 35

55. Which of the following characteristics describe *Legionella pneumophila?*
   a. Gram-positive, motile, facultative anaerobic rod
   b. Gram-negative, nonmotile, aerobic rod
   c. rod-shaped bacillus, motile by one polar flagellum with poor Gram stain reaction
   d. spirochete with polar flagella

   ans. c.   diff. 5   obj. 4   p. 36

56. Which of the following methods of transmission is not effective for the transfer of *L. pneumophila?*
   a. person-to-person spread
   b. contaminated aerosols
   c. nebulizers
   d. showers and water faucets

   ans. a.   diff. 6   obj. 4   p. 38

57. Influenza A and B are caused by the
   a. orthomyxoviruses
   b. paramyxoviruses
   c. cytomegaloviruses
   d. adenoviruses

   ans. a.   diff. 7   obj. 6   p. 41

58. Projecting from the outer envelope of the influenza A and B virus are two kinds of glycoprotein spikes—
   a. macrocin and lymphocidin
   b. hyaluronidase and coagulase
   c. leukocidin and hemolysin
   d. hemagglutinin and neuraminidase

   ans. d.   diff. 6   obj. 6   p. 42

59. Immunity to influenza is due to the formation of antibodies to
   a. neuraminidase
   b. hemagglutinin
   c. leukocidin
   d. protein M

   ans. b.   diff. 7   obj. 6   p. 43

60. In influenza, destruction of the _____ severely impairs one of the body's major defenses against infection.
   a. mucociliary escalator
   b. blood-brain barrier
   c. keratin layer of the skin
   d. tonsils

   ans. a.   diff. 6   obj. 6   p. 43

61. Which of the following organisms is not usually involved in the secondary infections of influenza?
   a. *Haemophilus influenzae*
   b. *Staphylococcus aureus*
   c. *Streptococcus pyogenes*
   d. *Mycobacterium tuberculosis*

   ans. d.   diff. 5   obj. 6   p. 44

62. Epidemics of influenza in the United States are associated with an estimated _____ deaths per year.
   a. 10,000 to 40,000
   b. 1 million
   c. 5,000 to 10,000
   d. 50,000 to 75,000

   ans. a.   diff. 7   obj. 6   p. 44

63. Antigenic shifts are represented by dramatic changes in the viral antigens. It is likely they arise from rare events in which
   a. conjugation of two viruses occurs
   b. there is lysogenic conversion of two separate viruses
   c. two different viruses infect a cell at the same time
   d. there is blending of several viral genomes

   ans. c.   diff. 7   obj. 6   p. 46

64. Major antigenic shifts are more common in influenza _____ viruses, and generally cause more severe and widespread diseases.
   a. D
   b. B
   c. A
   d. Bangkok

   ans. c.   diff. 5   obj. 6   p. 46

65. The curious affliction that occasionally occurs in association with influenza B infections is referred to as
   a. Guillain-Barré syndrome
   b. Reye's syndrome
   c. aspirin myopathy
   d. paralytic epidemic

   ans. b.   diff. 6   obj. 6   p. 47

66. Epidemiologic evidence suggests the risk of Reye's syndrome increases with the use of _____ for fever.
   a. aspirin
   b. ibuprofen
   c. penicillin
   d. cool baths

   ans. a.   diff. 7   obj. 6   p. 47

67. The syndrome that occurred in the U.S. during a nationwide immunization program with the swine influenza vaccine is called
   a. Reye's syndrome
   b. Guillain-Barré syndrome
   c. paralytic syndrome
   d. histoplasmic syndrome

   ans. b.   diff. 6   obj. 6   p. 48

68. Two widespread North American lung mycoses are
   a. coccidioidomycosis and histoplasmosis
   b. candidiasis and aspergillosis
   c. aspergillosis and histoplasmosis
   d. candidiasis and coccidioidomycosis

   ans. a.   diff. 7   obj. 7   p. 50

69. The causative agent of coccidioidomycosis is *Coccidioides immitis*, which can be described as a(n)
   a. dimorphic fungus living in the soil
   b. yeastlike fungus found on plant roots
   c. spherulic spore of the damp soil
   d. endospore, larger and less resistant than bacterial spores

   ans. a.   diff. 6   obj. 6   p. 51

70. The "tissue phase" of *C. immitis* can be identified in sputum or pus as a thick-walled, nonbudding _____.
   a. hypha
   b. spore
   c. spherule
   d. endospore

   ans. c.   diff. 7   obj. 7   p. 51

71. Transmission of coccidioidomycosis occurs by
   a. aerosol droplets of the infective agent
   b. ingestion of contaminated foods
   c. swallowing and coughing up spherules
   d. inhalation of spores

   ans. d.   diff. 6   obj. 7   p. 54

72. Coccidioidomycosis and histoplasmosis are usually benign but occasionally mimic
   a. diphtheria
   b. influenza
   c. pneumonia
   d. tuberculosis

   ans. d.   diff. 7   obj. 7   p. 54

73. The causative agent of histoplasmosis is found most frequently in soil contaminated by
   a. fecal material
   b. bat or bird droppings
   c. polluted water
   d. bovine excrement

   ans. b: diff. 6   obj. 7   p. 54

74. Another name for histoplasmosis is
   a. spelunkers disease
   b. mycopneumonia
   c. blastomycosis
   d. tuberculosis-like

   ans. a.   diff. 7   obj. 7   p. 55

75. The causative agent of histoplasmosis is the dimorphic fungus
    a. *Emmonsiella bovis*
    b. *Coccidioides immitis*
    c. *Histoplasma capsulatum*
    d. *Candida albicans*

    ans. c.   diff. 5   obj. 7   p. 57

True/False

1. Mucous membranes that cover the lungs and line the chest cavity provide protection to these organs.
   false   (pleura membranes provide frictionless movement)

2. Mucus is propelled out of the respiratory system by the columnar epithelial cells that line the nasopharynx and oropharynx.
   false   (ciliated pseudostratified columnar epithelium)

3. One of the most common causes of pneumonia is *Streptococcus pneumoniae*.
   true

4. The virulence of *K. pneumoniae* is due to its endotoxin production, which causes serious destruction to the lung.
   true

5. Treatment of infections with *Klebsiella pneumoniae* is extremely difficult due to the presence of F factor plasmids.
   false   (R factor plasmids)

6. The common name for the pneumonia caused by *Mycoplasma* sp. is mononucleosis.
   false   (walking pneumonia)

7. The sudden, violent, uncontrollable cough of pertussis is described as paroxysmal.
   true

8. Tuberculosis is a chronic bacterial infection of the lungs.
   true

9. The toxin of *M. tuberculosis* is responsible for its virulence.
   false   (virulence depends on survival within host macrophages)

10. A diagnostic feature of *M. tuberculosis* is its Gram-stain response.
    false   (acid-fast stain response)

11. The chronic aspects of tuberculosis are the results of anaphylaxis.
    false   (delayed hypersensitivity)

12. The positive tuberculin test shows immunity to tuberculosis.
    false   (infection with *M. tuberculosis*)

13. The diagnosis of Legionnaires' disease is easily demonstrated by Gram-stain and culture techniques.
    false   (special stains and special culture techniques)

14. The most serious influenza epidemics are caused by the influenza A virus.
    true

15. Pandemics of influenza occur approximately every 10 years.
    true

16. Deaths from influenza are caused by the viral destruction of pulmonary tissue.
    false   (secondary bacterial infections)

17. Reye's syndrome occasionally occurs in association with influenza B infection.
    true

18. Most cases of coccidioidomycosis occur in the western United States by inhaling airborne spores of the fungus.
    true

19. Coccidioidomycosis is a fungal disease that occurs in the arid regions of the western United States.
    true

20. Another name for Histoplasmosis is Guana disease.
    false   (spelunkers disease)

21. Tuberculosis is a modern disease of urban civilization.
    false   (ancient, found in Stone Age skeletons)

22. The normal flora of the lower respiratory tract are primarily viruses.
    false   (no normal flora of lower respiratory tract)

23. Over 60% of the bacterial pneumonias that require hospitalization are caused by *Klebsiella pneumoniae*.
    false   (*Streptococcus pneumoniae*)

24. Pneumococcal pneumonia is caused by encapsulated strains of *S. pneumoniae*.
    true

25. The vaccine that prevents pneumococcal pneumonia contains 23 capsular antigens.
true

26. A virulent pneumonia found among college students is caused by *Klebsiella pneumoniae*.
false (*Mycoplasma pneumoniae*)

27. The vaccine available for *K. pneumoniae* is highly effective in preventing infections in hospital personnel.
false (there is no vaccine available)

Chapter 22

Multiple Choice

1. The passageway that runs from the mouth to the anus is referred to as the
   a. oral cavity
   b. alimentary tract
   c. anal cavity
   d. gut canal

   ans. b.   diff. 5   obj. 1   p. 2

2. Why is the alimentary tract one of the major entry routes for invading infectious agents?
   a. tissues are exposed to organisms from food and drink
   b. dental caries interferes with mechanical breakdown of food
   c. system is always open to invasion through swallowing mucus
   d. highly processed foods are easily contaminated

   ans. a.   diff. 6   obj. 1   p. 2

3. The major purpose of the alimentary system is to
   a. filter out infectious agents
   b. create an exit for metabolic waste
   c. provide nourishment to the body
   d. provide the body with oxygen

   ans. c.   diff. 6   obj. 1   p. 3

4. The exceedingly widespread and economically important disease of the upper alimentary system is
   a. mumps
   b. herpes simplex
   c. dental caries
   d. acne

   ans. c.   diff. 6   obj. 1   p. 3

5. The portion of the tooth that provides a protective layer of densely packed crystals is
   a. enamel
   b. dentine
   c. pulp cavity
   d. fluoride treatment

   ans. a.   diff. 6   obj. 1   p. 4

6. If inflammation develops in the space between the tooth and the gum, the condition is referred to as
   a. pyorrhea
   b. periodontal disease
   c. cavities
   d. gingivitis

   ans. d.   diff. 5   obj. 1   p. 4

7. How does saliva aid in the breakdown of complex food molecules for absorption by the body cells?
   a. lubricates the mouth
   b. keeps the mouth clean
   c. helps maintain a neutral pH
   d. contains an enzyme that begins the breakdown of starch

   ans. d.   diff. 5   obj. 1   p. 4

8. The connecting structure between the oral cavity and the stomach cavity is the
   a. esophagus
   b. trachea
   c. nasopharynx
   d. small intestine

   ans. a.   diff. 5   obj. 1   p. 6

9. The condition that can result from food being inadequately chewed, infected with microbes, and then rotting due to prolonged stay in the stomach, is referred to as
   a. halitosis
   b. dental caries
   c. indigestion
   d. vomiting

   ans. a.   diff. 6   obj. 1   p. 7

10. Among the normal flora of the mouth is *Streptococcus sanguis*, which colonizes the
    a. upper part of the tongue
    b. teeth
    c. mucosa of the cheek
    d. floor of the mouth

    ans. b.   diff. 6   obj. 1   p. 8

11. Collections of bacteria that adhere to the smooth surfaces and crevices of the teeth are called
    a. tooth decay
    b. dental caries
    c. dental plaque
    d. halitosis

    ans. c.   diff. 6   obj. 1   p. 8

12. The gingival crevice characteristically is populated by strict anaerobes of the genera
    a. *Clostridium*
    b. *Streptococcus*
    c. *Bacteroides*
    d. *Pseudomonas*

    ans. c.   diff. 7   obj. 1   p. 8

13. The most common infectious disease of humans is
    a. the common cold
    b. dental caries
    c. hepatitis A
    d. athlete's foot

    ans. b.   diff. 7   obj. 2   p. 10

14. The principal cause of dental caries is *Streptococcus*
    a. *mutans*
    b. *salivarius*
    c. *mitis*
    d. *sanguis*

    ans. a.   diff. 7   obj. 3   p. 10

15. Which of the following genera are not ordinarily found as inhabitants of dental plaque?
    a. *Streptococcus*
    b. *Actinomyces*
    c. *Staphylococcus*
    d. *Neisseria*

    ans. c.   diff. 7   obj. 2   p. 13

16. Which of the following genera of organisms found in dental plaque does not prefer an anaerobic environment for growth?
    a. *Streptococcus*
    b. *Lactobacillus*
    c. *Neisseria*
    d. *Veillonella*

    ans. c.   diff. 8   obj. 2   p. 13

17. The production of dental caries by *S. mutans* is due in part to its ability to
    a. change the pH of the saliva in the presence of starches
    b. invade plaque and break it down to dissolve enamel
    c. convert sugar to an organic compound with calcium, which then erodes the enamel
    d. store an intracellular polysaccharide that is later metabolized to acid

    ans. d.   diff. 8   obj. 3   p. 14

18. The white or brown spot seen on the tooth under the plaque represents a(n)
    a. area of weakened enamel
    b. break in the enamel that exposes the dentin
    c. hole in the outer tooth structure and an opening into the pulp
    d. stain made by foodstuffs not adequately cleaned

    ans. a.   diff. 7   obj. 2   p. 14

19. The mineral, when taken in minute quantities in water, that makes enamel more resistant to dissolving in acid is
    a. fluoride
    b. calcium
    c. chlorine
    d. zinc

    ans. a.   diff. 7   obj. 2   p. 15

20. Root caries, rather than dental caries, is a problem in
    a. adolescents
    b. teething infants
    c. older people
    d. pregnant women

    ans. c.   diff. 6   obj. 3   p. 16

21. The source of the *S. mutans* in children is
    a. household pet saliva
    b. fruit sugar
    c. cow's milk
    d. mother's saliva

    ans. d.　　diff. 7　　obj. 3　　p. 17

22. The chronic inflammatory process involving the tissues around the roots of the teeth is referred to as
    a. root caries
    b. dental plaque
    c. periodontal disease
    d. calculus

    ans. c.　　diff. 6　　obj. 3　　p. 17

23. Gum disorders caused by calculus stimulate the inflammatory response in the gums known as
    a. gingivitis
    b. plaque
    c. dental caries
    d. root caries

    ans. a.　　diff. 6　　obj. 3　　p. 18

24. Acute necrotizing ulcerative gingivitis is also known as
    a. gingivitis
    b. root caries
    c. pyorrhea
    d. trenchmouth

    ans. d.　　diff. 7　　obj. 3　　p. 20

25. In individuals with abnormal heart valves, what treatment is used before dental procedures are begun?
    a. treatment with an antibiotic is begun several hours before treatment
    b. a gas anesthesia is used instead of a local anesthetic to reduce stress
    c. the patient is hospitalized as a precaution against severe problems
    d. the patient is given an EKG as a precaution

    ans. a.　　diff. 9　　obj. 3　　p. 21

26. Which of the following is mismatched?
    a. swollen salivary glands—herpes
    b. Koplik's spots—measles
    c. oral blisters—chicken pox
    d. oral ulcers—herpangina

    ans. a.　　diff. 7　　obj. 4　　p. 22

27. The disease that is characterized by a group of small blisters, surrounded by an area of redness, which rupture within a day or two and leave superficial, painful ulcers that heal without treatment within about ten days is
    a. measles
    b. chicken pox
    c. herpes simplex
    d. foot, hand, and mouth disease

    ans. c.　　diff. 5　　obj. 4　　p. 24

28. The virus family that includes the varicella virus and the Epstein-Barr virus is
    a. picovirus
    b. rhinovirus
    c. herpesvirus
    d. parvovirus

    ans. c.　　diff. 7　　obj. 4　　p. 24

29. Which of the following immune responses is not employed by the body in limiting a herpes virus infection?
    a. interferon
    b. cyst formation
    c. humoral antibodies
    d. cell-mediated immunity

    ans. b.　　diff. 7　　obj. 4　　p. 25

30. Where in the body do the noninfectious forms of the herpes simplex virus persist between eruptions?
    a. motor neurons
    b. sensory neurons
    c. interneurons
    d. cranial nerves

    ans. b.　　diff. 6　　obj. 4　　p. 26

31. A painful finger infection attributable to the herpes virus is known as a(n)
    a. herpetic whitlow
    b. finger-cold sore
    c. abrasion lesion
    d. viral impetigo

    ans. a.   diff. 5   obj. 4   p. 27

32. Recurrent infection with the _____ virus is the most common cause of corneal blindness in the United States.
    a. coxsackie
    b. varicella-zoster
    c. Epstein-Barr
    d. herpes simplex

    ans. d.   diff. 6   obj. 4   p. 27

33. The most frequently identified cause of nonepidemic viral encephalitis is _____ virus.
    a. coxsackie
    b. varicella-zoster
    c. Epstein-Barr
    d. herpes simplex

    ans. d.   diff. 6   obj. 4   p. 27

34. The viral disease that characteristically involves the parotid glands is
    a. Rocky Mountain spotted fever
    b. measles
    c. mumps
    d. mouth ulcers

    ans. c.   diff. 5   obj. 4   p. 28

35. The viral disease that frequently produces painful swelling of one or both parotid glands is
    a. mumps
    b. measles
    c. herpes simplex
    d. chicken pox

    ans. a.   diff. 5   obj. 4   p. 28

36. About 25% of cases in postpubertal men and boys are complicated by rapid swelling of one or both testicles and are accompanied by intense pain.
    a. herpes simplex
    b. chicken pox
    c. measles
    d. mumps

    ans. d.   diff. 6   obj. 4   p. 29

37. The mumps virus is an enveloped member of which viral group?
    a. parvovirus group
    b. paramyxovirus family
    c. herpes zoster group
    d. coxsackie virus group

    ans. b.   diff. 7   obj. 4   p. 29

38. Which of the following diseases is not among those caused by the paramyxovirus family?
    a. measles
    b. cold sores
    c. mumps
    d. respiratory syncytial disease

    ans. b.   diff. 8   obj. 4   p. 29

39. How many antigenic types of mumps virus are known today?
    a. two
    b. one
    c. several
    d. over fifty

    ans. b.   diff. 7   obj. 4   p. 29

40. The most frequent manner in which mumps is spread to nonimmunized hosts is by contact with the infected patient's _____.
    a. saliva
    b. blood
    c. semen
    d. stools

    ans. a.   diff. 6   obj. 4   p. 30

41. The incubation period of mumps is generally
    a. 1–7 days
    b. 7–14 days
    c. 15–21 days
    d. 4–6 weeks

    ans. c.   diff. 6   obj. 4   p. 30

42. The natural host of the mumps virus are
    a. rabbits
    b. dogs
    c. humans
    d. mice

    ans. c.   diff. 5   obj. 4   p. 31

43. The available vaccine for mumps is a(n) _____ vaccine.
    a. genetically engineered
    b. inactive
    c. toxoid
    d. live attenuated

    ans. d.   diff. 5   obj. 4   p. 31

44. The active attenuated mumps vaccine is grown in
    a. chicken embryos
    b. horses
    c. duck embryos
    d. humans

    ans. a.   diff. 6   obj. 4   p. 32

45. _____ should be immunized against mumps.
    a. people infected with the AIDS virus
    b. babies less than 1 year of age
    c. those with extreme allergy to egg protein
    d. pregnant women

    ans. a.   diff. 9   obj. 4   p. 33

46. Inflammation of the esophagus is usually the result of
    a. diet
    b. poorly chewed food
    c. careless use of endoscopy equipment
    d. acid reflux from the stomach

    ans. d.   diff. 6   obj. 5   p. 33

47. In addition to *Candida albicans,* which of the following is a major cause of infection of the esophagus?
    a. *Escherichia coli*
    b. bacteroides
    c. herpes simplex
    d. *Helicobacter pylori*

    ans. c.   diff. 7   obj. 5   p. 33

48. A common stomach ailment is gastritis. It can be caused by the bacterium
    a. *Escherichia coli*
    b. *Helicobacter pylori*
    c. *Bacillus acidophilus*
    d. *Lactobacillus rugae*

    ans. b.   diff. 7   obj. 5   p. 34

49. More than 75% of patients with gastric ulcers and 95% of patients with duodenal ulcers are infected with the bacterium
    a. *H. pylori*
    b. *E. coli*
    c. *B. acidophilus*
    d. *L. rugae*

    ans. a.   diff. 8   obj. 5   p. 35

50. Gastric infection with *H. pylori* is associated with a rise in stomach pH, probably as a result of
    a. a decrease in the production of stomach acid
    b. the organism converting urea to ammonia
    c. an increase in the amount of mucus secreted by the stomach lining
    d. the increase in ingestion of high pH medications

    ans. b.   diff. 7   obj. 5   p. 38

True/False

1. The upper alimentary system is composed of the mouth, salivary glands, esophagus, and stomach.
   true

2. Salivary amylase is produced by the salivary glands and begins the digestion of sugar to acetylCoA.
   false   (digestion of starch to maltose)

3. The acid produced by the stomach is hydrochloric acid. It maintains the acid environment of the stomach.
   true

4. The stomach enzyme renin begins the digestion of fats.
   false   (pepsin begins degradation of some proteins)

5. Dental plaque consists of large quantities of bacteria and yeast, which cause halitosis.
   false   (eventually allow for anaerobes to degrade enamel)

6. Plaque formation is initiated by species of streptococci and actinomycetes.
   true

7. Dental caries is caused primarily by *Streptococcus pyogenes*.
   false   (*Streptococcus mutans*)

8. The major cause of tooth loss in older people is periodontal disease.
   true

9. A complication of dental procedures in patients with abnormal heart valves is pericarditis.
   false   (endocarditis)

10. Herpes simplex is an RNA virus that cannot be cultivated in tissue cells.
    false   (DNA virus and can be readily cultivated in tissue cultures)

11. Herpes simplex viruses can persist in noninfectious forms in the epithelial cells.
    false   (in sensory neurons)

12. Herpes simplex virus produces an acute disease in the mucous membranes and is followed by permanent immunity.
    false   (can remain noninfectious for a time and then recur during periods of stress)

13. Another name for parotiditis is mumps.
    true

14. Mumps frequently causes birth defects in infected pregnant women.
    false   (may cause miscarriage but not birth defects)

15. The mumps vaccine is in the form of an active viral vaccine.
    true

16. Peptic ulcers frequently involve infections with *Candida albicans*.
    false   (*Helicobacter pylori*)

Chapter 23

Multiple Choice

1. Cholera first appeared in the United States in _____ and was described by Army surgeon _____.
   a. 1866, Dr. Escherich
   b. 1828, John Snow
   c. 1854, Robert Koch
   d. 1832, S.B. Smith

   ans. d.   diff. 7   obj. 1   p. 2

2. Which London physician in 1854 demonstrated that cholera was transmitted by contaminated water?
   a. John Snow
   b. Robert Koch
   c. S.B. Smith
   d. Louis Pasteur

   ans. a.   diff. 5   obj. 1   p. 2

3. The scientist in 1883 who isolated *Vibrio cholerae* was
   a. Robert Koch
   b. John Snow
   c. S.B. Smith
   d. Louis Pasteur

   ans. a.   diff. 6   obj. 1   p. 2

4. Which of the following is not a function of the lower alimentary tract?
   a. degrading macromolecules of food to simpler subunits
   b. manufacture of red blood cells
   c. absorbing nutrients, minerals, and vitamins into the blood stream
   d. recycling of water

   ans. b.   diff. 5   obj. 1   p. 4

5. If an obstruction of the bile ducts interferes with the excretion of the breakdown products of hemoglobin, the resulting condition is described as
   a. gallstones
   b. hepatitis
   c. jaundice
   d. food poisoning

   ans. c.   diff. 6   obj. 2   p. 4

6. Infection of the gallbladder is termed
   a. gallstones
   b. cholecystitis
   c. jaundice
   d. hepatitis

   ans. b.   diff. 6   obj. 2   p. 4

7. Which of the following pathogens can colonize the gallbladder for years without causing symptoms?
   a. *Proteus vulgaris*
   b. *Escherichia coli*
   c. *Shigella dysenteria*
   d. *Salmonella typhi*

   ans. d.   diff. 6   obj. 2   p. 4

8. The pancreas is involved in which two body systems?
   a. digestive and excretory
   b. circulatory and lymphatic
   c. circulatory and immune
   d. endocrine and digestive

   ans. d.   diff. 5   obj. 2   p. 5

9. The pancreas produces digestive enzymes and _____.
   a. insulin
   b. bile salts
   c. proteases
   d. thyroxin

   ans. a.   diff. 6   obj. 2   p. 5

10. As the food enters the small intestine, digestive enzymes and bile are added, which are produced by the
    a. small intestine and stomach
    b. gallbladder and small intestine
    c. pancreas and liver
    d. small intestine and liver

    ans. c.   diff. 6   obj. 2   p. 6

11. Which one of the following is not usually absorbed by the small intestine?
    a. vitamins
    b. amino acids
    c. fatty acids
    d. alcohol

    ans. d.   diff. 6   obj. 2   p. 7

12. Small patches of lymphoid tissue are located in the walls of the small intestine. These lymphoid cells are responsible for the immune response to invading microorganisms and are called
    a. Peyer's patches
    b. crypts of Lieberkuhn
    c. intestinal glands
    d. microvilli

    ans. a.   diff. 6   obj. 2   p. 7

13. A major function of the large intestine is to
    a. recycle intestinal water
    b. digest protein
    c. absorb minerals and vitamins
    d. regulate peristalsis

    ans. a.   diff. 7   obj. 2   p. 8

14. Substances found in the alimentary tract that disassociate into ions when dissolved are referred to as
    a. salts
    b. electrolytes
    c. acids
    d. minerals

    ans. b.   diff. 6   obj. 2   p. 8

15. Pathogens that infect the cells of the small intestine adhere to its epithelial cells by specialized structures called
    a. capsules
    b. pili
    c. flagella
    d. adhesins

    ans. d.   diff. 7   obj. 3   p. 9

16. Bacteria make up what percent of the fecal weight?
    a. one-fourth
    b. one-half
    c. one-third
    d. three-fourths

    ans. c.   diff. 7   obj. 3   p. 9

17. _____ species of anaerobes make up the bulk of the organisms in the colon.
    a. *Escherichia*
    b. *Bacteroides*
    c. *Clostridium*
    d. *Proteus*

    ans. b.   diff. 6   obj. 3   p. 9

18. Which of the following is not among the facultative anaerobes in feces?
    a. *Klebsiella*
    b. *Proteus*
    c. *Bacteroides*
    d. *Enterobacter*

    ans. c.   diff. 7   obj. 3   p. 9

19. Which of the following is not among the useful vitamins synthesized by the intestinal bacteria?
    a. niacin
    b. biotin
    c. vitamin D
    d. vitamin K

    ans. c.   diff. 8   obj. 3   p. 10

20. When an individual is poorly nourished, oral antibiotics can cause vitamin deficiency by
    a. causing constipation
    b. causing diarrhea
    c. suppressing growth of intestinal flora
    d. suppressing water reabsorption

    ans. c.   diff. 8   obj. 3   p. 10

21. Antibiotic-associated colitis can follow the use of some broad-spectrum antibiotics. The disease is caused by the growth of the toxin-producing anaerobe
    a. *Clostridium difficile*
    b. *Clostridium botulinum*
    c. *Salmonella enteritidis*
    d. *Bacteroides species*

    ans. a.   diff. 8   obj. 3   p. 10

22. Suppression of intestinal flora with antibacterial medications can increase the patient's susceptibility to pathogens such as
    a. *Salmonella enteritidis*
    b. *Shigella flexneri*
    c. *Escherichia coli*
    d. *Enterobacter aerogenes*

    ans. a.   diff. 7   obj. 3   p. 10

23. The condition in which large populations of intestinal flora build up in an isolated pocket of small intestine is called
    a. acites
    b. ulcerative colitis
    c. antibiotic-associated colitis
    d. blind loop syndrome

    ans. d.   diff. 6   obj. 3   p. 11

24. The condition referred to as blind loop syndrome is corrected by surgery or
    a. antibiotic treatment
    b. diet
    c. laxatives
    d. specific exercises

    ans. a.   diff. 6   obj. 3   p. 11

25. Conversion of various substances to carcinogens in the colon along with diet may be responsible for intestinal cancer. The responsible substances are
    a. bile
    b. antibiotics
    c. high-fiber diets
    d. bacterial enzymes

    ans. d.   diff. 8   obj. 3   p. 11

26. Each year, more than five million people die of _____, 80 percent of them less than one year of age.
    a. malnutrition
    b. malaria
    c. diarrhea
    d. colon cancer

    ans. c.   diff. 5   obj. 3   p. 12

27. Bloody, pus-filled diarrhea known as dysentery results from damage to the
    a. upper alimentary tract
    b. small intestine
    c. large intestine
    d. liver

    ans. c.   diff. 5   obj. 3   p. 12

28. Since the ability of the small intestine to absorb fluid remains intact in most diarrheal diseases, rehydration is possible either by using intravenous treatment or by the use of
    a. oral rehydration solution (ORS)
    b. oral antibiotic treatment
    c. anal suppositories to prevent vomiting
    d. lukewarm baths to maintain hydration

    ans. a.   diff. 6   obj. 2   p. 13

29. The oral rehydration solution used in developing countries to counteract the severe dehydration resulting from diarrhea contains water and
    a. essential amino acids and the essential fatty acid
    b. vitamins and minerals
    c. amino acids and fatty acids
    d. glucose and electrolytes

    ans. d.   diff. 6   obj. 2   p. 13

30. There have been seven cholera pandemics since the early 1800s. The latest epidemic appeared in _____ in ____.
    a. Peru, 1991
    b. Africa, 1989
    c. Asia, 1979
    d. Europe, 1959

    ans. a.   diff. 6   obj. 3   p. 14

31. Cholera is the classic example of a(n)
    a. zoonose
    b. food-borne illness
    c. very severe form of diarrhea
    d. normal flora turned virulent opportunist

    ans. c.   diff. 6   obj. 3   p. 14

32. The massive diarrhea of cholera has a peculiar appearance and has been described as
    a. low-viscosity feces
    b. "rice water stool"
    c. 20 liters of water loss
    d. cholera-like

    ans. b.   diff. 6   obj. 4   p. 15

33. The symptoms of cholera are related to the production of a(n)
    a. pyrogenic endotoxin
    b. potent enterotoxin, choleragen
    c. inflammation due to capsular antigens of the organism
    d. mucus, which destroys the epithelium of the small intestine

    ans. b.   diff. 7   obj. 4   p. 15

34. Which of the following is not a common source of cholera?
    a. water contaminated with feces
    b. unpasteurized milk
    c. vegetables fertilized with human feces
    d. seafood harvested from contaminated water

    ans. b.   diff. 6   obj. 4   p. 16

35. The primary concern of the medical personnel treating a cholera patient is the
    a. replacement of lost fluids and salts
    b. isolation of the patient to protect the personnel
    c. protection of the patient due to his immunosuppressed state
    d. proper antibiotic therapy to halt the damage of the infection

    ans. a.   diff. 8   obj. 4   p. 18

36. The incubation period of cholera is
    a. 2–5 days
    b. 7–14 days
    c. 14–21 days
    d. 12–48 hours

    ans. d.   diff. 6   obj. 4   p. 19

37. Epithelial invasion and destruction characterize dysentery caused by _____ species.
    a. *Salmonella*
    b. *Shigella*
    c. *Escherichia*
    d. *Proteus*

    ans. b.   diff. 6   obj. 4   p. 20

38. Which of the following is not an ordinary means of acquiring members of the *Shigella* species responsible for dysentery?
    a. unpasteurized milk
    b. person-to-person contact
    c. day care centers
    d. contaminated food and water

    ans. a.   diff. 7   obj. 5   p. 21

39. The most virulent species of *Shigella* is
    a. *flexneri*
    b. *dysenteriae*
    c. *bovis*
    d. *extensi*

    ans. b.   diff. 6   obj. 5   p. 21

40. The disease that results in fever, diarrhea, and vomiting, and pus and blood in the feces and is caused by an organism that multiplies within the intestinal epithelial cells is
    a. "rice water stool" disease
    b. amoebic dysentery
    c. bacterial dysentery
    d. salmonellosis

    ans. c.   diff. 6   obj. 5   p. 22

41. Which of the following animals is not considered a common source of the *Salmonella* strains that cause gastroenteritis?
    a. turtles
    b. chickens
    c. rats
    d. pigs

    ans. d.   diff. 7   obj. 6   p. 23

42. It is generally recognized that there are _____ species of *Salmonella* and over _____ serotypes.
   a. twenty, 5,000
   b. ten, 1,000
   c. five, 1,500
   d. one, 2,000

   ans. d.   diff. 6   obj. 6   p. 23

43. The food products most commonly contaminated with *Salmonella* strains are
   a. milk and cheese
   b. meat and seafood
   c. eggs and poultry
   d. medications

   ans. c.   diff. 5   obj. 6   p. 26

44. The common source of *Salmonella* gastroenteritis is
   a. plant
   b. animal
   c. human
   d. polluted water

   ans. b.   diff. 6   obj. 6   p. 26

45. Which of the following is not a major consideration in the control of *Salmonella* infections?
   a. tracing the sources and careful identification of the causative agents
   b. elimination of all seafood beds in the shipping lanes
   c. routine sampling of animal products for contamination
   d. adequate cooking, especially of frozen food

   ans. b.   diff. 8   obj. 6   p. 28

46. Most cases of salmonellosis
   a. recover spontaneously and do not require antimicrobial medicines
   b. require hospitalization and antibiotic treatment
   c. weaken the patient severely and require extensive rehabilitation
   d. are contagious via the genitourinary tract

   ans. a.   diff. 8   obj. 6   p. 28

47. Laboratory diagnosis of salmonellosis requires
   a. fecal culture and media selective for *Salmonella* sp.
   b. inoculation of a laboratory animal and recovery of the organism
   c. a patient with all the classic symptoms and at least one that is unique
   d. elimination of all other Gram-negative bacilli from the feces

   ans. a.   diff. 7   obj. 6   p. 29

48. Which of the following is an almost universal member of the normal intestinal flora of the human and a possible cause of gastrointestinal disease?
   a. *Candida albicans*
   b. *Shigella dysenteriae*
   c. *Salmonella typhi*
   d. *Escherichia coli*

   ans. d.   diff. 5   obj. 7   p. 29

49. Which of the following groups of *E. coli* has not been implicated in diarrhea?
   a. enteroinvasive
   b. enterotoxigenic
   c. enterocapsular
   d. enterohemorrhagic

   ans. c.   diff. 6   obj. 7   p. 30

50. Which strain of *E. coli* has been identified as the cause of diarrheal outbreaks in hospital nurseries?
   a. enterohemorrhagic
   b. enterotoxigenic
   c. enteroinvasive
   d. enteropathogenic

   ans. d.   diff. 8   obj. 7   p. 30

51. Which strain of *E. coli* produces toxins that are closely related to the Shiga toxin found in some *Shigella* strains?
   a. enteropathogenic
   b. enterohemorrhagic
   c. enterotoxigenic
   d. enteroinvasive

   ans. b.   diff. 8   obj. 7   p. 31

52. The 1993 Washington state outbreak of *E. coli* disease was traced to inadequately cooked hamburgers served by a fast-food restaurant chain. The epidemic was caused by which strain of *E. coli?*
    a. enterotoxigenic
    b. enteropathogenic
    c. enterohemorrhagic
    d. enteroinvasive

    ans. c.　　diff. 8　　obj. 7　　p. 31

53. Traveler's diarrhea caused by *E. coli* can be prevented by taking bismuth preparations such as _____.
    a. Pepto-Bismol
    b. sodium bichloride
    c. Rolaids
    d. Metamucil

    ans. a.　　diff. 6　　obj. 7　　p. 32

54. Which recently isolated infectious agent is responsible for gastroenteritis with symptoms similar to those caused by *Shigella* species?
    a. *Campylobacter jejuni*
    b. *Escherichia coli*
    c. *Salmonella enteriditis*
    d. *Shigella flexneri*

    ans. a.　　diff. 6　　obj. 7　　p. 34

55. A serious form of *Salmonella* infection occurring only in humans and referred to as an example of "enteric fever" is
    a. bacillary dysentery
    b. staphylococcal intoxication
    c. botulism
    d. typhoid fever

    ans. d.　　diff. 5　　obj. 8　　p. 36

56. *Salmonella typhi* is the causative agent of
    a. typhoid fever
    b. paratyphoid fever
    c. diarrhea
    d. hemorrhagic fever

    ans. a.　　diff. 5　　obj. 8　　p. 36

57. Typhoid fever is usually spread by which route?
    a. genitourinary
    b. fecal-oral
    c. respiratory
    d. droplet infection

    ans. b.　　diff. 5　　obj. 8　　p. 36

58. The difference among the strains of *Salmonella typhi* are made on the basis of their
    a. H_2S production
    b. production of gas from the fermentation of glucose
    c. susceptibility to various bacteriophages
    d. motility

    ans. c.　　diff. 6　　obj. 8　　p. 36

59. In which of the following organs are the typhoid bacilli able to multiply and remain for long periods?
    a. Peyer's patches
    b. liver
    c. gallbladder
    d. colon

    ans. c.　　diff. 5　　obj. 8　　p. 37

60. The most notorious typhoid carrier was "Typhoid Mary." She is said to have been directly responsible for at least ____ cases of typhoid fever over a 15-year period.
    a. 200
    b. 100
    c. 150
    d. 50

    ans. d.　　diff. 6　　obj. 8　　p. 38

61. The incubation period for typhoid fever is
    a. 1–2 months
    b. 1–7 days
    c. 14–30 days
    d. 7–21 days

    ans. d.　　diff. 7　　obj. 8　　p. 39

62. Food poisonings are defined as illnesses caused by
    a. Gram-positive rods in water
    b. infection with organisms that contaminate food
    c. microbial products rather than by ingestion of the organisms
    d. fecal contamination

    ans. c.　　diff. 7　　obj. 9　　p. 40

63. Which of the following organisms is not ordinarily associated with food intoxications?
   a. *Clostridium*
   b. *Salmonella*
   c. *Bacillus*
   d. *Staphylococcus*

   ans. b.   diff. 7   obj. 9   p. 40

64. One of the most common forms of food poisoning is due to
   a. *Escherichia coli*
   b. *Staphylococcus aureus*
   c. *Salmonella enteriditis*
   d. *Clostridium botulinum*

   ans. b.   diff. 7   obj. 9   p. 41

65. One of the unique characteristics of the *S. aureus* exotoxin is that it is
   a. heat-stable
   b. heat-sensitive
   c. a polysaccharide
   d. not recoverable from contaminated food samples

   ans. a.   diff. 6   obj. 9   p. 41

66. Food poisoning from _____ is usually mild, the toxin is similar to that of *S. aureus,* and it commonly contaminates rice dishes.
   a. *Escherichia coli*
   b. *Clostridium perfringens*
   c. *Clostridium botulinum*
   d. *Bacillus cereus*

   ans. d.   diff. 7   obj. 9   p. 42

67. The type of food poisoning that can result in paralysis and death is
   a. gastroenteritis
   b. botulism
   c. bacillary dysentery
   d. aflatoxication

   ans. b.   diff. 5   obj. 9   p. 42

68. The most common cause of death from botulism is
   a. potassium depletion
   b. blindness
   c. dehydration
   d. respiratory paralysis

   ans. d.   diff. 7   obj. 9   p. 44

69. The incubation period for botulism is approximately
   a. 0–12 hours
   b. 1–6 days
   c. 1–2 weeks
   d. 12–36 hours

   ans. d.   diff. 7   obj. 9   p. 45

70. Which of the following foods is an unlikely source of *Clostridium botulinum* endospores?
   a. milk
   b. vegetables
   c. meat
   d. seafood

   ans. a.   diff. 6   obj. 9   p. 45

71. Which of the following organisms is capable of producing one of the most powerful neurotoxins known?
   a. *Clostridium botulinum*
   b. *Staphylococcus aureus*
   c. *Bacillus* cereus
   d. *Clostridium perfringens*

   ans. a.   diff. 5   obj. 9   p. 46

72. Which of the following foods has been associated with botulism in infants?
   a. vegetables
   b. milk
   c. fruit
   d. honey

   ans. d.   diff. 5   obj. 9   p. 47

73. Another name for ergot poisoning is
   a. lichen disease
   b. intoxication
   c. St. Anthony's fire
   d. botulism

   ans. c.   diff. 7   obj. 9   p. 48

74. The type of food poisoning that can result from eating bread prepared from grain on which certain types of mold have grown is
   a. ergot poisoning
   b. penicillin toxicity
   c. candidiasis
   d. tiniasis

   ans. a.   diff. 6   obj. 9   p. 48

75. Which fungal poison has been associated with liver damage and possible liver cancer?
    a. aflatoxins
    b. hemagglutinins
    c. aggressins
    d. exotoxins

    ans. a.   diff. 7   obj. 9   p. 49

76. Two viral infections of the lower alimentary system are hepatitis and
    a. St. Anthony's fire
    b. pancreatitis
    c. typhoid fever
    d. gastroenteritis

    ans. d.   diff. 6   obj. 10   p. 50

77. Typical hepatitis cases have been associated with many causes. Which of the following is not an expected cause of hepatitis?
    a. allergic reactions to medicines
    b. toxic chemicals including anesthetic gases
    c. exercise and vitamins
    d. various microorganisms and viruses

    ans. c.   diff. 6   obj. 10   p. 50

78. Most cases of hepatitis are caused by
    a. hepatitis C virus and hepatitis D virus
    b. hepatitis A virus and hepatitis B virus
    c. alcoholism
    d. poor diet

    ans. b.   diff. 5   obj. 10   p. 50

79. The hepatitis virus that is described as a small RNA virus of the enterovirus group of the picornavirus family is responsible for hepatitis
    a. B
    b. A
    c. C
    d. D

    ans. b.   diff. 6   obj. 10   p. 51

80. The epidemiology of hepatitis A is via the
    a. fecal-oral route
    b. blood
    c. semen
    d. sex route

    ans. a.   diff. 5   obj. 10   p. 51

81. The vaccine used to prevent hepatitis A infections is described as
    a. an antitoxin to the hepatitis toxin
    b. an attenuated vaccine
    c. a formalin inactivated vaccine
    d. gamma globulin

    ans. c.   diff. 7   obj. 10   p. 52

82. The hepatitis virus that is described as an enveloped DNA virus of the hepadnavirus family is responsible for hepatitis
    a. A
    b. B
    c. C
    d. D

    ans. b.   diff. 5   obj. 10   p. 53

83. The incubation period for hepatitis B is
    a. 10–15 weeks
    b. 3–5 weeks
    c. 5–9 weeks
    d. indefinite

    ans. a.   diff. 7   obj. 10   p. 53

84. Liver cell destruction of HBV is probably the result of the
    a. predetermined factors associated with liver abuse due to malnutrition and alcoholism
    b. destruction of the enzyme-producing mechanisms
    c. interactions and by-products of the medicines used for control
    d. body's immune system attacking the infected liver cells

    ans. d.   diff. 8   obj. 10   p. 54

85. The disease that infects at least five percent of the world's population and is the ninth leading cause of death is
   a. hepatitis B
   b. mumps
   c. malaria
   d. heart disease

   ans. a.   diff. 6   obj. 10   p. 55

86. Which of the following is not normally associated with the spread of the hepatitis B virus?
   a. saliva transfer
   b. blood-to-blood transfer
   c. plasma-to-mouth transfer
   d. unsterile tattooing techniques

   ans. a.   diff. 7   obj. 10   p. 56

87. Hepatitis B can be spread from mother to offspring
   a. via the fecal-oral route
   b. across the placenta
   c. at the time of delivery
   d. by kissing

   ans. c.   diff. 8   obj. 10   p. 57

88. The current hepatitis B vaccine is produced by genetic engineering using
   a. *Saccharomyces cerevisiae*
   b. *Escherichia coli*
   c. *Candida albicans*
   d. herpes zoster

   ans. a.   diff. 8   obj. 10   p. 57

89. The major reservoir of the hepatitis B virus is the
   a. domestic dog
   b. human
   c. monkey
   d. chicken

   ans. b.   diff. 8   obj. 10   p. 57

90. Which of the following efforts has not been shown to be effective in limiting the transfer of hepatitis B virus?
   a. treating all human blood as though it were infectious
   b. antibiotic therapy for cure
   c. using gloves and protective clothing
   d. using condoms

   ans. b.   diff. 7   obj. 10   p. 58

91. The epidemiology of HBV is identical to that of
   a. malaria
   b. hepatitis A
   c. syphilis
   d. AIDS

   ans. d.   diff. 6   obj. 10   p. 58

92. Another name for the "transfusion-associated virus" is
   a. hepatitis C virus
   b. hepatitis B virus
   c. hepatitis D virus
   d. hepatitis A virus

   ans. a.   diff. 7   obj. 10   p. 58

93. Delta hepatitis virus is a defective RNA virus that depends on hepatitis ___ virus for replication.
   a. C
   b. A
   c. B
   d. D

   ans. c.   diff. 7   obj. 10   p. 59

94. Hepatitis D viral infections show a high frequency in
   a. Italy
   b. United States
   c. Canada
   d. Asia

   ans. a.   diff. 8   obj. 10   p. 59

95. Hepatitis E virus is generally transmitted by the
   a. blood
   b. fecal-oral route
   c. semen
   d. rupture of the placenta during the birth process

   ans. b.   diff. 7   obj. 10   p. 60

96. Viral gastroenteritis in children is most commonly caused by
   a. rotavirus
   b. herpes zoster
   c. hepatitis virus A
   d. rabies virus

   ans. a.   diff. 7   obj. 10   p. 61

97. About fifty percent of the viral gastroenteritis outbreaks in the United States are due to the
    a. Norwalk virus
    b. rotavirus
    c. herpes virus
    d. hepatitis virus A

    ans. a.   diff. 7   obj. 10   p. 62

98. The most commonly identified water-borne illness in the United States is
    a. cryptosporosis
    b. amebiasis
    c. giardiasis
    d. balantidiasis

    ans. c.   diff. 7   obj. 11   p. 63

99. The life stage of *Giardia lamblia* responsible for infection is the
    a. ciliated
    b. spore
    c. trophozoite
    d. cyst

    ans. d.   diff. 6   obj. 11   p. 65

100. Which of the following is not a suspected source of *G. lamblia*?
    a. cats
    b. humans
    c. beavers
    d. poultry

    ans. d.   diff. 7   obj. 11   p. 66

101. Which of the following means has not been associated with transfer of *Giardia lamblia*?
    a. person-to-person
    b. fecal-oral route
    c. male homosexuals
    d. breast milk

    ans. d.   diff. 6   obj. 11   p. 66

102. Although it is usually a mild disease, about 30,000 deaths per year in Mexico, South America, Asia, and Africa are caused by
    a. *Balantidium coli*
    b. *G. lamblia*
    c. *E. histolytica*
    d. *Trichomonas vaginalis*

    ans. c.   diff. 7   obj. 11   p. 69

103. Which of the following organisms has a characteristic structure called a chromatoidal body in its immature cyst?
    a. *E. histolytica*
    b. *G. lamblia*
    c. *Balantidium coli*
    d. *Trichomonas vaginalis*

    ans. a.   diff. 8   obj. 11   p. 69

104. The infectious form of *E. histolytica* is the
    a. ciliated stage
    b. immature cyst
    c. trophozoite
    d. mature cyst

    ans. d.   diff. 6   obj. 11   p. 70

True/False

1. The pancreas is involved in the production of digestive enzymes and insulin for breakdown of protein and glucose.
   false   (insulin converts glucose to glycogen)

2. The small intestine is necessary to begin the digestion of carbohydrates, fats, and proteins.
   false   (it completes their breakdown)

3. The major function of the large intestine is to recycle water.
   true

4. The sterile environment of the large intestine is maintained by its acidity.
   false   (environment is inhabited by large amounts of normal flora)

5. Microorganisms make up 90% of the weight of feces in the large intestines.
   false   (40% of the weight)

6. The most prevalent anaerobes of the large intestine are of the *Bacteroides* genera.
   true

312

7. The major cause of death worldwide is diarrhea.
   true

8. The disease that produces "rice water stools" is cholera.
   true

9. Species of *Salmonella* are common causes of dysentery because they invade the colon epithelium.
   false (*Shigella*)

10. Dysentery is a more severe form of water and blood loss than diarrhea.
    true

11. The only reservoir of *Salmonella* is the human.
    false (poultry and their products are frequent reservoirs)

12. The virulence factors of *Shigella* sp. often depend on plasmids.
    false (*Escherichia coli*)

13. A common cause of diarrhea in the United States is *Vibrio cholerae.*
    false (*Campylobacter jejuni*)

14. Typhoid fever is caused by *Salmonella paratyphi.*
    false (*Salmonella typhi*)

15. Typhoid fever, although characterized by high fever, headache, and abdominal pain, will rarely cause death. Complete recovery will follow even if untreated.
    false (high mortality if untreated)

16. Carriers of typhoid fever are humans.
    true

17. Staphylococcal food poisoning is produced by an accumulation of the toxin in high-carbohydrate food.
    true

18. The toxin of *S. aureus* is readily denatured by heat.
    false (is heat stable)

19. Botulism is a deadly form of food intoxication.
    true

20. Botulism can only be acquired as a food-borne illness.
    false (may occur in three forms: food-borne, intestinal, and wound)

21. Aflatoxins are fungal poisonings produced by molds of the genus *Saccharomyces.*
    false (*Aspergillus*)

22. Hepatitis A virus is spread by blood products from contaminated needles.
    false (spread by fecal-oral route)

23. Hepatitis B virus is spread only by blood and blood products.
    false (also by sexual intercourse and from mother to infant)

24. Viral gastroenteritis is caused by the hepatitis viruses.
    false (caused by rotaviruses and Norwalk viruses)

25. Norwalk viruses cause about one-half of the United States gastroenteritis outbreaks.
    true

Chapter 24

Multiple Choice

1. The disease that was first named the "French pox" is
   a. gonorrhea
   b. syphilis
   c. vaginitis
   d. herpes genitalia

   ans. b.   diff. 5   obj. 1   p. 2

2. The causative agent of syphilis was originally named *Spirochaeta pallida* by
   a. Boccacio
   b. Fracastoro
   c. Schaudinn
   d. Churchill

   ans. c.   diff. 6   obj. 1   p. 3

3. Which of the following organs is not considered part of the urinary tract?
   a. adrenals
   b. kidneys
   c. bladder
   d. urethra

   ans. a.   diff. 5   obj. 2   p. 4

4. The normal pH range of urine is
   a. 7.0 to 8.5
   b. 4.8 to 7.5
   c. 7.0 to 9.0
   d. 4.0 to 6.5

   ans. b.   diff. 5   obj. 2   p. 4

5. A consistently alkaline urine suggests infection with urease-producing bacteria that convert urea in urine to ammonia. Which of the following organisms is capable of this?
   a. *Proteus vulgaris*
   b. *Staphylococcus aureus*
   c. *Escherichia coli*
   d. *Enterobacter aerogenes*

   ans. a.   diff. 7   obj. 2   p. 4

6. The kidneys empty the produced urine into the
   a. urethra
   b. bladder
   c. ureters
   d. vagina

   ans. c.   diff. 4   obj. 2   p. 4

7. Which of the following is not a protective mechanism of the urinary tract during an infectious condition?
   a. inflammatory response results in phagocytosis of invading microorganisms
   b. presence of antibodies in urine
   c. antibody-forming lymphoid cells infiltrate kidneys to form antibodies locally
   d. increased secretion of sodium and hydrogen ions

   ans. d.   diff. 7   obj. 2   p. 5

8. The vagina is subject to colonization by potential pathogens from the
   a. urethra
   b. anus
   c. cervix
   d. penis

   ans. b.   diff. 6   obj. 2   p. 6

9. The normal flora of the urinary tract (above the bladder) is made up of
   a. *E. coli*
   b. no microorganisms
   c. *S. aureus*
   d. *P. vulgaris*

   ans. b.   diff. 6   obj. 2   p. 6

10. Which of the following is not normally among the normal flora of the lower urethra?
    a. *Escherichia*
    b. *Corynebacterium*
    c. *Lactobacillus*
    d. *Bacteroides*

    ans. a.   diff. 4   obj. 2   p. 6

11. The pH of postpubescent females is maintained in an acid state due to the presence of

a, estrogens
b. *Streptococcus*
c. *Neisseria*
d. progesterone

ans. a.   diff. 6   obj. 2   p. 7

12. Which group of people are in the greatest danger of urinary tract infections?

a. adolescent females
b. pregnant women
c. infants
d. paraplegics

ans. d.   diff. 6   obj. 2   p. 7

13. Which of the following treatments is responsible for the greatest number of bladder infections?

a. surgery
b. poor nutrition due to long hospital stay
c. catheterization
d. intravenous drug therapy

ans. c.   diff. 6   obj. 3   p. 8

14. Most kidney infections develop after

a. childbirth
b. sexual intercourse for the first time
c. a bladder infection
d. catheterization

ans. c.   diff. 7   obj. 3   p. 8

15. Bladder infections usually originate from

a. the environment
b. normal intestinal flora
c. the overuse of antibiotics
d. contaminated tampons

ans. b.   diff. 6   obj. 4   p. 9

16. The most common infectious agent causing urinary tract infections of young, otherwise healthy women is

a. *Escherichia coli*
b. *Proteus vulgaris*
c. *Enterobacter aerogenes*
d. *Staphylococcus aureus*

ans. a.   diff. 6   obj. 4   p. 9

17. Concentrations of bacteria greater than _____ organisms per milliliter indicate an infection of the urinary tract.

a. 5,000,000
b. 10,000
c. 1,000
d. 100,000

ans. d.   diff. 6   obj. 4   p. 10

18. Which of the following is not normally considered a means of increasing the spread of sexually transmitted diseases?

a. male or female prostitution
b. condoms
c. use of oral contraceptives
d. decrease in funding for tracking contacts of diagnosed cases

ans. b.   diff. 7   obj. 4   p. 12

19. Which of the following methods has not been suggested as effective in preventing a sexually transmitted disease?

a. proper use of latex condoms
b. proper use of an IUD
c. monogamy with a noninfected person
d. abstinence

ans. b.   diff. 6   obj. 4   p. 13

20. The sexually transmitted disease that leads the list of reportable bacterial diseases in the United States is

a. herpes
b. syphilis
c. gonorrhea
d. AIDS

ans. c.   diff. 6   obj. 5   p. 14

21. The incubation period of gonorrhea is approximately

a. 2–5 days
b. 1–2 weeks
c. 14–21 days
d. several months

ans. a.   diff. 6   obj. 5   p. 15

22. If gonorrhea goes untreated in males, all of the following complications can result except which one?
    a. urinary tract infections
    b. prostate cancer
    c. prostatitis
    d. sterility

    ans. b.   diff. 7   obj. 5   p. 15

23. A frequent complication of untreated gonorrhea in women is
    a. pelvic inflammatory disease
    b. syphilis
    c. dysuria
    d. vaginal discharge

    ans. a.   diff. 6   obj. 5   p. 15

24. Scar tissue can form in the fallopian tubes as a result of untreated gonorrhea, leading to the dangerous complication of
    a. arthritis
    b. death
    c. ectopic pregnancy
    d. cervicitis

    ans. c.   diff. 7   obj. 5   p. 16

25. Disseminated gonococcal infections can be characterized by all of the following symptoms except which one?
    a. infection of heart valves
    b. rash
    c. toxemia
    d. urogenital infection

    ans. d.   diff. 8   obj. 5   p. 16

26. The causative agent of gonorrhea is described as a
    a. Gram-negative diplococcus that shows flattening of adjacent sides
    b. flagellated spirochete that is difficult to stain but easily visible when using dark-field microscopy
    c. Gram-positive cocci in clusters that produces several agressins and a heat-stable toxin
    d. small bacterium that lacks a cell wall but shows a variety of intracellular stages

    ans. a.   diff. 8   obj. 5   p. 16

27. A vaccine for gonorrhea has not been developed due to the ability of the organism to
    a. traverse the uterus to reach the fallopian tube even though it is nonmotile
    b. attach to several different types of cells
    c. interfere with phagocytosis to macrophages
    d. express different surface antigens

    ans. d.   diff. 7   obj. 5   p. 18

28. Which of the following antibiotics is currently effective in treating gonorrhea?
    a. penicillin
    b. spectinomycin
    c. ceftrixone
    d. tetracycline

    ans. c.   diff. 7   obj. 5   p. 19

29. The increased resistance of *N. gonorrhoeae* to penicillin is due to
    a. chromosomal mutations that have altered penicillin binding sites on the cell wall
    b. the development of R factors in the organism
    c. strains appearing with plasmid-mediated resistance
    d. the development of an effective vaccine for historical gonorrhea

    ans. a.   diff. 7   obj. 5   p. 19

30. A contributing factor of the virulence of *N. gonorrhoeae* is its ability to destroy the protective nature of the _____ antibody found on the mucosal surfaces.
    a. IgE
    b. IgA
    c. IgD
    d. IgM

    ans. b.   diff. 5   obj. 5   p. 20

31. It is believed that the penicillinase-producing *N. gonorrhoeae* introduced into the United States was brought here from
    a. Europe
    b. outer space
    c. southeast Asia
    d. South Africa

    ans. c.   diff. 7   obj. 5   p. 21

316

32. Which of the following reportable bacterial diseases shows the highest incidence rate?
   a. gonorrhea
   b. syphilis
   c. AIDS
   d. PID

   ans. a.   diff. 6   obj. 5   p. 22

33. How many cases of gonorrhea have been reported in the U.S. from 1976 to 1980?
   a. over 100,000
   b. over 1,000,000
   c. less than 5,000
   d. about 200,000

   ans. b.   diff. 6   obj. 5   p. 22

34. Which of the following conditions of women taking oral contraceptives are not significant in the formation of serious complications of gonorrhea?
   a. migration of gonorrhea-susceptible cells to the outer cervix
   b. decreased pH of the vagina
   c. increased pH of the vagina
   d. increased moisture of the vagina

   ans. b.   diff. 6   obj. 5   p. 22

35. Following an attack of gonorrhea, immunity is
   a. permanent
   b. lacking
   c. temporary
   d. slow to develop

   ans. b.   diff. 4   obj. 5   p. 22

36. Which of the following conditions is probably responsible for the decreased incidence of gonorrhea since 1988?
   a. use of condoms
   b. use of the pill
   c. celibacy
   d. use of antibiotics

   ans. a.   diff. 6   obj. 5   p. 23

37. A serious, nonsexually transmitted gonococcal disease is
   a. ophthalmia neonatorum
   b. secondary syphilis
   c. NGU
   d. PID

   ans. a.   diff. 6   obj. 5   p. 23

38. The treatment of neonates with drops of erythromycin directly into the eyes is to prevent the development of
   a. herpes
   b. conjunctivitis
   c. cataracts
   d. ophthalmia neonatorum

   ans. d.   diff. 7   obj. 5   p. 23

39. The incubation period for gonorrhea is
   a. several months
   b. 7 to 14 days
   c. 2 to 3 weeks
   d. 3 to 5 days

   ans. d.   diff. 6   obj. 5   p. 24

40. Which of the following sexually transmitted organisms mimics the infection caused by *Neisseria gonorrhoeae*?
   a. *Treponema pallidum*
   b. *Mycoplasma pneumoniae*
   c. *Chlamydia trachomatis*
   d. *Escherichia coli*

   ans. c.   diff. 6   obj. 5   p. 25

41. The disease that results from *C. trachomatis* attaching to sperm and invading the female reproductive organs is
   a. pelvic inflammatory disease
   b. conjunctivitis
   c. nongonococcal urethritis
   d. trachoma

   ans. a.   diff. 6   obj. 5   p. 25

42. The incubation period of *C. trachomatis* is
   a. 3 to 5 days
   b. 7 to 14 days
   c. 2 to 3 weeks
   d. 1 month

   ans. b.   diff. 7   obj. 5   p. 25

43. The current method of diagnosing *Chlamydia trachomatis* infections is by using
   a. complement fixation
   b. fluorescent antibodies
   c. the precipitation test
   d. monoclonal antibodies

   ans. d.   diff. 7   obj. 5   p. 26

44. Infections caused by *C. trachomatis* can usually be treated with
   a. tetracycline
   b. penicillin
   c. cephalothin
   d. flagyl

   ans. a.   diff. 8   obj. 5   p. 27

45. The causative agent of syphilis is
   a. *Treponema pallidum*
   b. *Neisseria gonorrhoeae*
   c. *Chlamydia trachomatis*
   d. papillomavirus

   ans. a.   diff. 4   obj. 5   p. 27

46. The current high incidence of syphilis has been linked to many factors except which one of the following?
   a. inner city poverty
   b. use of birth control pills
   c. drug use
   d. prostitution

   ans. b.   diff. 4   obj. 5   p. 27

47. About three weeks after infection with _____ a painless red ulcer with a hard rim called a _____ appears at the site of infection.
   a. Herpes simplex, vesicle
   b. *Neisseria gonorrhoeae,* gumma
   c. *Chlamydia trachomatis,* boil
   d. *Treponema pallidum,* chancre

   ans. d.   diff. 8   obj. 5   p. 27

48. Which of the following diseases manifests itself in three clinical stages?
   a. pelvic inflammatory disease
   b. nongonococcal urethritis
   c. syphilis
   d. gonorrhea

   ans. c.   diff. 6   obj. 5   p. 27

49. The presence of the hard chancre in primary syphilis is representative of the body's
   a. intense inflammatory response
   b. antibody response
   c. rejection of invaded tissue
   d. septicemia due to infiltration of the organism

   ans. a.   diff. 7   obj. 5   p. 28

50. The reactions typical of secondary syphilis are due to
   a. reinfection with the disease
   b. delayed hypersensitivity
   c. circulatory antibodies
   d. immediate hypersensitivity

   ans. c.   diff. 6   obj. 5   p. 28

51. The secondary stage of syphilis has been known to last as long as
   a. 3 weeks
   b. 2 years
   c. 1 year
   d. a lifetime

   ans. c.   diff. 7   obj. 5   p. 28

52. During which of the following stages of syphilis is the patient non-infectious?
   a. second
   b. third
   c. first
   d. none, the patient remains infectious until cured

   ans. b.   diff. 7   obj. 5   p. 29

53. The characteristic pattern of symptoms in tertiary syphilis is referred to as
   a. general paresis
   b. blindness
   c. stroke
   d. personality change

   ans. a.   diff. 7   obj. 5   p. 29

54. Which stage of syphilis is characterized by rash, aches, pain, and mucous membrane lesions?
   a. initial
   b. primary
   c. tertiary
   d. secondary

   ans. d.   diff. 5   obj. 5   p. 30

55. Which stage of syphilis is characterized by damage to blood vessels, eyes, and nervous system?
   a. secondary
   b. tertiary
   c. primary
   d. initial

   ans. b.   diff. 5   obj. 5   p. 30

56. *T. pallidum* can cross the placenta and possibly infect the fetus during which stage of pregnancy?
   a. any stage
   b. first trimester
   c. second trimester
   d. third trimester

   ans. a.   diff. 6   obj. 5   p. 30

57. During which period of the pregnancy can the mother's syphilis be diagnosed and treated to prevent the disease from developing in the unborn child?
   a. after the fifth month
   b. until birth of the child
   c. before the fourth month
   d. only during the fifth month

   ans. c.   diff. 7   obj. 5   p. 30

58. The greatest risk to the unborn child occurs if the mother is in the _____ stage(s) of syphilis.
   a. primary, secondary, or tertiary
   b. secondary or early tertiary
   c. primary or early secondary
   d. tertiary

   ans. c.   diff. 8   obj. 5   p. 30

59. The causative agent of syphilis is described morphologically as a
   a. spirochete
   b. bacillus
   c. virus
   d. cocco-bacillus

   ans. a.   diff. 5   obj. 5   p. 31

60. The *T. pallidum* organism of syphilis can be transmitted by
   a. sexual or oral contact
   b. contact with contaminated objects
   c. clothing
   d. the fecal-oral route

   ans. a.   diff. 4   obj. 5   p. 31

61. The reservoir of *T. pallidum* typically is the
   a. guinea pig
   b. rabbit
   c. monkey
   d. human

   ans. d.   diff. 6   obj. 5   p. 32

62. The highly specific diagnostic test for syphilis, which is based on serum antibodies reacting with antigens specific to *T. pallidum*, is the
   a. non-treponemal test
   b. IgM capture method
   c. immunoblot technique
   d. indirect immunofluorescence test

   ans. c.   diff. 8   obj. 5   p. 36

63. The test that is especially helpful in diagnosing congenital syphilis is the
   a. non-treponemal test
   b. IgM capture method
   c. indirect immunofluorescence test
   d. immunoblot technique

   ans. b.   diff. 8   obj. 5   p. 36

64. The incubation period of syphilis is
   a. 1 month
   b. 2 weeks
   c. 10–90 days
   d. 1 year

   ans. c.   diff. 6   obj. 5   p. 38

65. The bacterial sexually transmitted disease that is characterized by painful genital ulcers is
    a. chancroid
    b. syphilis
    c. gonorrhea
    d. yaws

    ans. a.   diff. 7   obj. 5   p. 38

66. The causative agent of the bacterial sexually transmitted disease characterized by painful genital ulcers is
    a. *Chlamydia trachomatis*
    b. *Treponema pallidum*
    c. *Neisseria gonorrhoeae*
    d. *Haemophilus ducreyi*

    ans. d.   diff. 7   obj. 5   p. 39

67. Which of the following statements concerning the sexually transmitted diseases caused by viruses is not true?
    a. HIV infections probably cause AIDS
    b. genital herpes can give recurrent symptoms for years
    c. papillomavirus infections can lead to cancer
    d. STDs caused by viruses are far less common than those caused by bacteria

    ans. d.   diff. 7   obj. 6   p. 40

68. Which of the following sexually transmitted diseases caused by viruses is characterized by painful genital blisters and ulcers?
    a. yaws
    b. AIDS
    c. genital warts
    d. genital herpes simplex

    ans. d.   diff. 6   obj. 6   p. 40

69. The usual cause of genital herpes is
    a. herpes simplex virus types 2 and 3
    b. herpes simplex virus type 1
    c. herpes simplex virus type 3
    d. herpes simplex virus type 2

    ans. d.   diff. 5   obj. 6   p. 41

70. Which of the following statements about genital herpes is true?
    a. herpes simplex virus type 1 can infect only mucous membranes above the waist
    b. the virus can be transmitted at any time, even if symptoms are absent
    c. the virus can be treated successfully and recurrent infections can be prevented
    d. genital herpes rarely is seen in patients with gonorrhea

    ans. b.   diff. 7   obj. 6   p. 41

71. Symptoms of genital herpes usually occur about _____ after exposure.
    a. 6 days
    b. 3 weeks
    c. several hours
    d. 1 month

    ans. a.   diff. 7   obj. 6   p. 41

72. The average number of recurrences of genital herpes within a year is about
    a. six
    b. one
    c. four
    d. twelve

    ans. c.   diff. 7   obj. 6   p. 42

73. The drug that is used to decrease the severity of the first attack and the incidence of recurrences of genital herpes is
    a. acyclovir
    b. penicillin
    c. erythromycin
    d. AZT

    ans. a.   diff. 6   obj. 6   p. 42

74. If a primary infection with genital herpes occurs near the time of delivery, the baby may
    a. be late in being delivered
    b. be born prematurely
    c. acquire the infection and die of it
    d. be injured due to placental damage

    ans. c.   diff. 7   obj. 6   p. 43

75. The most common of the sexually transmitted viral disease agents is the
    a. herpes virus
    b. human papillomavirus
    c. human immunodeficiency virus
    d. *Gordnerella vaginalis*

    ans. b.   diff. 6   obj. 6   p. 44

76. Dysplasias of the cervix can be detected with which of the following tests?
    a. Papanicolaou test
    b. ELISA test
    c. fluorescent antibody test
    d. precipitation test

    ans. a.   diff. 6   obj. 6   p. 45

77. Which of the following populations is not considered at high risk of contracting AIDS?
    a. monogamous males
    b. intravenous drug users
    c. sexually promiscuous bisexuals
    d. male homosexuals

    ans. a.   diff. 5   obj. 6   p. 46

78. Most infected children contract HIV from their mothers
    a. early in pregnancy via the placenta
    b. while nursing
    c. around the time of birth
    d. during the second trimester

    ans. c.   diff. 7   obj. 6   p. 46

79. Most of the college students infected with HIV are
    a. celibate males
    b. homosexual females
    c. homosexual males
    d. twenty years of age or older

    ans. c.   diff. 5   obj. 6   p. 48

80. Which of the following genital tract infections is caused by a protozoan?
    a. candidiasis
    b. trichomoniasis
    c. puerperal sepsis
    d. toxic shock syndrome

    ans. b.   diff. 6   obj. 7   p. 49

81. The genital tract infection often associated with menstruation is
    a. candidiasis
    b. trichomoniasis
    c. toxic shock syndrome
    d. puerperal sepsis

    ans. c.   diff. 6   obj. 7   p. 49

82. Pregnant women with bacterial vaginosis are at risk of
    a. being unable to nurse
    b. having a premature baby
    c. being unable to deliver vaginally
    d. placental insufficiency

    ans. b.   diff. 6   obj. 7   p. 49

83. The genital tract infection characterized by an unpleasant odor and the presence of clue cells is
    a. candidiasis
    b. trichomoniasis
    c. bacterial vaginosis
    d. amebiasis

    ans. c.   diff. 7   obj. 7   p. 50

84. Although the cause of bacterial vaginosis is unknown, which of the following organisms has been shown to be consistently present in large numbers?
    a. *Gardnerella vaginalis*
    b. lactobacilli
    c. staphylococci
    d. streptococci

    ans. a.   diff. 6   obj. 7   p. 51

85. Bacterial vaginosis has not been connected epidemiologically to which of the following?
    a. use of vaginal tampons
    b. sexual abuse of children
    c. promiscuity
    d. sexually active women

    ans. a.   diff. 7   obj. 7   p. 51

86. Which of the following antimicrobial drugs has been effective in treating bacterial vaginosis?
    a.   erythromycin
    b.   penicillin
    c.   flagyl
    d.   metronidazole

    ans. d.   diff. 7   obj. 7   p. 52

87. The genital tract infection of women that is characterized by itching and a whitish, curdlike vaginal discharge is
    a.   trichomoniasis
    b.   vulvovaginal candidiasis
    c.   bacterial vaginosis
    d.   gonorrhea

    ans. b.   diff. 7   obj. 7   p. 52

88. One of the major causes of vulvovaginal candidiasis is
    a.   intensive treatment with antibiotics
    b.   sexual promiscuity
    c.   homosexuality
    d.   continuous tampon use during menstruation

    ans. a.   diff. 7   obj. 7   p. 53

89. The vaginitis that produces itching of the vulva and inner thighs among its symptoms and is asymptomatic in males is
    a.   bacterial vaginosis
    b.   candidiasis
    c.   gonorrhea
    d.   trichomoniasis

    ans. d.   diff. 6   obj. 7   p. 54

90. The vaginal infection that is protozoan in origin with no reservoir other than the human is caused by
    a.   *Giardia lamblia*
    b.   *Candida albicans*
    c.   *Trichomonas vaginalis*
    d.   *Escherichia coli*

    ans. c.   diff. 6   obj. 7   p. 56

91. The major cause of puerperal fever is
    a.   *Streptococcus pyogenes*
    b.   *Staphylococcus aureus*
    c.   *Pseudomonas aeruginosa*
    d.   *Clostridium perfringens*

    ans. a.   diff. 7   obj. 7   p. 57

92. The organism responsible for many fatalities following abortions induced under unclean conditions is
    a.   *Clostridium perfringens*
    b.   *Streptococcus pyogenes*
    c.   *Escherichia coli*
    d.   *Giardia lamblia*

    ans. a.   diff. 7   obj. 7   p. 57

93. *Staphylococcus aureus* is the causative agent of
    a.   puerperal fever
    b.   toxic shock syndrome
    c.   bacterial vaginosis
    d.   gas gangrene after abortions

    ans. b.   diff. 6   obj. 7   p. 58

94. Toxic shock syndrome caused by staphylococcal exotoxins is analogous to another staphylococcal disease referred to as
    a.   scalded skin syndrome
    b.   puerperal fever
    c.   pneumonia
    d.   glomerulonephritis

    ans. a.   diff. 6   obj. 7   p. 58

95. The reduction of which of the following ions appears to be involved in the development of toxic shock syndrome?
    a.   sodium
    b.   calcium
    c.   magnesium
    d.   chlorine

    ans. c.   diff. 8   obj. 7   p. 58

True/False

1. One of the key defense mechanisms against bladder infections is the ciliated epithelium of the urethra.
   false   (flushing action of urination)

2. The pH of the adult female vagina is maintained at a slightly acidic pH by the presence of estrogen.
   true

3. Syphilis was brought from the New World to Europe by Columbus's men.
   false   (strong evidence that syphilis existed in the Old World long before Columbus returned from New World)

4. "French pox" is the slang name for gonorrhea.
false (syphilis)

5. Kidneys empty their newly formed urine into the urinary bladder through the urethra.
false (through the ureter)

6. The female urinary system is more susceptible to urinary tract infections than is the male system due to the short length of the urethra.
true

7. Most kidney infections are easily treated with sulfa drugs and antibiotics.
false (very difficult to eradicate)

8. The normal flora population of the female vagina is maintained by its hormones.
true

9. Paraplegics are the population with the greatest susceptibility to urinary tract infections.
true

10. Gonorrhea in women can lead to sterility by scar tissue blocking the fallopian tubes.
true

11. Gonorrhea can be transmitted *in utero* by the organism crossing the placenta.
false (syphilis)

12. The only reservoir for *Neisseria gonorrhoeae* is the human.
true

13. The first case of gonorrhea due to a penicillinase-producing gonococcus was reported in 1986.
false (March 1976)

14. The sexually transmitted disease that can affect the eyes of the newborn is ophthalmia neonatorum.
true

15. Monoclonal antibodies can be used to detect *C. trachomatis* from pus or other clinical specimens.
true

16. The sexually transmitted disease that can occur in three clinical stages is syphilis.
true

17. General paresis of syphilis occurs in the second stage of the disease.
false (third)

18. Syphilis is contagious only during stages one and two of the disease.
true

19. The ulcer typically associated with syphilis is usually described as painful and oozing.
false (painless with a hard rim)

20. The characteristic pattern of general paresis is typical of tertiary syphilis.
true

21. The permanent cure for genital herpes is treatment with acyclovir.
false (no cure, only treatment)

22. Chlamydia infections are very similar to those of syphilis.
false (gonorrhea)

23. The most common viral genital disease is venereal warts.
false (genital herpes simplex)

24. The genital disease frequently associated with cervical cancer is genital warts.
true

25. The STD that is currently pandemic and almost universally fatal is HIV infection.
true

26. Vulvovaginitis is most commonly caused by *Gardnerella vaginalis*.
false (*Candida albicans*)

27. The protozoan *Trichomonas vaginatis* is usually asymptomatic in males.
true

28. The common cause of puerperal fever is *Staphylococcus aureus*.
false (*Streptococcus pyogenes*)

29. Toxic shock syndrome is caused by absorption of the endotoxin of *Staphylococcus aureus*.
false (exotoxin)

30. Dysplasia of the cervix is easily detected with the
     Papanicolaou test.
     true

Matching

Match each of the following organisms with the disease it causes.

1. *Treponema pallidum*   a. syphilis
2. *Candida albicans*   b. gonorrhea
3. *Haemophilus ducreyi*   c. chancroid
4. *Neisseria gonorrhoeae*   d. vulvovaginitis

1 - a; 2 - d; 3 - c; 4 - b

Chapter 25

Multiple Choice

1. The first polio epidemic of modern times was reported in
   _____ in ____.
   a. Egypt, 596 B.C.
   b. Sweden, 1887
   c. America, 1776
   d. China, 1826

   ans. b.   diff. 6   obj. 1   p. 1

2. The common name for the National Foundation for
   Infantile Paralysis is the
   a. March of Dimes
   b. Jimmy Fund
   c. Jerry Lewis Foundation
   d. Muscular Dystrophy Fund

   ans. a.   diff. 5   obj. 1   p. 2

3. Which of the following researchers received the Nobel prize
   for their work in polio research?
   a. Salk, Sabin, and Pauling
   b. Roosevelt, Salk, and Enders
   c. Enders, Weller, and Robbins
   d. Salk and Sabin

   ans. c.   diff. 6   obj. 1   p. 2

4. An inactivated polio vaccine was developed in the 1950s by
   a. John Enders
   b. Albert Sabin
   c. Linus Pauling
   d. Jonas Salk

   ans. d.   diff. 7   obj. 1   p. 2

5. An effective attenuated polio vaccine was developed by
   a. Frederick Robbins
   b. Jonas Salk
   c. John Enders
   d. Albert Sabin

   ans. d.   diff. 6   obj. 1   p. 2

6. The results of polio vaccination programs have been so
   effective that the number of reported cases of polio has been
   reduced from 57,879 in 1952 to ___ in 1986.
   a. 3
   b. 72
   c. 910
   d. 485

   ans. a.   diff. 8   obj. 1   p. 2

7. Which of the following groups of organisms does not
   commonly cause central nervous system diseases?
   a. protozoa
   b. worms
   c. fungi
   d. viruses

   ans. b.   diff. 7   obj. 2   p. 3

8. The central nervous system is made up of the
   a. brain and spinal cord
   b. spinal cord and spinal nerves
   c. sympathetic nervous system and parasympathetic
      system
   d. brain and endocrine organs

   ans. a.   diff. 6   obj. 2   p. 3

9. The nerves that carry information into the CNS are referred
   to as
   a. efferent (motor)
   b. afferent (sensory)
   c. efferent (sensory)
   d. afferent (motor)

   ans. b.   diff. 8   obj. 2   p. 3

10. The nerves that transmit information from the CNS to the periphery are known as
    a. afferent (motor)
    b. afferent (sensory)
    c. efferent (sensory)
    d. efferent (motor)

    ans. d.   diff. 8   obj. 2   p. 4

11. Nerves cannot regenerate if
    a. the nerve cell of which they are a part is killed
    b. their distal portions are killed
    c. their proximal fibers are severed
    d. their neurolemma is intact

    ans. a.   diff. 7   obj. 2   p. 4

12. The cavities inside the brain are known as
    a. cavities
    b. sinuses
    c. ventricles
    d. sulci

    ans. c.   diff. 6   obj. 2   p. 4

13. To determine the causative agent of a CNS infection, a sample is taken from the
    a. venous blood
    b. arterial blood
    c. cerebrospinal fluid
    d. urine

    ans. c.   diff. 7   obj. 2   p. 4

14. The membranes that cover the surface of the brain and spinal cord are known as the
    a. neurolemma
    b. dura mater
    c. meninges
    d. collagen

    ans. c.   diff. 6   obj. 2   p. 5

15. An infection that occurs in the area between the arachnoid layer and the pia mater of the brain covering is
    a. meningitis
    b. encephalitis
    c. arachnitis
    d. materitis

    ans. a.   diff. 5   obj. 2   p. 5

16. Which of the following routes is usually not involved in the entrance of infectious agents to the CNS?
    a. nerves
    b. genitals
    c. bone
    d. blood

    ans. b.   diff. 5   obj. 3   p. 5

17. The chief route of CNS infections is by way of the
    a. skin
    b. blood stream
    c. genitourinary system
    d. nerves

    ans. b.   diff. 6   obj. 3   p. 6

18. The blood-brain barrier functions to prevent infectious agents from traveling to the CNS from the blood stream due to
    a. the high concentration of sodium ions
    b. the impenetrability of the capillaries
    c. the lack of connection between capillaries and the CNS
    d. special cells lining the CNS capillaries

    ans. d.   diff. 7   obj. 3   p. 6

19. Which of the following diseases has a causative agent that penetrates the CNS by traveling up the nerve bundles?
    a. polio
    b. viral encephalitis
    c. bacterial meningitis
    d. rabies

    ans. d.   diff. 8   obj. 3   p. 6

20. CNS infections do not normally occur through bone. However, if they do, the more commonly used entrance site is the
    a. compound fracture
    b. middle ear
    c. skull fracture
    d. joint replacement surgery

    ans. b.   diff. 7   obj. 5   p. 7

21. Which of the following is true of the normal flora of the CNS?
    a. there are none; it is a sterile environment
    b. the staphylococci and streptococci are commensals
    c. only viruses are found there
    d. they rarely are responsible for infectious conditions

    ans. a.   diff. 7   obj. 3   p. 7

22. The group most frequently affected by bacterial meningitis is
    a. the elderly
    b. pubescent males between eleven and fourteen
    c. children under five years old
    d. hospitalized persons

    ans. c.   diff. 7   obj. 4   p. 7

23. The leading cause of bacterial meningitis in children is
    a. *Streptococcus pneumoniae*
    b. *Escherichia coli*
    c. *Haemophilus influenzae*
    d. *Neisseria meningitidis*

    ans. c.   diff. 6   obj. 4   p. 8

24. Which of the following organisms is least frequently involved in bacterial meningitis of adults?
    a. *Neisseria meningitidis*
    b. *Escherichia coli*
    c. *Streptococcus pneumoniae*
    d. *Haemophilus influenzae*

    ans. d.   diff. 6   obj. 4   p. 8

25. The age group that is most frequently affected by meningococcal meningitis is
    a. children aged six to eleven months
    b. adults in a hospital setting
    c. nursing home residents
    d. neonates

    ans. a.   diff. 7   obj. 4   p. 8

26. The process of identification of the different strains of *Neisseria meningitidis* is
    a. serotyping
    b. Gram staining
    c. titer measure
    d. fluorescent antibody technique

    ans. a.   diff. 6   obj. 4   p. 9

27. The causative agent of meningococcal meningitis is
    a. *Haemophilus influenzae*
    b. *Neisseria meningitidis*
    c. *Escherichia coli*
    d. *Streptococcus pneumoniae*

    ans. b.   diff. 5   obj. 4   p. 9

28. Acquisition of the infectious agent of meningococcal pneumonia is by
    a. inhaling airborne droplets from an infected person
    b. contact with materials contaminated by the organism
    c. ingestion of food products containing the infectious agent
    d. genital contact with an infected individual

    ans. a.   diff. 6   obj. 4   p. 10

29. The development of shock in meningococcal meningitis is due to the release of _____ by the infecting agent.
    a. exotoxin
    b. endotoxin
    c. agressins
    d. pus

    ans. b.   diff. 7   obj. 4   p. 11

30. The Shwartzman phenomenon is the result of increased sensitivity to repeated exposures to the _____ of the infectious agent of meningococcal meningitis.
    a. agressins
    b. exotoxin
    c. endotoxin
    d. pus

    ans. c.   diff. 7   obj. 4   p. 11

31. The mortality rate of treated cases of meningococcal meningitis is
    a. extremely low (less than 1 percent)
    b. greater than 50 percent
    c. less than 10 percent
    d. extremely high (greater than 74 percent)

    ans. c.   diff. 8   obj. 4   p. 13

32. The incubation period of meningococcal meningitis is
    a. 1–7 days
    b. 7–14 days
    c. 14–21 days
    d. 1 month

    ans. a.   diff. 6   obj. 4   p. 14

33. Most serious cases of bacterial meningitis are caused by *Haemophilus influenzae* type __.
    a. a
    b. c
    c. d
    d. b

    ans. d.   diff. 7   obj. 4   p. 15

34. *Streptococcus pneumoniae* is a leading cause of
    a. viral encephalitis in infants
    b. childhood meningitis
    c. fatal meningitis in the newborn
    d. adult meningitis

    ans. d.   diff. 7   obj. 4   p. 16

35. Many cases of pneumococcal meningitis are preceded by
    a. otitis media
    b. septicemia
    c. meningococcal meningitis
    d. tuberculosis

    ans. a.   diff. 6   obj. 4   p. 16

36. Which of the following organisms is most likely the cause of meningitis in the newborn?
    a. *Neisseria meningitidis*
    b. *Streptococcus pneumoniae*
    c. *Haemophilus influenzae*
    d. *Escherichia coli*

    ans. d.   diff. 6   obj. 4   p. 17

37. The causative agent of leprosy was discovered in ____ by _____.
    a. 1794, Dr. Joseph Lister
    b. 1868, Dr. Armauer Hansen
    c. 1845, Louis Pasteur
    d. 1900, Hans Krebs

    ans. b.   diff. 7   obj. 5   p. 20

38. The only known human pathogen that preferentially attacks the peripheral nerves is
    a. poliomyelitis virus
    b. *M. leprae*
    c. *E. coli*
    d. *N. meningitidis*

    ans. b.   diff. 7   obj. 5   p. 21

39. The limited type of leprosy in which cell-mediated immunity suppresses proliferation of the bacilli is called the _____ type.
    a. tuberculoid
    b. lepromatous
    c. limited
    d. immune

    ans. a.   diff. 7   obj. 5   p. 21

40. The limited growth of *M. leprae* in the tissues is due to the development of
    a. functional antibodies
    b. humoral immunity
    c. cellular immunity
    d. opsonins

    ans. c.   diff. 7   obj. 5   p. 21

41. The most contagious form of leprosy is the _____ type.
    a. peripheral
    b. tuberculoid
    c. lepromatous
    d. limited

    ans. c.   diff. 6   obj. 5   p. 22

42. The incubation period of leprosy is
    a. 1–2 months
    b. often ten years or longer
    c. 7–14 days
    d. 6–9 months

    ans. b.   diff. 5   obj. 5   p. 22

43. The definitive staining technique used to identify *M. leprae* from a lesion is the _____ stain.
    a. negative
    b. Gram
    c. capsule
    d. acid-fast

    ans. d.   diff. 5   obj. 5   p. 22

44. The transmission of leprosy is by
    a. direct human-to-human contact
    b. contact with materials used by the leper
    c. the fecal-oral route
    d. contact with infected blood products

    ans. a.   diff. 6   obj. 5   p. 23

45. Which of the following has not been associated with the transmission of leprosy?
    a. skin abrasions
    b. nasal secretions of a lepromatous case
    c. mucous membranes
    d. the armadillo

    ans. d.   diff. 7   obj. 5   p. 24

46. The symptoms of fever, headache, stiff neck, disorientation, seizures, coma, and impairment of one or more motor or sensory nerves are typical of
    a. meningococcal meningitis
    b. viral encephalitis
    c. leprosy
    d. polio

    ans. b.   diff. 6   obj. 6   p. 27

47. The major cause of epidemic encephalitis in the United States is the
    a. sporadic encephalitis virus
    b. adenovirus
    c. LaCrosse encephalitis virus
    d. *Aedes* mosquito

    ans. c.   diff. 6   obj. 6   p. 27

48. Humans acquire the arbovirus that causes epidemic encephalitis by
    a. the bite of an infected mosquito
    b. the fecal-oral route
    c. mucous membrane contact with an infected person
    d. human-to-human contact

    ans. a.   diff. 7   obj. 6   p. 27

49. Which of the following vectors is responsible for transmission of epidemic encephalitis?
    a. sentinel chickens
    b. *Culex* mosquito
    c. tsetse fly
    d. *Aedes* mosquito

    ans. d.   diff. 8   obj. 6   p. 27

50. The St. Louis encephalitis virus is maintained in nature in birds and the _____.
    a. tsetse fly
    b. *Culex* mosquito
    c. *Aedes* mosquito
    d. sentinel chickens

    ans. b.   diff. 8   obj. 6   p. 27

51. The poliomyelitis virus appears to selectively destroy the _____ nerve cells of the spinal cord.
    a. sensory
    b. motor
    c. mixed
    d. autonomic

    ans. b.   diff. 7   obj. 6   p. 29

52. Which of the following does not describe the poliomyelitis virus?
    a. enterovirus
    b. RNA virus
    c. DNA virus
    d. picornavirus

    ans. c.   diff. 6   obj. 7   p. 30

53. The polio viruses usually enter the body by the _____ route.
    a. parenteral
    b. blood stream
    c. genital
    d. oral

    ans. d.   diff. 6   obj. 7   p. 30

54. The incubation period of poliomyelitis is
    a. 3–4 weeks
    b. 1–6 days
    c. 7–14 days
    d. 1–2 months

    ans. c.   diff. 6   obj. 7   p. 33

55. Poliomyelitis virus is included within the group recognized as the
    a. arboviruses
    b. adenoviruses
    c. enteroviruses
    d. dermotropic viruses

    ans. c.   diff. 5   obj. 7   p. 34

56. Vaccine-related paralytic poliomyelitis is rare but could occur with the _____ dose of the vaccine.
    a. first
    b. second
    c. third
    d. fourth

    ans. a.   diff. 5   obj. 7   p. 34

57. Polio survivors who develop muscle pain, increased weakness, and muscle degeneration years after they suffered from the disease have a condition described as
    a. relapsing polio
    b. post-polio syndrome
    c. recurring polio
    d. secondary polio

    ans. b.   diff. 6   obj. 7   p. 35

58. The symptom at the site of the animal bite that strongly suggests rabies is
    a. festering without fever
    b. foaming pus
    c. inability to heal
    d. tingling or twitching

    ans. d.   diff. 7   obj. 8   p. 36

59. Painful spasms of the throat and respiratory muscles provoked by swallowing or even seeing liquids is
    a. hydrophobia
    b. anaphylaxis
    c. tetany
    d. paralysis

    ans. a.   diff. 7   obj. 8   p. 36

60. The disease that is caused by an enveloped, bullet-shaped RNA virus is
    a. polio
    b. herpes
    c. rabies
    d. encephalitis

    ans. c.   diff. 6   obj. 8   p. 38

61. In rabies, the virus multiplies in the cells at the site of infection. The virus then attaches to receptors in the
    a. region where the nerve joins the muscle
    b. spinal cord
    c. area affected by rabies
    d. abdomen most susceptible

    ans. a.   diff. 7   obj. 8   p. 38

62. The incubation period of rabies is partly determined by the
    a. season of the year that the disease is acquired
    b. length of the journey from the bite site to the brain
    c. age of the patient
    d. physical condition of the infected animal

    ans. b.   diff. 6   obj. 8   p. 38

63. Sites of rabies viral replication in the brain are called
    a. included bodies
    b. Koplik spots
    c. foamy microcytes
    d. Negri bodies

    ans. d.   diff. 5   obj. 8   p. 39

64. Rare cases in which the rabies virus can be identified in the _____ allow diagnosis to be made before the death of the victim.
   a. eye
   b. heart
   c. salivary glands
   d. fatty tissue under the skin

   ans. a.   diff. 7   obj. 8   p. 39

65. Seller's stain is frequently used for diagnosing
   a. rabies
   b. polio
   c. herpes
   d. meningitis

   ans. a.   diff. 7   obj. 8   p. 40

66. Which of the following is not included in the list of animals considered chief reservoirs of rabies?
   a. raccoons
   b. skunks
   c. squirrels
   d. bats

   ans. c.   diff. 6   obj. 8   p. 40

67. Pasteur's rabies vaccine was made from
   a. chick embryos
   b. dried rabbit spinal cords
   c. formalin-treated dog brains
   d. heat-treated dog spinal cords

   ans. b.   diff. 6   obj. 8   p. 42

68. A complication in the recipient from the use of Pasteur's rabies vaccine is referred to as
   a. allergic encephalitis
   b. rabies meningitis
   c. paralytic rabies
   d. chronic rabies

   ans. a.   diff. 8   obj. 8   p. 42

69. The rabies vaccine that has replaced the Pasteur rabies vaccine in the prevention of the disease is the
   a. human diploid cell vaccine
   b. armadillo foot pad vaccine
   c. nonallergic serum
   d. attenuated rabies virus vaccine

   ans. a.   diff. 6   obj. 8   p. 42

70. How many people have recovered from rabies?
   a. one
   b. two
   c. three
   d. none

   ans. c.   diff. 6   obj. 9   p. 43

71. CNS diseases of fungi occur rarely in
   a. children
   b. cancer patients
   c. AIDS victims
   d. transplant recipients

   ans. a.   diff. 7   obj. 9   p. 44

72. Crytococcal meningoencephalitis differs from the general pattern of CNS fungal diseases in that it can occur in
   a. children
   b. normal people
   c. AIDS victims
   d. debilitated persons

   ans. b.   diff. 7   obj. 9   p. 45

73. The causative agent of cryptococcal meningoencephalitis belongs to the group of organisms known as
   a. protozoa
   b. bacteria
   c. yeasts
   d. worms

   ans. c.   diff. 8   obj. 9   p. 46

74. The causative agent of cryptococcal meningoencephalitis is currently known as
   a. *Cryptococcus neoformans*
   b. *Filobasidiella neoformans*
   c. *Balantidium coli*
   d. *Enterobacter meningitis*

   ans. b.   diff. 7   obj. 9   p. 46

75. In the pathology of cryptococcal meningoencephalitis, the infection is established first in the
   a. oral cavity
   b. skin
   c. brain
   d. lung

   ans. d.   diff. 7   obj. 9   p. 46

76. Cryptococcal meningoencephalitis is frequently associated with
    a. medical treatment centers
    b. mushrooms
    c. dairies
    d. pigeon droppings

    ans. d.   diff. 8   obj. 9   p. 48

77. The protozoan CNS disease caused by *Naegleria fowleri* is
    a. meningoencephalitis
    b. meningitis
    c. African sleeping sickness
    d. amebiasis

    ans. a.   diff. 5   obj. 10   p. 50

78. African sleeping sickness is caused by a bite of an infected
    a. *Aedes* mosquito
    b. tsetse fly
    c. bird
    d. *Culex* mosquito

    ans. b.   diff. 6   obj. 10   p. 51

79. Another name for African trypanosomiasis is
    a. sleeping sickness
    b. malaria
    c. ornithosis
    d. candidiasis

    ans. a.   diff. 5   obj. 10   p. 51

80. The causative agent for African sleeping sickness is
    a. *Naegleria fowleri*
    b. *Trypanosoma gambiense*
    c. *Trichomonas vaginalis*
    d. *Trypanasoma brucei*

    ans. d.   diff. 6   obj. 10   p. 51

81. The *rhodesiense* subspecies of *T. brucei* occurs mainly in the cattle raising areas of
    a. West Africa
    b. Central Africa
    c. East Africa
    d. southeast Asia

    ans. c.   diff. 5   obj. 1   p. 51

82. There is more than one form of sleeping sickness, the most severe form is the _____ form.
    a. Rhodesian
    b. Gambian
    c. Kenyan
    d. Southern

    ans. a.   diff. 6   obj. 10   p. 53

83. The main reservoir for the milder form of African sleeping sickness are
    a. humans
    b. tsetse flies
    c. wild animals
    d. *Culex* mosquitoes

    ans. a.   diff. 7   obj. 10   p. 53

84. The group of chronic degenerative brain diseases are known collectively as
    a. brain abscesses
    b. brain inflammations
    c. spongiform encephalopathies
    d. delayed hypersensitivities

    ans. c.   diff. 6   obj.   p. 55

85. The chronic degenerative brain disease that is associated with cannibalism is
    a. scrapie
    b. Creutzfeldt-Jakob disease
    c. kuru
    d. Tay-Sachs disease

    ans. c.   diff. 6   obj.   p. 56

True/False

1. Infectious diseases of the CNS are uncommon because pathogens cannot penetrate the myelin sheath.
   false   (blood-brain barrier)

2. The nervous system is infected primarily from the skin.
   false   (blood stream)

3. The first evidence of poliomyelitis was described by ancient Egyptian inscriptions.
   true

4. The first polio vaccine was an active oral vaccine introduced by Dr. Albert Sabin.
false (inactive vaccine introduced by Dr. Jonas Salk)

5. The use of the polio vaccine reduced the incidence of cases from 5,000 in 1952 to none in 1986.
false (57,879 in 1952 to three in 1986)

6. Infections of the nervous system are common but rarely serious.
false (uncommon and serious)

7. The central nervous system is composed of the brain, spinal cord, and peripheral nerves.
false (brain and spinal cord only)

8. The most common means by which infectious agents enter the CNS is via the middle ear.
true

9. The normal flora of the CNS are similar to those found in the respiratory system.
false (no normal flora, sterile environment)

10. Bacterial meningitis is an infection of the brain coverings.
true

11. Meningococcal meningitis frequently develops into a rapidly fatal condition.
true

12. Among the first symptoms of meningococcal meningitis are those of a mild cold.
true

13. The causative organism of meningococcal meningitis is *Neisseria gonorrhea.*
false (*Neisseria meningitidis*)

14. Humans are the only reservoir for meningococcal meningitis.
true

15. The leading cause of bacterial meningitis is *Streptococcus pneumoniae.*
false (*Haemophilus influenzae*)

16. The major cause of bacterial meningitis in the newborn is *Escherichia coli.*
true

17. The causative agent of leprosy is *Mycobacterium leprae.* It is grown in the lab in tissue culture.
false (has not been cultivated *in vitro*)

18. In the areas of the body affected by leprosy, sensation is always decreased.
false (increased or decreased sensation)

19. The common name for leprosy is Hansen's disease.
true

20. The major reservoir of human leprosy is the armadillo.
false (no animal or environmental reservoir)

21. The leading cause of encephalitis in humans is the epidemic encephalitis virus.
false (LaCrosse encephalitis virus)

22. Encephalitis is spread by bites of an infected tsetse fly.
false (*Aedes* mosquito)

23. The characteristic feature of poliomyelitis is the selective destruction of motor nerve cells.
true

24. The World Health Organization has effectively eradicated polio as an infectious disease.
false (goal is to eliminate the disease by the year 2000)

25. The rabies virus multiplies in the muscle cells and then the toxin enters an axon.
true

26. Post-polio syndrome occurs due to a reinfection of the same virus type as the initial infectious agent.
false (death of nerve cells that had taken over for those killed initially)

27. Cryptococcal meningoencephalitis originates in the intestines after ingestion of the fungi.
false (originates in the lung after inhalation of the fungi)

Chapter 26

Multiple Choice

1. The normal habitat of *Clostridium tetani* is
   a. unpasteurized milk
   b. the human intestine
   c. the lungs
   d. soil and dust

   e. animal intestine

   ans. d.   diff. 5   obj. 1   p. 1

2. The researcher who paved the way for the development of immunization against tetanus was
   a. Von Behring
   b. Koch
   c. Kitasato
   d. Pasteur

   ans. c.   diff. 6   obj. 1   p. 3

3. The discovery that tetanus is caused by the toxin of the organism rather than the organism itself was made by
   a. Kitasato
   b. Koch
   c. Von Behring
   d. Pasteur

   ans. a.   diff. 6   obj. 1   p. 2

4. Wounds that introduce potentially pathogenic organisms into the tissue are often complicated by the presence of
   a. dead tissue
   b. normal flora
   c. antiseptic ointments
   d. a good blood supply

   ans. a.   diff. 5   obj. 1   p. 4

5. A localized collection of pus in a wound is termed a(n)
   a. leukocyte
   b. abscess
   c. mass of dead tissue
   d. colony of normal flora

   ans. b.   diff. 5   obj. 2   p. 4

6. Factor(s) not found in abscesses is (are)
   a. tissue remnants
   b. pus
   c. dead leukocytes
   d. blood vessels

   ans. d.   diff. 6   obj. 2   p. 5

7. An important feature of wounds, which can lead to serious complications, is that they are
   a. well aerated
   b. walled off from the surrounding area
   c. relatively anaerobic
   d. sterile

   ans. c.   diff. 6   obj. 2   p. 5

8. Wounds caused by projectiles are especially dangerous because they
   a. rupture blood vessels and may cause severe bleeding
   b. do serious damage to the surrounding area
   c. disturb the homeostasis of the body
   d. can carry contaminated fragments into the tissue

   ans. d.   diff. 7   obj. 2   p. 5

9. The Gram-negative rod that lives in soil and water, and can infect a minor wound but not cause serious damage unless the patient is immunodeficient or the wound is severely contaminated is
   a. *Staphylococcus aureus*
   b. *Aeromonas hydrophila*
   c. *Sporothrix schenckii*
   d. *Pseudomonas aeruginosa*

   ans. b.   diff. 6   obj. 2   p. 6

10. The most frequent cause of wound infections in healthy people is
    a. *S. aureus*
    b. *P. aeruginosa*
    c. *Pasteurella multocida*
    d. *Rochalimea henselae*

    ans. a.   diff. 6   obj. 2   p. 6

11. Anaerobic bacteria are responsible for about ___ percent of wound infections.
    a. 20
    b. 10
    c. 30
    d. 40

    ans. b.   diff. 7   obj. 2   p. 7

12. Which of the following genera is not among the anaerobic opportunistic pathogens?
    a. *Bacteroides*
    b. *Clostridium*
    c. *Pseudomonas*
    d. *Peptostreptococcus*

    ans. c.   diff. 8   obj. 2   p. 7

13. Which of the following is not significant when considering the factors involved in the development of harmful wound infections?
    a. the skin type of the host
    b. virulence of the invading organisms
    c. the number of invading pathogens
    d. the natural defenses of the host

    ans. a.   diff. 6   obj. 2   p. 8

14. The most common source of the *S. aureus* responsible for wound infections is the
    a. soil in the area
    b. patient's own nose
    c. food eaten by the host
    d. contact with contaminated objects

    ans. b.   diff. 6   obj. 2   p. 8

15. Some strains of *S. aureus* are resistant to penicillin because of the production of _____ by _____.
    a. antimicrobics, the patient
    b. methicillinase, the genome
    c. coagulase, *S. aureus*
    d. penicillinase, plasmids

    ans. d.   diff. 7   obj. 2   p. 9

16. The antibiotic generally used to treat infections that are caused by penicillinase-producing staphylococci is
    a. tetracycline
    b. acyclovir
    c. streptomycin
    d. methicillin

    ans. d.   diff. 6   obj. 2   p. 9

17. Which of the following organisms, although less invasive than *S. aureus*, are more difficult to destroy because of their antibiotic resistance?
    a. species of *Clostridium*
    b. coagulase-positive staphylococci
    c. species of *Pseudomonas*
    d. *Shigella* species

    ans. c.   diff. 8   obj. 2   p. 10

18. Among the newly recognized causes of wound infections are species of
    a. *Legionella*
    b. *Candida*
    c. *Serratia*
    d. *Shigella*

    ans. a.   diff. 8   obj. 2   p. 11

19. Ideal sites of infection by environmental bacteria or normal flora are
    a. in the genitourinary tract
    b. mucous membranes
    c. burned areas of the skin
    d. foods being prepared in the kitchen

    ans. c.   diff. 7   obj. 3   p. 12

20. One of the most common and hard to treat opportunistic pathogens of burns is the Gram-negative rod
    a. *Pseudomonas aeruginosa*
    b. *Enterobacter aerogenes*
    c. *Serratia marscesens*
    d. *Escherichia coli*

    ans. a.   diff. 7   obj. 3   p. 13

21. The Gram-negative opportunistic rod that grows aerobically in a variety of aqueous solutions, distilled water, and even some disinfectants is
    a. *Escherichia coli*
    b. *Enterobacter aerogenes*
    c. *Serratia marscesens*
    d. *Pseudomonas aeruginosa*

    ans. d.   diff. 6   obj. 3   p. 13

22. One successful method of killing *Pseudomonas aeruginosa* is by using
    a. hexachlorophene
    b. phenolic disinfectants
    c. quanternary ammonia
    d. alcohol

    ans. b.   diff. 7   obj. 3   p. 14

23. The disease that has a high mortality rate but is easily prevented by immunization is
    a. tetanus
    b. the common cold
    c. gas gangrene
    d. tuberculosis

    ans. a.   diff. 5   obj. 4   p. 14

24. Although anaerobic, *Clostridium* species can grow on wound surfaces because
    a. the surface is coated with anaerobic cells
    b. the organisms can use the oxygen available under these conditions
    c. the by-products of the wound supply oxygen
    d. pus and dead tissue cells consume available oxygen

    ans. d.   diff. 6   obj. 4   p. 14

25. The popular name for tetanus is
    a. Hansen's disease
    b. hydrophobia
    c. "lockjaw"
    d. consumption

    ans. c.   diff. 4   obj. 4   p. 15

26. The disease that characteristically shows uncontrollable contractions of the body muscles often beginning with the jaw is
    a. tetanus
    b. rabies
    c. polio
    d. toxic shock syndrome

    ans. a.   diff. 5   obj. 4   p. 15

27. The average incubation period for tetanus is
    a. 24 hours
    b. 2 weeks
    c. 8 days
    d. 1 month

    ans. c.   diff. 5   obj. 4   p. 16

28. Tetanus is caused solely by the action of a powerful _____ produced by the vegetative cells of *Clostridium tetani.*
    a. endotoxin
    b. exotoxin
    c. enzyme
    d. plasmid

    ans. b.   diff. 5   obj. 4   p. 16

29. A striking feature of *Clostridium tetani* that contrasts it with other pathogenic species of *Clostridium* is its ability to
    a. divide anaerobically
    b. sporulate
    c. produce flagellae
    d. form a terminal endospore

    ans. d.   diff. 5   obj. 4   p. 17

30. The exotoxin of tetanus is effective in producing paralytic muscle spasms by preventing the release of _____.
    a. acetylcholine, an inhibitory chemical
    b. cholinesterase
    c. seratonin
    d. tyrosine

    ans. a.   diff. 5   obj. 4   p. 18

31. Why is no immunity present after recovery from tetanus?
    a. the organism does not migrate from the wound site
    b. the quantity of toxin is too small to elicit the immune response
    c. the phagocytes are not attracted to the area, and no antigens are released
    d. the toxin is not antigenic

    ans. b.   diff. 8   obj. 4   p. 18

32. The early stages of tetanus are generally treated by administering
    a. tetanus toxoid
    b. tetanus antitoxin
    c. hyperbarometric oxygen
    d. penicillin

    ans. b.   diff. 5   obj. 4   p. 19

33. _____ percent of the tetanus cases in recent years have occurred in people who have never been immunized with the toxoid.
    a. 10
    b. 51
    c. 28
    d. 97

    ans. d.   diff. 7   obj. 4   p. 21

34. Life-threatening gas gangrene is most commonly caused by
    a. *Clostridium perfringens*, type B
    b. *Clostridium perfringens*, type A
    c. *Clostridium tetani*
    d. *Streptococcus pyogenes*

    ans. b.   diff. 5   obj. 4   p. 23

35. The tissue-attacking enzymes produced by *Clostridium tetani* that probably promote the process of infection are
    a. amino acids and glycogen
    b. streptokinase and leukocidins
    c. collagenase and hyaluronidase
    d. hydrogen and carbon dioxide

    ans. c.   diff. 6   obj. 4   p. 24

36. The spread of the infection by *Clostridium tetani* is enhanced by the
    a. accumulated pressure produced by hydrogen gas and carbon dioxide
    b. presence of amino acids and glycogen
    c. invasion of surrounding connective tissue
    d. lymphocytes and macrophages

    ans. a.   diff. 6   obj. 4   p. 24

37. Which of the following situations would not result in the development of gas gangrene by *Clostridium perfringens?*
    a. accidents
    b. battlefield wounds
    c. nonsterile abortions
    d. swimming in a freshwater lake

    ans. d.   diff. 6   obj. 4   p. 26

38. The vaccine for prevention of gas gangrene can be best described as a(n)
    a. there is no vaccine
    b. toxoid
    c. attenuated bacterium
    d. antitoxin

    ans. a.   diff. 5   obj. 5   p. 28

39. Which of the following anaerobes is not responsible for the development of wound infections?
    a. *Fusobacterium*
    b. *Bacteroides*
    c. *Neisseria*
    d. *Peptostreptococcus*

    ans. c.   diff. 6   obj. 5   p. 29

40. The organism that grows anaerobically in dead or damaged tissue and produces dense colonies called sulfur granules is
    a. *Peptostreptococcus*
    b. *Actinomyces israelii*
    c. *Bacteroides*
    d. *Neisseria*

    ans. b.   diff. 7   obj. 5   p. 29

41. The most feared infection transmitted by animal bites is
    a. rabies
    b. tetanus
    c. gas gangrene
    d. cat scratch disease

    ans. a.   diff. 5   obj. 6   p. 30

42. The most frequent infectious agent acquired from the bite wounds of cats, dogs, and humans is
    a. *Peptostreptococcus*
    b. *Actinomyces israelii*
    c. *Pasteurella multocida*
    d. *Bacteroides*

    ans. c.   diff. 6   obj. 6   p. 30

43. Human infections with *P. multocida* usually respond to treatment with soaking to drain pus and the antibiotic

    _____
    a. amphotericin B
    b. tetracycline
    c. penicillin
    d. streptomycin

    ans. c.   diff. 6   obj. 6   p. 31

44. The most common cause of chronic lymph node enlargement at localized body sites in young children is
    a. sporotrichosis
    b. rat bite fever
    c. Haverhill fever
    d. cat scratch disease

    ans. d.   diff. 6   obj. 6   p. 31

45. _____ percent of annual cases of cat scratch fever result in hospitalization of the victim.
    a. 10
    b. 16
    c. 8
    d. 0

    ans. c.   diff. 8   obj. 6   p. 31

46. Most cases of cat scratch disease are caused by a bite that becomes infected with
    a. *Rochalimea henselae*
    b. *Afipia felis*
    c. *P. multocida*
    d. *Peptostreptococcus*

    ans. a.   diff. 8   obj. 6   p. 32

47. The rat bite fever that is characterized by relapsing fever and rash is caused by
    a. *R. henselae*
    b. *Streptobacillus moniliformis*
    c. *Afipia felis*
    d. *Spirillum minus*

    ans. d.   diff. 7   obj. 6   p. 33

48. In Japan, rat bite fever is known as
    a. sodoku
    b. Haverhill fever
    c. cat scratch disease
    d. Kitasato disease

    ans. a.   diff. 6   obj. 6   p. 33

49. The rat bite fever that is characterized by fever, rash, and arthritis is caused by
    a. *Afipia felis*
    b. *Streptobacillus moniliformis*
    c. *Rochalimea henselae*
    d. *P. multocida*

    ans. b.   diff. 7   obj. 6   p. 33

50. When *Streptobacillus moniliformis* infects nonpasteurized milk and causes disease in humans, it is called
    a. Haverhill fever
    b. cat scratch fever
    c. rat bite fever
    d. actinomycosis

    ans. a.   diff. 8   obj. 6   p. 33

51. Human bites can occasionally be responsible for the unusual entry of human pathogens. Which of the following has not been associated with human bites?
   a. Haverhill fever
   b. syphilis
   c. tuberculosis
   d. hepatitis B

   ans. a.   diff. 7   obj. 6   p. 34

52. Although fungal infections of wounds are unusual in the United States, which of the following infections occasionally causes small epidemics?
   a. candidiasis
   b. tineasis
   c. sporotrichosis
   d. trichomoniasis

   ans. c.   diff. 6   obj. 7   p. 35

53. Frequently sporotrichosis is caused when the infectious agent is introduced into wounds by
   a. food preparation
   b. artificial nails
   c. thorns
   d. lotions

   ans. c.   diff. 7   obj. 7   p. 35

54. In the United States, most cases of sporotrichosis occur in the
   a. Indian reservations of Oklahoma
   b. ski areas of Colorado
   c. logging camps of Oregon and Montana
   d. Mississippi and Missouri river valleys

   ans. d.   diff. 8   obj. 7   p. 37

55. A potential source of *Sporothrix schenkii* is
   a. evergreen seedlings
   b. sphagnum moss
   c. cow manure
   d. mushrooms

   ans. b.   diff. 6   obj. 7   p. 37

True/False

1. Wound abscesses are localized pus pockets invaded by numerous capillaries.
   false (lack blood vessels)

2. Surface skin abrasions are likely areas of anaerobic infection.
   false (puncture wounds)

3. Serious infection can result if tissue is crushed and the host is immunodeficient.
   true

4. The most common cause of surgical wound infections is *S. aureus*.
   true

5. The most common cause of burn infections is *S. aureus*.
   false (*Pseudomonas aeruginosa*)

6. Tetanus is caused by an exotoxin of *Clostridium tetani*.
   true

7. The toxin of *Clostridium tetani* is referred to as an endotoxin.
   false (exotoxin)

8. Tetanus can be treated successfully at any stage with its antitoxin.
   false (only works very early in the disease)

9. Tetanus can be prevented by immunization with the toxoid.
   true

10. Gas gangrene can be characterized by the necrosis that occurs in connective tissue.
   false (muscle tissue)

11. Anaerobic infections of wounds are frequently caused by *Clostridium perfringens*, a normal flora of the mouth and intestine.
   false (*Actinomyces israelii*)

12. The organism that commonly complicates animal bite wounds is *Pasteurella multocida*.
   true

13. Cat scratch disease is the most common cause of chronic lymph node enlargement at one body site in children.
   true

14. Cat scratch disease requires aggressive treatment with very specific antimicrobials or fatality can result.
   false (most patients get well in time without treatment)

15. Rat bite fever is common in Japan and rare in the United States.
    true

16. In Japan, rat bite fever is known as Haverhill fever.
    false (sodoku)

17. Human bite wounds are rarely if ever contaminated with human pathogens.
    false (can be infected with causative agents of hepatitis B, syphilis, or tuberculosis)

18. A common cause of human bite wound infections is *Streptococcus pyogenes*.
    false (*S. aureus*)

19. Sporotrichosis is a chronic fungal disease of people whose work exposes them to splinters and thorns.
    true

20. A 1988 epidemic of sporotrichosis is known to have been caused by patients handling spruce seedlings.
    false (infected sphagnum moss)

Chapter 27

Multiple Choice

1. Who was the researcher who influenced Roux to study diphtheria and tuberculosis?
   a. Louis Pasteur
   b. Joseph Marburg
   c. Alexandre Yersin
   d. Robert Koch

   ans. c.   diff. 5   obj. 1   p. 2

2. The scientist responsible for the development of the first anti-plague vaccine in 1896 was
   a. Alexandre Yersin
   b. Joseph Marburg
   c. Louis Pasteur
   d. Robert Koch

   ans. a.   diff. 6   obj. 1   p. 4

3. Which of the following is not included in the list of functions of the circulatory system?
   a. fighting of diseases
   b. transportation of oxygen to the cells
   c. manufacture of blood cells
   d. supply of nutrients to the body

   ans. c.   diff. 5   obj. 2   p. 4

4. The normal flora of the circulatory system are
   a. always commensalistic
   b. absent
   c. opportunistic
   d. always aerobic

   ans. b.   diff. 6   obj. 2   p. 5

5. The sac that surrounds the heart is called the
   a. myocardium
   b. endocardium
   c. atrium
   d. pericardium

   ans. d.   diff. 4   obj. 2   p. 5

6. The heart chamber that discharges blood to the lungs is the
   a. right ventricle
   b. left ventricle
   c. right atrium
   d. left atrium

   ans. a.   diff. 5   obj. 2   p. 6

7. The vessels that return deoxygenated blood to the _____ are the _____.
   a. left ventricle, arteries
   b. left atrium, arteries
   c. right ventricle, veins
   d. right atrium, veins

   ans. d.   diff. 4   obj. 2   p. 6

8. The major systemic artery that carries blood throughout the body is the
   a. subclavian vein
   b. superior vena cava
   c. carotid artery
   d. aorta

   ans. d.   diff. 4   obj. 2   p. 6

9. The function of the fixed phagocyte system is to
   a. distribute oxygen to the cells
   b. remove cellular carbon dioxide from the system
   c. cleanse the blood of foreign material
   d. produce antibodies

   ans. c.   diff. 6   obj. 2   p. 6

10. The _____ fluid bathes and nourishes the tissue cells.
   a. cytoplasm
   b. lymph
   c. blood
   d. interstitial

   ans. d.   diff. 6   obj. 2   p. 7

11. The small bean-shaped bodies into which the lymphatic vessels drain are the
    a. lymph nodes
    b. adrenals
    c. subclavian veins
    d. valves

    ans. a.   diff. 5   obj. 2   p. 7

12. The condition that is described as a red streak spreading up from an infected limb toward the trunk is referred to as
    a. septicemia
    b. lymphangitis
    c. bacteremia
    d. edema

    ans. b.   diff. 6   obj. 2   p. 7

13. The condition that develops as a result of infection in a heart valve previously damaged by rheumatic fever or a birth defect is
    a. subacute bacterial endocarditis
    b. acute bacterial endonyocarditis
    c. bacterial myocarditis
    d. perieorditis

    ans. a.   diff. 6   obj. 3   p. 9

14. The most common cause of infection involving heart valves previously damaged by disease or a congenital defect is(are)
    a. *Staphylococcus aureus*
    b. *Streptococcus pyogenes*
    c. members of the normal skin or mouth flora
    d. members of the genus *Bacteroides*

    ans. c.   diff. 6   obj. 3   p. 9

15. Bacteria that cause SBE rarely gain entrance to the blood stream by
    a. trauma
    b. dental procedures
    c. ingestion of the infectious agent
    d. brushing and flossing teeth

    ans. c.   diff. 7   obj. 3   p. 10

16. Although patients with SBE may have high levels of antibodies, they are of little value in eliminating the bacteria because the
    a. antibodies are nonfunctional
    b. surface antigens change so rapidly the antibodies are ineffective
    c. bacteria are inaccessible to phagocytosis
    d. antibodies are too few

    ans. c.   diff. 5   obj. 3   p. 10

17. The inflammatory effect of immune complexes lodging in the kidney is described as
    a. renal failure
    b. glomerulonephritis
    c. urethritis
    d. renal inflammation

    ans. b.   diff. 5   obj. 3   p. 10

18. If the SBE is on the right side of the heart, the blood from the arm vein will be free of the infective agent because
    a. blood must pass through several areas where phagocytosis will occur
    b. the agent is seen only in arterial samples
    c. the bacteria can localize in the right kidney
    d. the valves of the veins will retard the progress of the agents

    ans. a.   diff. 7   obj. 3   p. 11

19. Currently, the target population of SBE cases has shifted. Which of the following populations has not shown an increase in the frequency of infection?
    a. patients with artificial heart valves
    b. drug abusers
    c. patients with long-term intravenous catheters
    d. dental patients

    ans. d.   diff. 7   obj. 3   p. 11

20. The newer populations affected with SBE show infection more frequently with _____ than any other organism.
    a. *Streptococcus pyogenes*
    b. *S. aureus*
    c. *B. cereus*
    d. *S. epidermidis*

    ans. d.   diff. 6   obj. 3   p. 12

21. Acute bacterial endocarditis differs from subacute bacterial endocarditis in the
   a. suddenness and severity of onset
   b. population affected
   c. resultant damage
   d. development of exotoxin shock

   ans. a.   diff. 6   obj. 3   p. 13

22. Acute bacterial endocarditis is most frequently caused by
   a. *Neisseria gonorrhoeae* or *Escherichia coli*
   b. *Staphylococcus aureus* or *Streptococcus pneumoniae*
   c. *Enterobacter aerogenes* or *Streptococcus viridans*
   d. *Staphylococcus epidermidis* or *Proteus vulgaris*

   ans. b.   diff. 7   obj. 3   p. 13

23. Which of the following conditions can not ordinarily be complicated by ABE?
   a. gonorrhea
   b. pneumonia
   c. pregnancy
   d. bacteremia

   ans. c.   diff. 7   obj. 3   p. 13

24. The acute illness caused by infectious agents or their products circulating in the blood stream is described as
   a. bacteremia
   b. septicemia
   c. parasitemia
   d. systememia

   ans. b.   diff. 7   obj. 4   p. 14

25. _____ percent of the cases of septicemia in the United States are caused by Gram-negative bacteria.
   a. 30
   b. 40
   c. 20
   d. 10

   ans. a.   diff. 8   obj. 4   p. 14

26. The symptoms of violent chills and fever often accompanied by anxiety and rapid breathing, which is be followed by death in severe cases, are indications of
   a. septicemia
   b. subacute endocarditis
   c. acute bacterial endocarditis
   d. bacteremia

   ans. a.   diff. 7   obj. 4   p. 15

27. Which of the following groups of organisms is not responsible for cases of septicemia?
   a. fungi
   b. bacteria
   c. worms
   d. viruses

   ans. c.   diff. 6   obj. 4   p. 15

28. Among the aerobic Gram-negative, rod-shaped bacteria encountered in septicemia are organisms such as
   a. *Bacteroides sp.*
   b. *Pseudomonas aeruginosa*
   c. *Escherichia coli*
   d. *Staphylococcus aureus*

   ans. b.   diff. 7   obj. 4   p. 15

29. The cytokine released from macrophages that is believed to be responsible for causing septic shock is referred to as
   a. bacteremic factor
   b. macrophage shock factor
   c. septicemia
   d. tumor necrosis factor

   ans. d.   diff. 6   obj. 4   p. 18

30. Disseminated intravascular coagulation occurs in septicemia and is often accompanied by hemorrhage due to the
   a. platelets being destroyed by the toxin
   b. absence of clotting proteins in the injured tissues
   c. liver enzymes being destroyed by the endotoxin
   d. accumulation of blood in the small veins

   ans. b.   diff. 8   obj. 4   p. 19

31. Low blood pressure is clinically described by the term
    a. hypotension
    b. hypertension
    c. dystension
    d. pulmonary edema

    ans. a.   diff. 5   obj. 4   p. 19

32. Although multiple organs are affected by endotoxemia, which organ is particularly vulnerable to serious irreversible damage?
    a. lung
    b. heart
    c. liver
    d. bladder

    ans. a.   diff. 8   obj. 4   p. 20

33. Infections that involve the mononuclear phagocyte system show as a prominent feature, the enlargement of
    a. kidneys and bladder
    b. heart and aorta
    c. spleen and lymph nodes
    d. liver and gallbladder

    ans. c.   diff. 7   obj. 5   p. 21

34. Which of the following diseases does not involve the mononuclear phagocyte system?
    a. endocarditis
    b. plague
    c. brucellosis
    d. tularemia

    ans. a.   diff. 8   obj. 5   p. 21

35. Although the bacterial diseases of tularemia, plague, and brucellosis are uncommon in the United States, they are still a threat because they _____.
    a. exist as zoonoses
    b. are responsible for endotoxin shock
    c. complicate surgical procedures
    d. may be brought into the country with imported goods

    ans. a.   diff. 7   obj. 5   p. 21

36. Tularemia can be acquired by humans by
    a. sharpening a knife used in skinning
    b. removing infected fish from the hook on which it was caught
    c. skinning infected animals
    d. being immunized against plague

    ans. c.   diff. 7   obj. 5   p. 22

37. The incubation period of tularemia is usually
    a. several months
    b. 7–14 days
    c. 3–4 weeks
    d. 2–5 days

    ans. d.   diff. 5   obj. 5   p. 23

38. The development of lymph node enlargement in the area of a skin ulceration as a result of a tick or insect bite two to five days after the incident is indicative of
    a. tularemia
    b. brucellosis
    c. endocarditis
    d. septicemia

    ans. a.   diff. 6   obj. 5   p. 23

39. The pleomorphic, nonmotile, aerobic, Gram-negative rod that is the causative agent of tularemia is
    a. *Brucella abortis*
    b. *Francisella tularensis*
    c. *Staphylococcus aureus*
    d. *Shigella flexneri*

    ans. b.   diff. 4   obj. 5   p. 24

40. The observation that *Francisella tularensis* is ingested by phagocytic cells and grows within them suggests that this organism is similar to _____ in its growth and reproductive patterns.
    a. a virus
    b. *Rickettsia rickettsii*
    c. *Trachoma trachactis*
    d. *Mycobacterium tuberculosis*

    ans. d.   diff. 8   obj. 5   p. 24

41. Since isolation of *F. tularensis* can be dangerous to the laboratory technicians, most cases of tularemia are diagnosed by the
   a. characteristic fever, chills, and achiness
   b. rise in antibody titer to the organisms
   c. identification of the animal vector
   d. skin ulceration

   ans. b.   diff. 7   obj. 5   p. 25

42. Which of the following states has not reported cases of tularemia?
   a. Iowa
   b. Alaska
   c. California
   d. Hawaii

   ans. d.   diff. 7   obj. 5   p. 26

43. The common name for tularemia is
   a. rabbit fever
   b. Bang's disease
   c. Hansen's disease
   d. ulcerative colitis

   ans. a.   diff. 6   obj. 5   p. 26

44. Another name for brucellosis is
   a. ulcerative colitis
   b. rabbit fever
   c. undulant fever
   d. Hansen's disease

   ans. c.   diff. 5   obj. 6   p. 26

45. The Danish veterinarian who discovered the cause of cattle brucellosis was
   a. Robert Koch
   b. Frederik Bang
   c. Carl von Linné
   d. George Rat

   ans. b.   diff. 6   obj. 6   p. 26

46. The aerobic, nonmotile, pleomorphic, short, Gram-negative rod that causes undulant fever in humans is
   a. *Francisella tularensis*
   b. *Brucella abortus*
   c. *Yersinia pestis*
   d. *Rickettsia rickettsii*

   ans. b.   diff. 6   obj. 6   p. 28

47. The most important control measure(s) against brucellosis is(are)
   a. elimination of dairy products in the diet
   b. immunization of milk-drinking children
   c. pasteurization of dairy products and inspection of domestic animals for evidence of the disease
   d. use of antibiotics in animal feed and immunization of dairy herds

   ans. c.   diff. 7   obj. 6   p. 29

48. The disease known as the "black death," which was epidemic in the Middle Ages in Europe, is also called
   a. plague
   b. typhus
   c. tularemia
   d. brucellosis

   ans. a.   diff. 5   obj. 7   p. 29

49. In Europe between 1346 and 1350 an epidemic disease that was responsible for the death of approximately one-third of the population was
   a. plague
   b. typhus
   c. tularemia
   d. brucellosis

   ans. a.   diff. 6   obj. 7   p. 29

50. The causative agent of plague is
   a. *Vibrio cholerae*
   b. *Francisella tularensis*
   c. *Brucella abortis*
   d. *Yersinia pestis*

   ans. d.   diff. 5   obj. 7   p. 30

51. The causative agent of plague is transmitted by the bite of an infected _____ carried on a _____.
   a. mite, human
   b. tick, deer
   c. louse, cattle
   d. flea, rat

   ans. d.   diff. 5   obj. 7   p. 30

52. The important virulence factor of *Yersinia pestis*, which contributes to the development of plague in its victims, appears to be its
    a. capsule
    b. A antigen
    c. C antigen
    d. V antigen

    ans. d.   diff. 7   obj. 7   p. 31

53. Dark hemorrhages into the skin and lack of oxygen from lung involvement give the body the physical appearance that probably inspired the name "_____" for the plague.
    a. hemorrhagic fever
    b. black death
    c. pneumonia
    d. congestive heart failure

    ans. b.   diff. 6   obj. 7   p. 31

54. Plague is endemic in rodent populations on all continents except
    a. New Zealand
    b. Asia
    c. Australia
    d. South America

    ans. c.   diff. 7   obj. 7   p. 32

55. The mortality rate for persons with untreated bubonic plague is between _____ and _____ percent.
    a. 50, 80
    b. 10, 30
    c. 70, 90
    d. 20, 50

    ans. a.   diff. 7   obj. 7   p. 33

56. The type of immunization that gives short-term, partial protection against plague is described as a
    a. pooled human serum
    b. attenuated vaccine
    c. gamma globulin
    d. killed vaccine

    ans. d.   diff. 6   obj. 7   p. 33

True/False

1. The cardiovascular system is made up of the heart and lungs.
   false   (heart, arteries, and veins)

2. The function of the lymphatic system is to manufacture white blood cells.
   false   (manufacture many products that contribute to immunity)

3. The function of the arteries is to distribute oxygenated blood to all areas of the body.
   true

4. The lymphatic system is composed of the lymph nodes and the veins.
   true

5. The anatomy of the lymph vessels is closer to that of the arteries than the veins.
   false   (anatomy is much more like veins)

6. Subacute bacterial endocarditis is diagnosed by its typical sudden and severe onset.
   false   (SBE has a gradual onset, ABE has the symptoms described)

7. The most common cause of SBE is *S. aureus*.
   false   (*S. epidermidis*)

8. SBE usually affects healthy individuals who have recently recovered from an acute infection.
   false   (affects valves previously damaged by disease or a congenital defect)

9. Acute bacterial endocarditis is usually caused by the virulent strains of *S. aureus*.
   true

10. ABE is characteristically slow to develop and difficult to diagnose.
    false   (onset is sudden, acute, and violent)

11. Septicemia usually results from bacteremia.
    true

12. Septicemia is most frequently caused by Gram-positive rods and anaerobes of the genus *Clostridium*.
    false   (Gram-negative rods and anaerobes of the genus *Bacteroides*)

13. Bacterial diseases that involve the lymph nodes inhibit red blood cell production.
false   (interfere with mononuclear phagocyte system)

14. In the western United States, tularemia is most often transmitted to humans by infected ticks or infected flies.
true

15. The cause of tularemia is the Gram-negative aerobe *Francisella tularensis.*
true

16. Brucellosis can be transferred from cattle or domestic animals, and is generally caused by species of *Francisella.*
false   (species of *Brucella*)

17. Another name by which tularemia is known is "rabbit fever."
true

18. Another name by which brucellosis is identified is "rabbit fever."
false   (undulant fever or Bang's disease)

19. An important control measure for brucellosis is the elimination of dairy products from the diet.
false   (pasteurization of milk)

20. Plague has also been called the black death.
true

21. Plague was epidemic in Europe in the Middle Ages but is virtually unknown today.
false   (is endemic in rodent populations)

22. Plague can be transferred to the human by the bite of an infected flea.
true

23. Plague is currently endemic on all continents except Australia.
true

24. There is no reservoir for plague other than the human.
false   (many rodents as well as dogs and cats)

25. Aerosols produced by coughing patients can transfer bubonic plague.
false   (pneumonia)

Chapter 28

Multiple Choice

1. In what century is it thought that yellow fever was introduced to the Americas?
   a. 17th
   b. 18th
   c. 16th
   d. 19th

   ans. a.   diff. 5   obj. 1   p. 1

2. Authorities suggest that yellow fever originated on which of the following continents?
   a. Asia
   b. Africa
   c. Europe
   d. Australia

   ans. b.   diff. 6   obj. 1   p. 1

3. Yellow fever is transmitted by which of the following infected vectors?
   a. *Aedes aegypti* mosquito
   b. *Anopheles* mosquito (female)
   c. *Anopheles* mosquito (male)
   d. tsetse fly

   ans. a.   diff. 5   obj. 1   p. 1

4. The person most often identified with the conquest of yellow fever is
   a. Patrick Manson
   b. John Snow
   c. Carlos Finlay
   d. Walter Reed

   ans. d.   diff. 5   obj. 1   p. 2

5. Who successfully eradicated typhoid fever outbreaks at army installations during the Spanish American War?
   a. Battista Grassi
   b. Ronald Ross
   c. Walter Reed
   d. Theobald Smith

   ans. c.   diff. 6   obj. 1   p. 2

6. As early as 1881, it was suggested that an insect vector was responsible for the transmission of yellow fever by
   a. Carlos Finlay
   b. Walter Reed
   c. Theobald Smith
   d. Patrick Manson

   ans. a.   diff. 6   obj. 1   p. 2

7. Ronald Ross and Battista Grassi showed in the latter part of the nineteenth century that the mosquito could carry the
   a. rickettsia of rocky mountain spotted fever
   b. virus of yellow fever
   c. malaria protozoa
   d. relapsing fever spirochete

   ans. c.   diff. 6   obj. 1   p. 2

8. A living organism in which a disease agent multiplies before transmission to a new host is referred to as a(n)
   a. insect
   b. mechanical vector
   c. biological vector
   d. vector

   ans. c.   diff. 5   obj. 1   p. 2

9. The Yellow Fever Commission of the United States Army was established in about 1900 to investigate the causes of yellow fever. During its investigations, a member of the commission who succumbed to the disease was
   a. Jesse Lazear
   b. Walter Reed
   c. James Carroll
   d. Ronald Ross

   ans. a.   diff. 7   obj. 1   p. 2

10. The scientist responsible for establishing mosquito control in the Panama Canal Zone and therefore allowing construction of the canal was
    a. James Carrol
    b. Walter Reed
    c. Jesse Lazear
    d. William Gorgas

    ans. d.   diff. 6   obj. 1   p. 4

11. An increase in the number of mononuclear leukocytes in the blood stream is an indication of which of the following conditions?
    a. leukemia
    b. infectious mononucleosis
    c. viral hepatitis
    d. AIDS

    ans. b.   diff. 5   obj. 2   p. 5

12. After an incubation period of 30 to 60 days, the symptoms of fever, sore throat covered with pus, and enlargement of the spleen and lymph nodes of the neck is indicative of
    a. infectious mononucleosis
    b. leukemia
    c. viral hepatitis
    d. AIDS

    ans. a.   diff. 6   obj. 2   p. 5

13. The DNA virus of the herpesvirus family that is the causative agent of infectious mononucleosis is called the
    a. yellow fever virus
    b. Herpes virus A
    c. HIV-1
    d. Epstein-Barr virus

    ans. d.   diff. 6   obj. 2   p. 5

14. The Epstein-Barr virus was isolated in the early 1960s from
    a. an infected spleen
    b. Burkitt's lymphoma
    c. pus pockets in the throat
    d. fecal samples

    ans. b.   diff. 7   obj. 2   p. 5

15. The cells that have specific receptor sites for the Epstein-Barr virus and are infected by it are the
    a. neutrophils
    b. T lymphocytes
    c. B lymphocytes
    d. monocytes

    ans. c.   diff. 7   obj. 2   p. 6

16. An antibody (IgM) that will react with the antigens of another animal is called a _____ antibody.
    a. heterophile
    b. non-productive
    c. cross-species
    d. Burkitt's

    ans. a.   diff. 6   obj. 2   p. 7

17. The malignancies most closely related to Epstein-Barr virus infections are
    a. myeloma and neuroma
    b. leukemia and infectious mononucleosis
    c. Burkitt's lymphoma and nasopharyngeal carcinoma
    d. sarcoma and adenocarcinoma

    ans. c.   diff. 7   obj. 2   p. 7

18. The EB viral genome is frequently found in specific malignancies of children of East Africa and New Guinea. The genome is found in _____ percent of these cases.
    a. 90–100%
    b. 50–75%
    c. 60%
    d. 30–60%

    ans. a.   diff. 6   obj. 2   p. 7

19. In the United States, evidence suggests that the Epstein-Barr virus may be a factor in causing some malignancies that develop in patients with
    a. polyglomerulonephritis
    b. cardiovascular disease
    c. hypertension
    d. defective immunity

    ans. d.   diff. 7   obj. 2   p. 8

20. Precise diagnostic techniques for demonstration of specific IgM antibody against the Epstein-Barr virus include the use of
    a. precipitation
    b. complement fixation
    c. fluorescent antibody
    d. nucleic acid probes

    ans. d.   diff. 6   obj. 2   p. 8

21. Immune compromised persons who have previously been infected with Epstein-Barr virus as shown by the presence of antibodies in their blood frequently shed viruses in their _____.
    a. feces
    b. urine
    c. saliva
    d. sweat

    ans. c.   diff. 7   obj. 2   p. 9

22. An important mode of transmission of infectious mononucleosis in young adults is by
    a. kissing
    b. sexual intercourse
    c. fecal-oral route
    d. use of contaminated cosmetics

    ans. a.   diff. 6   obj. 2   p. 9

23. Infections such as those caused by *Pneumocystis carinii* occur in individuals with
    a. other fungal infections
    b. severely impaired immune systems
    c. congenital heart disease
    d. multi-ethnic backgrounds

    ans. b.   diff. 5   obj. 3   p. 11

24. In 1981, a number of cases of *P. carinii* pneumonia in homosexual men led to the recognition of the disease known as
    a. Legionnaires' disease
    b. HIV
    c. AIDS
    d. penicillin-resistant gonorrhea

    ans. c.   diff. 6   obj. 3   p. 11

25. The first symptoms of HIV disease become apparent after an incubation period of
    a. 6 months to 6 years
    b. 6 days to 6 weeks
    c. 1 year to 10 years
    d. an undetermined period (as yet)

    ans. b.   diff. 7   obj. 3   p. 12

26. The symptoms of fever, headache, sore throat, muscle aches, enlarged lymph nodes, and a generalized rash appear after a period of six days to six weeks in people with
    a. brucellosis
    b. mumps
    c. HIV disease
    d. yellow fever

    ans. c.   diff. 7   obj. 3   p. 12

27. The end stage of HIV disease is referred to as
    a. leprosy
    b. AIDS
    c. *P. carinii* pneumonia
    d. a fatal complication

    ans. b.   diff. 6   obj. 3   p. 13

28. The fuzzy white patch frequently seen on the tongues of AIDS patients is thought to be the result of
    a. latent Epstein-Barr virus reactivation
    b. candidiasis
    c. gingivitis
    d. trench mouth

    ans. a.   diff. 7   obj. 3   p. 13

29. The fuzzy white patch frequently seen on the tongues of AIDS patients is referred to as
    a. hairy leukoplakia
    b. thrush
    c. opsonized glottis
    d. blanco glottitis

    ans. a.   diff. 7   obj. 3   p. 13

30. Which of the following often presents itself during the AIDS stage of HIV disease as purplish nodular skin swellings that often resemble bruises?
    a. exoerythrocytic cycle
    b. Burkitt's lymphoma
    c. hairy leukoplakia
    d. Kaposi's sarcoma

    ans. d.   diff. 7   obj. 3   p. 13

31. The human immunodeficiency virus type-1 is recognized as being a(n) _____ virus of the _____ family.
    a. RNA, retrovirus
    b. DNA, adenovirus
    c. RNA, neurotropic
    d. DNA, herpes virus

    ans. a.   diff. 6   obj. 3   p. 13

32. The virus that belongs to the lentivirus subgroup of the retrovirus family causes
    a. infectious mononucleosis
    b. recurrent herpes infection
    c. yellow fever
    d. human immunodeficiency virus disease

    ans. d.   diff. 6   obj. 3   p. 13

33. It is believed that the human immunodeficiency virus type 1 appeared in the United States in the
    a. early 1980s
    b. early 1950s
    c. late 1970s
    d. decade of the 60s

    ans. c.   diff. 6   obj. 3   p. 14

34. Outside the body, HIV-1 is easily inactivated by
    a. hexachlorophene
    b. several disinfectants and by moderate heat
    c. drying
    d. alcohol

    ans. b.   diff. 7   obj. 3   p. 14

35. In what year(s) was enough HIV-1 antigen produced to do serological studies?
    a. 1970s
    b. 1983–1984
    c. 1981–1982
    d. 1956–1957

    ans. b.   diff. 8   obj. 3   p. 15

36. Which of the following is an effective inactivating agent that has not proved to be successful in eliminating HIV-1?
    a. moist heat
    b. dry heat
    c. alcohols
    d. formaldehyde

    ans. b.   diff. 7   obj. 3   p. 15

37. What is the estimated percent of sexually active adult homosexual men in the United States?
    a. 10 to 15
    b. 1.4 to 10
    c. 25
    d. 1

    ans. b.   diff. 7   obj. 3   p. 16

38. Which of the following is one of the many ways in which HIV disease has spread throughout the human population?
    a. monogamous sexual conduct
    b. vaccinating young children against HIV
    c. using gamma globulin to prevent hepatitis outbreaks
    d. use of donated blood products to treat hemophiliacs

    ans. d.   diff. 6   obj. 3   p. 16

39. What is the projected figure for the HIV-infected population in the United States by the year 2000? _____ million.
    a. 40
    b. 10
    c. 20
    d. 30

    ans. a.   diff. 7   obj. 3   p. 16

40. Which of the following is an unlikely manner in which the transmission of HIV can occur among the human population?
    a. exposure to blood through contaminated needles
    b. sexual contact
    c. nonintimate contact with infected persons
    d. mother to baby transmission during pregnancy or around the time of childbirth

    ans. c.   diff. 6   obj. 3   p. 17

41. The subpopulations of the body's lymphocytes are identified by the
    a. antigens on their surfaces
    b. percentages in the population
    c. differences in staining results
    d. locations in which each is found

    ans. a.   diff. 6   obj. 3   p. 18

42. The prime target of the HIV are the lymphocytes designated as
    a. T-3
    b. B-a
    c. T-4
    d. T-2

    ans. c.   diff. 6   obj. 3   p. 18

43. The lymphocytes that are responsible for the effective immune activity of macrophages, B lymphocytes, and other T-lymphocyte populations are the cells designated as
    a. T-3
    b. B-a
    c. T-4
    d. T-2

    ans. c.   diff. 6   obj. 3   p. 18

44. Besides the T-4 lymphocytes, what other body cell types are susceptible to HIV-infection?
    a. chondrocytes
    b. erythrocytes
    c. collagen-producing cells
    d. mononuclear phagocytes

    ans. d.   diff. 7   obj. 3   p. 18

45. Which of the HIV-infected cells are capable of crossing the blood-brain barrier and infecting the brain with HIV?
    a. macrophages
    b. epithelial cells
    c. lymphocytes
    d. bone marrow cells

    ans. a.   diff. 6   obj. 3   p. 18

46. The presence of pneumocystosis in a patient suggests AIDS since
    a. over 80 percent of AIDS patients eventually develop the disease
    b. as a protozoan, it is highly infectious
    c. it shows T-4 cell preference
    d. this condition is latent in patients with activated T cells

    ans. a.   diff. 6   obj. 3   p. 23

47. The most common opportunistic brain infection in AIDS patients is
    a. tuberculosis
    b. toxoplasmosis
    c. pneumocystosis
    d. herpes

    ans. b.   diff. 8   obj. 3   p. 24

48. A major cause of disability and ultimate death in AIDS patients is
    a. herpes
    b. tuberculosis
    c. pneumocystosis
    d. toxoplasmosis

    ans. d.   diff. 8   obj. 3   p. 24

49. Repeated infections with different strains of _____ is thought to contribute to the susceptibility to HIV infection of promiscuous male homosexuals.
    a. cytomegalovirus
    b. *Toxoplasma gondii*
    c. herpes virus
    d. *Mycobacterium tuberculosis*

    ans. a.   diff. 7   obj. 3   p. 26

50. Groups of gay men have lowered the incidence of new HIV infections from 10 to 20 percent per year to only _____ percent per year by avoiding unsafe sexual practices.
   a. 1–5
   b. 5–10
   c. 1–2
   d. zero

   ans. c.   diff. 7   obj. 3   p. 28

51. Which of the following approaches to the treatment of HIV disease prevents attachment of the HIV to the host's cells?
   a. pseudomonas toxin
   b. zidovudine
   c. dextran sulfate
   d. alpha interferon

   ans. c.   diff. 8   obj. 3   p. 32

52. Yellow fever was first recognized in _____ in 1648; it was probably introduced there from _____.
   a. United States, Europe
   b. South America, Asia
   c. North America, Australia
   d. Central America, Africa

   ans. d.   diff. 7   obj. 4   p. 32

53. One of the worst epidemics of yellow fever occurred in the 1960s in
   a. Central America
   b. Ethiopia
   c. Argentina
   d. Brazil

   ans. b.   diff. 8   obj. 4   p. 33

54. The term "black vomit" used to describe one of the classic symptoms of yellow fever arises from
   a. gastrointestinal bleeding
   b. bleeding from the nose
   c. bleeding into the skin
   d. high fever

   ans. a.   diff. 6   obj. 4   p. 34

55. The disease caused by an enveloped RNA arbovirus of the flavivirus family is
   a. malaria
   b. yellow fever
   c. AIDS
   d. herpes

   ans. b.   diff. 6   obj. 4   p. 34

56. Yellow fever is transmitted to the susceptible population by
   a. human to human contact
   b. an infected *Aedes* mosquito
   c. contact with contaminated articles
   d. blood products

   ans. b.   diff. 6   obj. 4   p. 34

57. The yellow fever virus affects the circulatory system. Which of the following is not involved in damage to this system?
   a. directly damaging heart muscle
   b. lysing of red blood cells
   c. causing bleeding from blood vessel injury in various tissues
   d. causing disseminated intravascular coagulation

   ans. b.   diff. 7   obj. 4   p. 35

58. The reservoir of yellow fever in the tropical jungles of Central and South America is infected mosquitoes and
   a. soil
   b. humans
   c. flies
   d. primates

   ans. d.   diff. 6   obj. 4   p. 35

59. The injection used to immunize foreign travelers and people who might become exposed to yellow fever is described as a(n)
   a. active attenuated vaccine
   b. genetically engineered inactive vaccine
   c. inactive viral product
   d. passive serum

   ans. a.   diff. 7   obj. 4   p. 35

60. One of the most important and widespread of the world's protozoan diseases is
    a. malaria
    b. amebiasis
    c. giardiasis
    d. balantidiasis

    ans. a.   diff. 6   obj. 5   p. 36

61. What group gave the disease associated with swamps its name *malaria* in the 17th century?
    a. Greeks
    b. Chinese
    c. Italians
    d. Romans

    ans. c.   diff. 5   obj. 5   p. 36

62. In 1902, the scientist _____ received the Nobel Prize for demonstrating the life cycle of the malaria protozoan.
    a. G. Burkitt
    b. M.A. Epstein
    c. Ronald Ross
    d. J. Lazear

    ans. c.   diff. 7   obj. 5   p. 36

63. Approximately how many deaths worldwide are attributable to malaria annually?
    a. 100,000
    b. one million
    c. 650,000
    d. about ten million

    ans. b.   diff. 6   obj. 5   p. 36

64. How many species of *Plasmodium* cause human malaria?
    a. two
    b. three
    c. four
    d. one

    ans. c.   diff. 6   obj. 5   p. 36

65. With which species of *Plasmodium* have the majority of patients in the United States been infected?
    a. *P. vivax*
    b. *P. falciparum*
    c. *P. malariae*
    d. *P. ovale*

    ans. a.   diff. 7   obj. 5   p. 37

66. In which phase of the life cycle is the malaria parasite able to infect its vector?
    a. sporozoites
    b. merozoites
    c. zygotes
    d. gametocytes

    ans. d.   diff. 7   obj. 5   p. 38

67. The specialized sexual form of the circulating plasmodia in the human host is the
    a. merozoite
    b. gametocyte
    c. cyst
    d. sporozoite

    ans. b.   diff. 6   obj. 5   p. 37

68. Which phase of the malaria parasite is responsible for the transmission of the disease from one person to another?
    a. merozoite
    b. gametocyte
    c. cyst
    d. sporozoite

    ans. b.   diff. 6   obj. 5   p. 37

69. Which of the developmental stages of the malaria parasite is injected into the new human host by the mosquito?
    a. sporozoite
    b. gametocyte
    c. cyst
    d. zygote

    ans. a.   diff. 7   obj. 5   p. 38

70. In the human host, which organ is infected by the malaria parasite, where it enlarges and subdivides, producing thousands of daughter cells that are released into the blood stream?
   a. spleen
   b. heart
   c. liver
   d. lymph nodes

   ans. c.   diff. 8   obj. 5   p. 38

71. The fever of malaria is due to the
   a. release of the merozoites from the erythrocytes
   b. accumulation of toxins from the merozoites
   c. secondary infections that follow malaria
   d. disturbances of the hypothalamic settings for temperature

   ans. a.   diff. 7   obj. 5   p. 38

72. For *P. malariae,* the growth cycle of the parasite takes 72 hours, so that fever recurs every _____ day.
   a. fifth
   b. third
   c. second
   d. fourth

   ans. d.   diff. 5   obj. 5   p. 38

73. Infections by *P. falciparum* tend to be very severe because all circulating erythrocytes are susceptible to infections. Thus, high levels of _____ can develop.
   a. carbon dioxide
   b. septicemia
   c. toxin
   d. parasitemia

   ans. d.   diff. 7   obj. 5   p. 39

74. Malaria caused by *Plasmodium vivax* and *P. ovale* often relapse because the organisms continue to infect the liver. This is called the _____ cycle.
   a. malarial
   b. hepatic
   c. exoerythrocytic
   d. cellular

   ans. c.   diff. 7   obj. 5   p. 39

75. In what decade was endemic malaria eliminated from the continental United States?
   a. 1940s
   b. 1930s
   c. 1950s
   d. 1960s

   ans. a.   diff. 6   obj. 5   p. 40

76. In what way can malaria be contracted so that it only affects the red blood cells, not the liver?
   a. via blood transfusions
   b. bite of an infected vector other than the *Anopheles* mosquito
   c. contact with water in which the mosquitoes breed
   d. if cirrhosis of the liver is a problem

   ans. a.   diff. 6   obj. 5   p. 41

77. Some black people are genetically resistant to *P. vivax* malaria because
   a. their red blood cells lack receptors for the parasite
   b. the mosquitoes do not bite them
   c. their liver cells cannot support the parasite's growth
   d. anemia is a constant problem

   ans. a.   diff. 5   obj. 5   p. 41

78. Present hopes for malaria control depend on
   a. detecting infected patients by obtaining blood smears
   b. employing insecticides against the mosquito vector
   c. vaccines currently under development and evaluation
   d. providing treatment of infected cases

   ans. c.   diff. 6   obj. 5   p. 42

79. Malaria is presently responsible for the death of _____ percent of African children under five years of age.
   a. 10–20
   b. 20–30
   c. 30–40
   d. 40–50

   ans. b.   diff. 6   obj. 5   p. 43

True/False

1. Viral diseases of the lymphatic system include herpes simplex diseases and infectious mononucleosis.
   false   ("mono" and AIDS)

2. Infectious mononucleosis is a viral disease principally involving B lymphocytes.
   true

3. The cause of infectious mononucleosis is the Epstein-Barr virus.
   true

4. The virus responsible for Burkitt's lymphoma is related to the HIV-1.
   false   (related to the Epstein-Barr virus)

5. Acquired immunodeficiency syndrome is a late manifestation of the Epstein-Barr virus.
   false   (human immunodeficiency virus)

6. Victims of infections with *Pneumocystis carinii* suggest an immune deficiency.
   true

7. In 1991, an estimated 200,000 patients with AIDS were diagnosed. Approximately 100,000 deaths from AIDS were reported in the same time period.
   false   (270,000 cases of AIDS, with 179,000 deaths before the year 2000)

8. The first symptoms of HIV disease appear after several months of incubation.
   false   (incubation of six days to six weeks)

9. One of the major concerns of patients with HIV disease is the moodiness and confusion associated with early manifestations of the disease.
   false   (persistent enlargement of lymph nodes)

10. AIDS represents the early stage of HIV disease, during which the patient can be treated and cured.
    false   (the end stage during which the patient is highly susceptible to infections)

11. The fuzzy white patch on the tongue in AIDS patients is due to latent Epstein-Barr virus reactivation.
    true

12. Hairy leukoplakia is a condition of abnormal leukocytes seen in AIDS patients.
    false   (fuzzy white patch on the tongue)

13. Kaposi's sarcoma as seen in AIDS patients often occurs as purplish nodular skin swellings.
    true

14. The HIV-1 virus is a DNA virus of the adenovirus family.
    false   (RNA virus of the retrovirus family)

15. It is believed that HIV-1 first appeared in the United States in the late 1970s.
    true

16. After extensive research it appears that the animal virus that is related to HIV-1 is the rabies virus.
    false   (there has been no closely related animal virus found)

17. Outside the body, HIV-1 can survive for a reasonable length of time on contaminated fomites.
    false   (is fragile; cannot survive longer than a few minutes)

18. HIV-1 is inactivated by several disinfectants and moderate heat.
    true

19. Epidemiologic evidence indicates that HIV is transmitted with great frequency by intimate contact with infected persons.
    true

20. Patients with HIV disease can develop several diseases, among them are toxoplasmosis and yellow fever.
    false   (not yellow fever but many others)

21. Abstaining from sexual intercourse will not protect the average monogamous couple from AIDS.
    false   (abstaining and monogamy both will eliminate the risk)

22. People at high risk of HIV infection are homosexual and bisexual men.
    true

23. It is believed that yellow fever was introduced in Central America in the late 1700s from Asia.
    false   (introduced in 1648 from Africa)

24. Yellow fever is caused by an enveloped RNA virus of the flavivirus family.
    true

25. Yellow fever is transmitted to the human by the bite of an infected *Anopheles* mosquito.
    false   (*Aedes* mosquito)

26. One of the most important and widespread of the protozoan diseases is toxoplasmosis.
    false   (malaria)

27. Malaria is a modern disease of urban populations.
   false   (an ancient scourge as evidenced by the early Chinese and Hindu writings)

28. The disease *mal aria,* or bad air, was named by the Italians in the seventeenth century.
   true

29. The sexual reproductive phase of the malaria protozoan takes place in the human host.
   true

30. The febrile phase of malaria is due to the release of gametocytes into the plasma of the patient.
   false   (merozoites)

31. The majority of malaria patients in the United States are infected with *Plasmodium falciparum.*
   false   (*P. vivax*)

Chapter 29

Multiple Choice

1. The search for safe medicines that would kill microbial pathogens was conducted in earnest by _____ in the late 1800s.
   a. Robert Hooke
   b. Robert Koch
   c. Paul Ehrlich
   d. Louis Pasteur

   ans. c.   diff. 5   obj. 1   p. 2

2. The arsenic compound that proved highly effective in treating syphilis was called arsphenamine or
   a. penicillin
   b. Salvarsan
   c. triple sulfa
   d. erythromycin

   ans. b.   diff. 5   obj. 1   p. 2

3. For their work with antisera, two of the following scientists received the Nobel Prize in 1908. They were
   a. Koch and Pasteur
   b. Ehrlich and Metchnikoff
   c. Hooke and Gordon
   d. Fleming and Sulfa

   ans. b.   diff. 6   obj. 1   p. 2

4. In 1910, Ehrlich's success with Salvarsan gave strong support to the concept of
   a. selective toxicity
   b. arsenic therapy
   c. human testing
   d. experimental drug use

   ans. a.   diff. 6   obj. 1   p. 2

5. The premise that substances that poison one kind of cell are generally poisonous to others is the central theme in the treatment of infectious disease. This is referred to as the
   a. developmental theme of antibiotic therapy
   b. minimal infectious concentration
   c. unity of biochemical principle
   d. minimum lethal dose

   ans. c.   diff. 5   obj. 2   p. 3

6. Medicine and/or disinfectants that kill microorganisms have the suffix "_____" in their description.
   a. -anti
   b. -static
   c. -cidal
   d. -genic

   ans. c.   diff. 5   obj. 2   p. 3

7. The term "static" is used to describe medicines and disinfectants that _____ the growth of microbes.
   a. inhibit
   b. kill
   c. mutate
   d. change

   ans. a.   diff. 5   obj. 2   p. 3

8. Prontosil, a red dye, effectively acted on streptococci when it was split by enzymes in the body to produce
   a. penicillin
   b. sulfanilamide
   c. streptomycin
   d. eosin

   ans. b.   diff. 6   obj. 2   p. 4

9. The highest dose of a medication a patient can tolerate without toxic effects, divided by the dose required to control a microbial infection is referred to as its
   a. minimal inhibitory amount
   b. therapeutic ratio
   c. resistance coefficient
   d. concentration equation

   ans. b.   diff. 7   obj. 2   p. 5

10. The group of diverse chemicals that are produced by the natural biosynthetic processes of some molds and bacteria are defined as
   a. antiseptics
   b. insecticides
   c. disinfectants
   d. antibiotics

   ans. d.   diff. 6   obj. 2   p. 5

11. Antibiotics are examples of _____ of microorganisms. They are not directly involved in the growth of the organism that produces them but many show a selective toxicity to other unrelated microbes.
   a. mutated enzymes
   b. waste material
   c. secondary metabolites
   d. DNA plasmids

   ans. c.   diff. 6   obj. 2   p. 6

12. An antibiotic that is made first by the microorganism and then by chemists is referred to as
   a. synthesized
   b. antimetabolite
   c. semisynthetic
   d. genetically engineered

   ans. c.   diff. 5   obj. 2   p. 6

13. Which of the following groups of microorganisms has not been involved in the manufacture of antibiotics?
   a. yeasts
   b. actinomycetes
   c. *Bacillus*
   d. molds

   ans. a.   diff. 6   obj. 2   p. 6

14. Which of the following is not included in the lists of synthetic antibacterial medicines?
   a. isoniazide
   b. sulfa drugs
   c. trimethoprim
   d. dicloxacillin

   ans. d.   diff. 6   obj. 2   p. 9

15. The first sulfa drug to be introduced for widespread medical use was
   a. sulfanilamide
   b. prontosil
   c. trimethoprim
   d. rifampin

   ans. a.   diff. 6   obj. 3   p. 9

16. Which of the following drugs lowered the death rates for pneumonia and meningitis by half within four years of its introduction?
   a. erythromycin
   b. rifampin
   c. ceftriaxone
   d. sulfanilamide

   ans. d.   diff. 6   obj. 3   p. 9

17. Which of the following facts about sulfa drugs is incorrect?
   a. active against Gram-positive and Gram-negative bacteria
   b. bacteriostatic
   c. absorbed well from gastrointestinal tract
   d. have serious side effects in twenty-five percent of patients

   ans. d.   diff. 7   obj. 3   p. 10

18. Trimethoprim functions to alter bacterial metabolism by
   a. inhibiting cell wall development
   b. blocking folic acid synthesis
   c. altering DNA sequences
   d. inhibiting bacterial enzyme production

   ans. b.   diff. 6   obj. 3   p. 11

19. *Pneumocystis carinii*, a fungus that frequently infects AIDS patients, has been found to be susceptible to a combination of
    a. penicillin and tetracycline
    b. trimethoprim and sulfa drugs
    c. tetracycline and trimethoprim
    d. penicillin and sulfa drugs

    ans. b.   diff. 6   obj. 3   p. 12

20. The family of synthetic bactericidal compounds that can be used in patients who are allergic to other antimicrobials or are infected with resistant bacteria is(are)
    a. trimethoprims
    b. sulfa drugs
    c. quinolones
    d. penicillins

    ans. c.   diff. 6   obj. 3   p. 12

21. The synthetic antibacterial medicines ethambutol, ethionamide, and dapsone function in the bacterial cells to
    a. prevent attachment to cell receptors
    b. block cell wall synthesis
    c. interfere with DNA replication
    d. block enzymatic reactions

    ans. d.   diff. 7   obj. 3   p. 12

22. The bactericidal that is particularly useful in treating bacterial vaginosis and some protozoan infections is
    a. metronidazole
    b. ethambutol
    c. cephalosporin
    d. trimethoprim

    ans. a.   diff. 7   obj. 3   p. 13

23. Several different penicillins have been found in mold cultures, and were designated by letters. Which form has been most suitable for treating infections?
    a. X
    b. F
    c. G
    d. K

    ans. c.   diff. 5   obj. 3   p. 13

24. The antibacterial activity of all members of the penicillin family depends on the presence of a(n)
    a. β-(beta) lactam ring
    b. α-(alpha)-reciprocal
    c. resistance to penicillinase
    d. total chemical structure

    ans. a.   diff. 5   obj. 3   p. 14

25. Many penicillins are inactivated by β-(beta) lactamase. How many different enzymes of this category are known?
    a. 30
    b. 40
    c. 10
    d. 20

    ans. a.   diff. 6   obj. 3   p. 15

26. Which antibiotic substances are somewhat similar in structure to penicillin and have the same mode of action?
    a. clavulanic acid
    b. cephalosporins
    c. sulbactom
    d. tetracycline

    ans. b.   diff. 6   obj. 3   p. 15

27. The antibacterial compound that is considered a β-lactam antibiotic but is resistant to destruction by most β-lactamases because of its chemical differences is
    a. trimethoprim
    b. bacitracin
    c. ampicillin
    d. aztreonam

    ans. d.   diff. 7   obj. 3   p. 18

28. Which of the following chemical substances have no antibacterial activity themselves but will enhance the effectiveness of many β-lactam antibiotics by protecting them from β-lactamase?
    a. phenylalanine and methionine
    b. linoleic acid and triminic
    c. clavulanic acid and sulbactam
    d. stomach acid and bile salts

    ans. c.   diff. 6   obj. 9   p. 16

29. Which of the following antimicrobial compounds is not derived from eukaryotes but is produced by certain prokaryotes of the actinomycetes?
    a.   ampicillin
    b.   aztreonam
    c.   bacitracin
    d.   tetracycline

    ans. d.   diff. 6   obj. 3   p. 20

30. Which of the following antibiotics is bacteriostatic rather than bacteriocidal?
    a.   cephalosporin
    b.   penicillin
    c.   tetracycline
    d.   aztreonam

    ans. c.   diff. 7   obj. 3   p. 20

31. The favorable therapeutic ratio of the tetracyclines is due to their active transport into _____ cells.
    a.   mammalian
    b.   bacterial
    c.   eukaryotic
    d.   living

    ans. b.   diff. 6   obj. 3   p. 20

32. Antibiotics that show activity against some strains of Gram-positive and Gram-negative strains of bacteria are referred to as
    a.   broad-spectrum
    b.   isolate usable
    c.   stress-induced
    d.   generally effective

    ans. a.   diff. 5   obj. 3   p. 20

33. Which of the following is not a common side effect of tetracycline use?
    a.   discoloration of teeth in infants and children
    b.   secondary infections
    c.   temporary sterility
    d.   nausea and diarrhea

    ans. c.   diff. 6   obj. 3   p. 21

34. Which of the following antibiotics can readily diffuse into cells and spinal fluid to act on bacterial pathogens beyond the reach of other antibiotics?
    a.   streptomycin
    b.   tetracycline
    c.   chloramphenicol
    d.   aminoglycosides

    ans. c.   diff. 7   obj. 3   p. 22

35. Which of the following is a serious, though infrequent, complication of chloramphenicol use?
    a.   Burkitt's lymphoma
    b.   malabsorption syndrome
    c.   aplastic anemia
    d.   allergic reaction

    ans. c.   diff. 6   obj. 3   p. 22

36. The first antibiotic to be found active against *Mycobacterium tuberculosis* was
    a.   streptomycin
    b.   aminoglycosides
    c.   neomycin
    d.   tobramycin

    ans. a.   diff. 6   obj. 3   p. 23

37. The most feared toxic effect of streptomycin is
    a.   aplastic anemia
    b.   damage to the kidneys
    c.   impairment of the sense of balance
    d.   irreversible deafness

    ans. d.   diff. 7   obj. 3   p. 23

38. Which of the following facts concerning neomycin cannot be corroborated experimentally?
    a.   reduces the number of intestinal bacteria
    b.   toxic when given orally
    c.   can lower blood cholesterol
    d.   decreases toxic levels of blood ammonia

    ans. b.   diff. 8   obj. 3   p. 24

361

39. Which of the following antibiotics is produced by *Streptomyces erythreus,* is active against Gram-positive bacteria, is generally bacteriostatic, and is generally administered orally?
   a. erythromycin
   b. neomycin
   c. streptomycin
   d. tetracycline

   ans. a.   diff. 5   obj. 3   p. 24

40. The major use of erythromycin is to treat infections in patients who are
   a. in the second trimester of pregnancy
   b. suffering endotoxin shock
   c. suffering from HIV
   d. allergic to penicillin

   ans. d.   diff. 7   obj. 3   p. 24

41. Which of the following is not a member of the macrolide group of antibiotics?
   a. azithromycin
   b. erythromycin
   c. neomycin
   d. clarithromycin

   ans. c.   diff. 7   obj. 3   p. 25

42. Which of the derivatives of lincomycin has an antimicrobial spectrum similar to that of erythromycin and is the form in general use?
   a. clindamycin
   b. streptomycin
   c. neomycin
   d. primaquine

   ans. a.   diff. 7   obj. 3   p. 25

43. An uncommon but potentially serious side effect of treatment with lincomycin is the development of
   a. pseudomembranous colitis
   b. irreversible deafness
   c. bacterial vaginosis
   d. toxic shock syndrome

   ans. a.   diff. 6   obj. 3   p. 25

44. Which of the following antibiotics is bactericidal against Gram-positive organisms and is frequently used as a topical antibiotic ointment?
   a. lincomycin
   b. erythromycin
   c. bacitracin
   d. streptomycin

   ans. c.   diff. 6   obj. 3   p. 26

45. Which of the following are considered polypeptide antibiotics?
   a. polymyxin and bacitracin
   b. penicillin V and rifampin
   c. rifampin and polymyxin
   d. erythromycin and penicillin G

   ans. a.   diff. 7   obj. 3   p. 26

46. Which of the following diseases is not generally treated with rifampin?
   a. leprosy
   b. gastroenteritis
   c. tuberculosis
   d. meningitis

   ans. b.   diff. 5   obj. 3   p. 27

47. The mechanism of action for the family of _____ is the competitive inhibition of an enzyme's activity necessary for the organism's metabolism and reproduction.
   a. aminoglycosides
   b. penicillins
   c. sulfa drugs
   d. cephalosporins

   ans. c.   diff. 7   obj. 4   p. 28

48. Which of the following antibiotics impairs cell wall synthesis?
   a. penicillin
   b. isoniazid
   c. sulfones
   d. polymyxin

   ans. a.   diff. 5   obj. 4   p. 29

49. The antibiotic polymyxin functions against pathogenic prokaryotes to
   a. prevent formation of ribosomes
   b. interfere with protein synthesis
   c. damage cytoplasmic membranes
   d. impair cell wall synthesis

   ans. c.   diff. 6   obj. 4   p. 31

50. The antimicrobial that inhibits mRNA synthesis by binding strongly to bacterial DNA-dependent RNA polymerase is
   a. sulfa drugs
   b. quinolones
   c. isoniazid
   d. rifampin

   ans. d.   diff. 7   obj. 4   p. 32

51. The lowest concentration of an antimicrobic capable of preventing growth of the microorganism is referred to as its
   a. least toxic concentration
   b. minimum lethal dose
   c. minimum lethal dose, 50 percent
   d. minimal inhibitory concentration

   ans. d.   diff. 6   obj. 5   p. 35

52. Which of the following statements describes the use of the term "sensitive" in a qualitative rather than a quantitative sense?
   a. no differentiation is made in defining the term "sensitive"
   b. the concentration of an antimicrobial necessary to inhibit bacterial growth *in vitro*
   c. the amount of an antimicrobial that will bring about complete destruction of the infective agent in twenty-four hours
   d. the concentration of the antimicrobial drugs known to occur in the blood of patients under treatment

   ans. d.   diff. 5   obj. 6   p. 37

53. Most medical laboratories determine qualitative susceptibilities of bacterial pathogens by using
   a. automated equipment
   b. filter paper disks impregnated with antimicrobial agents
   c. miniaturized computer-enhanced techniques
   d. laboratory animals injected with both the infectious agent and the antimicrobial being tested

   ans. b.   diff. 6   obj. 6   p. 37

54. The usefulness of all known antimicrobial medications has been limited by the
   a. discovery of new and totally different infectious disease agents
   b. increasing number of allergies that develop to the antimicrobials
   c. discovery of new diseases
   d. development of resistant strains of infectious agents

   ans. d.   diff. 6   obj. 6   p. 39

55. Between the years 1940 and 1993, the percent of penicillin-resistant strains of infectious agents increased from less than 3 percent to greater than
   a. 85 percent
   b. 14 percent
   c. 58 percent
   d. 95 percent

   ans. a.   diff. 6   obj. 6   p. 40

56. The genes for antibacterial resistance are least likely to be located on the
   a. chromosomes
   b. plasmids
   c. cell membrane
   d. transposons

   ans. c.   diff. 5   obj. 7   p. 40

57. The transfer of genetic information from resistant to sensitive strains of bacteria is least likely to take place by
   a. binary fission
   b. transduction
   c. conjugation
   d. transformation
   e. recombination

   ans. a.   diff. 5   obj. 7   p. 40

58. In what year did penicillin-resistant plasmids appear in *N. gonorrhoeae?*
   a. 1982
   b. 1989
   c. 1976
   d. 1949

   ans. c.   diff. 6   obj. 7   p. 41

59. Almost half of the antibiotic production in the United States is
    a. used experimentally in laboratories
    b. given to farm animals in their feed
    c. converted to injectable doses for treatment
    d. used as oral medications for the treatment of childhood diseases

    ans. b.   diff. 6   obj. 7   p. 42

60. The establishment of resistant strains of an organism in the environment is dependent upon all of the factors listed below except which one?
    a. virulence of the pathogen
    b. antibiotic usage
    c. human genetics
    d. mobility of the human population

    ans. c.   diff. 5   obj. 7   p. 43

61. Antimicrobial medicines function in many ways. Which of the following is not an action of these medicines?
    a. neutralization of bacterial toxins
    b. destroy actively growing organisms
    c. prevent the growth of microbes in the system
    d. increase the death rate of the microorganisms

    ans. a.   diff. 5   obj. 7   p. 45

62. The antifungal antibiotic produced by *Streptomyces nodosus* that is directed at the destruction of the cell membrane, is used against fungi of the genera *Aspergillus, Blastomyces, Candida* and others, and can be toxic to the kidneys is
    a. amphotericin B
    b. chloroquine
    c. flucytosine
    d. melarsoprol

    ans. a.   diff. 6   obj. 8   p. 49

63. The synthetic derivative of cytosine that frequently is used in combination with amphotericin B to treat deep tissue infections of fungi is
    a. flucytosine
    b. chloroquine
    c. ketoconazole
    d. fluconazole

    ans. a.   diff. 7   obj. 8   p. 49

64. The oral fungal antibiotic used to treat deep fungal infections by effectively interfering with synthesis of sterol components of fungal cytoplasmic membranes is
    a. flucytosine
    b. ketoconazole
    c. chloroquine
    d. primaquine

    ans. b.   diff. 7   obj. 8   p. 50

65. Fluconazole is an antifungal antibiotic that can be given orally or intravenously and can diffuse into the cerebrospinal fluid to treat
    a. coccidioidomycosis
    b. systemic candidiasis
    c. vaginosis
    d. cryptococcal meningitis

    ans. d.   diff. 7   obj. 8   p. 50

66. The iodine compound potassium iodide, when taken orally, is effective in treating the fungal infection
    a. vaginosis
    b. cryptococcal meningitis
    c. sporotrichosis
    d. pneumonia

    ans. c.   diff. 7   obj. 8   p. 50

67. All substances used for deep fungal infections are also effective for treating superficial fungal infections except
    a. potassium iodide
    b. fluconazole
    c. griseofulvin
    d. nystatin

    ans. a.   diff. 7   obj. 8   p. 51

68. The fungal antibiotic that is useful exclusively against the dermatophytes is
    a. potassium iodide
    b. griseofulvin
    c. ketoconazole
    d. hydroxystilbamadine

    ans. b.   diff. 6   obj. 8   p. 51

69. The fungal antibiotic that is chemically related to amphotericin but because of its toxicity can be used only on body surfaces to treat superficial infections is
    a. nystatin
    b. ketoconazole
    c. clotrimazole
    d. miconazole

    ans. a.   diff. 7   obj. 8   p. 52

70. Dehydroemetine and iodoquinol have similar modes of action and are used in treating diseases caused by members of the genus
    a. *Trichomonas*
    b. *Trypanosoma*
    c. *Naegleria*
    d. *Entamoeba*

    ans. d.   diff. 7   obj. 9   p. 54

71. Which of the following drugs is effective in treating the liver stage of malaria?
    a. primaquine
    b. quinine
    c. chloroquine
    d. iodoquinol

    ans. a.   diff. 7   obj. 9   p. 54

72. The antiviral medication that prevents assembly of the viral core protein of influenza A is
    a. ribavarin
    b. α interferon
    c. amantadine
    d. foscarnet

    ans. c.   diff. 6   obj. 10   p. 57

73. Use of amantadine to prevent influenza A has reduced the number of people who become ill with the disease by _____ percent.
    a. 10 to 60
    b. 25 to 50
    c. 50 to 100
    d. 15 to 70

    ans. c.   diff. 7   obj. 10   p. 57

74. The antiviral medicine that inhibits the function of viral mRNA or degrades mRNA, and can also enhance cellular immunity is
    a. foscarnet
    b. α-interferon
    c. acyclovir
    d. amantadine

    ans. b.   diff. 6   obj. 10   p. 56

75. Although amantadine is effective in preventing influenza A in a certain percentage of patients, it has also shown to be helpful in reducing the symptoms of
    a. AIDS
    b. Parkinson's disease
    c. hepatitis B
    d. herpes

    ans. b.   diff. 7   obj. 10   p. 58

76. Many antiviral medications are effective in treating human diseases because they are structurally
    a. able to replace essential elements of the viral enzymes
    b. capable of blocking cell receptor sites
    c. similar to viral enzymes
    d. similar to the purines and pyrimidines of human DNA and RNA

    ans. d.   diff. 8   obj. 10   p. 58

77. The antiviral medication that has been used to treat AIDS patients by interfering with viral nucleic acid production is
    a. azidothymidine
    b. amantadine
    c. interferon
    d. foscarnet

    ans. a.   diff. 6   obj. 10   p. 58

78. The antiviral substance that is made by human cells in response to invasion by viruses or other intracellular foreign substances is
    a. interferon
    b. trifluridine
    c. amantadine
    d. aidovudine

    ans. a.   diff. 6   obj. 10   p. 59

79. Interferon has been helpful in treating all of the following viral diseases except which one?

    a.   AIDS, in combination with zidovudine

    b.   chronic hepatitis B

    c.   herpes

    d.   Kaposi's sarcoma

    ans. c.   diff. 7   obj. 10   p. 60

True/False

1. The scientist who spent a great deal of his research time searching for the "magic bullet" to fight disease was Ehrlich.
    true

2. Due to the similarity in biochemical makeup of all living things, substances that poison one kind of cell are generously poisonous to others. This is referred to as the "similarity principle."
    false (unity of biochemical principle)

3. Medicines that inhibit microbial activity have the suffix "-cidal" attached to their description.
    false (-static or -stasis)

4. Antimicrobics that kill microorganisms are referred to as "-static."
    false ("-cidal")

5. Penicillin was among the first antibiotics discovered in 1929.
    true

6. Sulfanilamide is an arsenic compound discovered by Salvarsan to treat syphilis.
    true

7. In relation to treatment of diseases with antimicrobials, a low therapeutic ratio is desirable.
    false (high ratio is desirable)

8. The term "antiseptic" is used to define a group of diverse chemicals that are produced by the natural biosynthetic processes of some molds and bacteria.
    false (antibiotic)

9. If an antibiotic is made in part by the microorganisms and in part by the chemist it is referred to as synthetic.
    false (semisynthetic)

10. The usual medication effective in treating strep throat is penicillin V.
    true

11. The first sulfa drug to be introduced for widespread use was sulfanilamide.
    true

12. One of the major complications in the use of sulfanilamide is its possible damage to the kidneys.
    true

13. Trimethoprim and sulfa drugs block folic acid metabolism in the same way so either is effective.
    false (at a different step so they are used together)

14. Quinolones are highly effective oral antibacterial medicines used in treating resistant strains of *Pseudomonas aeruginosa*.
    true

15. Penicillin G is the drug of choice for susceptible organisms, and its development has overcome the problem of penicillin V, which is unstable in acids.
    false (penicillin V is a relatively acid-stable compound)

16. All the penicillins have in common the β-(beta) lactam ring, which can degrade the penicillinase produced by some bacteria.
    false (the bacteria that produce penicillinase inactivate the β-lactam ring)

17. A small percentage of people who are allergic to penicillin are also allergic to the cephalosporins.
    true

18. Tetracyclines are produced by certain fungi and are therefore eukaryotic in origin.
    false (produced by prokaryotes of the actinomycete genus *Streptomyces*)

19. A serious complication in the use of chloramphenicol is aplastic anemia.
    true

20. Neomycin is an aminoglycoside antibiotic too toxic to administer by injection so is always given by mouth.
    true

21. Neomycin treatment has been known to increase the toxic levels of blood ammonia in people with severe liver disease.
    false (decrease toxic levels of blood ammonia)

22. An uncommon but potentially serious side effect of treatment with lincomycins is the development of pseudomembranous colitis.
true

23. Penicillin interferes with the synthesis of the cell membrane in the infectious agents.
false (synthesis of cell wall components)

24. The sulfa drugs function to block folic acid synthesis.
true

25. Susceptibility of microorganisms to antimicrobial medicines can be quantified by determining the minimum lethal dose (MLD).
false (minimal inhibitory concentration [MIC])

26. The terms "resistant" and "sensitive" refer to whether a patient is allergic to a medication or not.
false (the treatability of the microorganism)

27. The percentage of penicillin G resistant strains of *S. aureus* have increased from less than three percent in 1940 to more than 85 percent in 1993.
true

28. Penicillin and tetracycline resistance plasmids appeared in *N. gonorrhoeae* in 1982.
false (penicillin resistance in 1976, tetracycline resistance in 1982)

29. Resistance plasmids are frequently referred to as P factors.
false (R factors)

30. Amphotericin B. is used principally against the genus *Phialophora*.
false (*Aspergillus* etc.)

31. Amphotericin B is a commonly used medicine for deep tissue infections.
true

Matching

1. Match the type of antibacterial with its mode of action:

1.	polymyxins	a)	block folic acid synthesis
2.	β-lactams	b)	prevent cell wall synthesis
3.	sulfa drugs	c)	inhibit ribosome function
4.	aminoglycosides	d)	damage cytoplasmic membrane

1 - d; 2 - b; 3 - a; 4 - c

2. Match the microbial resistance mechanism with the medication affected:

Resistance Mechanism   *Medication Affected*

1.	change in DNA gyrase	a)	aminoglycosides
2.	change in RNA polymerase	b)	quinolones
3.	change in ribosome structure	c)	rifampin
4.	transport of medication out of cell	d)	tetracycline

1 - b; 2 - c; 3 - a; 4 - d

3. Match the anti protozoa medication with the disease it is generally used to treat:

1.	iodoquinol	a)	amebic dysentery
2.	melarsoprol	b)	African sleeping sickness
3.	dehydroemetine	c)	malaria (liver stage)
4.	primaquine	d)	amebic abscess

1 - a; 2 - b; 3 - d; 4 - c

4. Match the antiviral medicines with their mode of action:

1. acyclovir
2. foscarnet
3. α-interferon
4. amantadine

a) prevents assembly of viral core protein
b) inhibits function of viral mRNA or degrades mRNA
c) inhibits reverse transcriptase by attaching to its phosphate receptors
d) inhibits viral DNA synthesis by reacting with DNA polymerases

1 - d; 2 - c; 3 - b; 4 - a

Chapter 30

Multiple Choice

1. An early contributor to the understanding of nitrogen fixation was
    a. Robert Hooke
    b. Sergei Winogradsky
    c. Anton von Leeuwenhoek
    d. Martinus Beijerinck

    ans. d.   diff. 6   obj. 1   p. 2

2. In order to sustain the productivity of a section of farm-land, early farmers would frequently
    a. allow a field to remain fallow for one or more seasons
    b. rejuvenate the soil with chemicals
    c. follow the Farmer's Almanac
    d. move his farm to another area

    ans. a.   diff. 6   obj. 1   p. 2

3. The process by which nitrogen is transformed into forms that are usable by plants and then released to the atmosphere is referred to as
    a. water cycle
    b. nodule building
    c. carbon dioxide exchange
    d. nitrogen fixation

    ans. d.   diff. 5   obj. 1   p. 2

4. The name given to identify the living organisms in a given area is referred to as a(n)
    a. ecosystem
    b. community
    c. niche
    d. genera

    ans. b.   diff. 5   obj. 2   p. 3

5. The term used to describe the interrelationship between the living organisms and their nonliving environment in an area is called a(n)
    a. group
    b. community
    c. ecosystem
    d. town

    ans. c.   diff. 5   obj. 2   p. 3

6. The role an organism plays in its environment as well as the physical space it occupies is called its
    a. community
    b. role
    c. system of environment
    d. ecological niche

    ans. d.   diff. 6   obj. 2   p. 3

7. The study of the relationship of organisms to each other and to their environment is referred to as
    a. ecology
    b. land study
    c. niche and community
    d. recycling

    ans. a.   diff. 6   obj. 2   p. 3

8. The region of the earth inhabited by living organisms is called its
    a. role
    b. biosphere
    c. niche
    d. ecosystem

    ans. b.   diff. 5   obj. 2   p. 4

9. The fundamental requisite for any organism in its environment is that it be capable of
    a. producing
    b. predation
    c. multiplying
    d. cooperation

    ans. c.   diff. 5   obj. 2   p. 4

10. What is the estimated weight per acre of bacteria and fungi in the top six inches of fertile soil?
    a. 10 pounds
    b. 1 ton
    c. 500 tons
    d. 2 tons

    ans. d.   diff. 6   obj. 2   p. 14

11. In harsh environments relatively few species are capable of existing for a significant period. Microorganisms survive by producing
    a. endospores
    b. zygotes
    c. capsules
    d. plasmids

    ans. a.   diff. 6   obj. 2   p. 5

12. The environment immediately surrounding an individual cell is referred to as its
    a. microenvironment
    b. microniche
    c. small community
    d. colony type

    ans. a.   diff. 6   obj. 2   p. 5

13. The species that is best adapted to live in a particular environment is the one that inhabits its _____ to the exclusion of all others.
    a. colony
    b. microniche
    c. specific microenvironment
    d. family

    ans. c.   diff. 6   obj. 2   p. 6

14. According to the basics of ecology, the natural defense mechanism against the establishment of intestinal pathogens is
    a. the environment within the cells of the organisms
    b. the homeostasis between nutrition and digestion
    c. its resident microbial population
    d. the macroenvironment of the body

    ans. c.   diff. 6   obj. 2   p. 7

15. The practice of cells manufacturing enzymes they can use and to cease synthesizing others that have lost their usefulness is referred to as a type of
    a. compromise
    b. adaptation
    c. utility metabolism
    d. nutritional changes

    ans. b.   diff. 6   obj. 2   p. 8

16. If a mutant within a pre-existing population becomes especially well adapted to a new environment where it has less competition, it can indeed become the
    a. dominant organism
    b. mutated species
    c. altered phenotype
    d. pathogen of the group

    ans. a.   diff. 7   obj. 2   p. 8

17. The emergence of antibiotic-resistant organisms in hospital environments is an example of
    a. mutation
    b. survival
    c. adaptation
    d. phenotype

    ans. c.   diff. 8   obj. 2   p. 9

18. The complete breakdown of protein that yields a completely clear and very odorous product is known as
    a. catabolism
    b. digestion
    c. anabolism
    d. putrefaction

    ans. d.   diff. 7   obj. 2   p. 10

19. Which of the following artificially created environments is considered oligotrophic?
    a. distilled H_2O
    b. polluted water
    c. sewers
    d. warm lakes

    ans. a.   diff. 7   obj. 2   p. 10

20. The genus of microorganism that is able to fix atmospheric nitrogen by reducing it to ammonia, which is then available for further conversion is
    a. *Myxobacteria* sp.
    b. *Bacteroides*
    c. *Streptomyces aureofaciens*
    d. *Rhizobium*

    ans. d.   diff. 7   obj. 3   p. 11

21. The atmospheric nitrogen that is reduced to ammonia is then available to the plants, which use it to
   a. remove waste products from the cell
   b. synthesize amino acids and eventually protein
   c. rejuvenate the soil
   d. stimulate the synthesis of cellulose

   ans. b.   diff. 6   obj. 3   p. 11

22. The decomposition of plant, animal, and microbial matter into complex molecules that are degraded and used as carbon and energy sources is referred to as
   a. compost
   b. humus
   c. fertilizer
   d. loam

   ans. b.   diff. 7   obj. 3   p. 12

23. The organism primarily responsible for the degradation of humus to CO_2 is
   a. *Streptomyces*
   b. Penicillium
   c. Actinomycetes
   d. myxobacteria

   ans. c.   diff. 75   obj. 36   p. 13

24. The average turnover rate for humus is approximately _____ year(s).
   a. 40
   b. 1
   c. 5
   d. 20

   ans. a.   diff. 6   obj. 3   p. 13

25. Among the variety of life forms supported by soil are viruses and bacteria, which are referred to as _____ inhabitants.
   a. submicroscopic
   b. macroscopic
   c. inorganic
   d. minerals

   ans. a.   diff. 5   obj. 3   p. 13

26. In desert or arctic soils, which of the following resistant forms of bacteria will be present?
   a. cysts or capsules
   b. zygotes or mycelium
   c. eggs or seeds
   d. endospores or cysts

   ans. d.   diff. 5   obj. 3   p. 13

27. The largest group of bacteria in the soil are the
   a. *Clostridium* sp.
   b. *Bacillus* sp.
   c. actinomycetes
   d. myxomyces

   ans. c.   diff. 6   obj. 3   p. 13

28. Which of the following members of the actinomycetes are not among the common genera found in soil?
   a. *Arthrobacter*
   b. *Nocardia*
   c. *Bacteroides*
   d. *Streptomyces*

   ans. c.   diff. 6   obj. 3   p. 13

29. Which of the following genera of soil bacteria are producers of useful antibiotics?
   a. *Streptomyces*
   b. *Nocardia*
   c. *Arthrobacter*
   d. *Myxobacteria*

   ans. a.   diff. 6   obj. 3   p. 14

30. Which of the following common groups of heterotrophic soil bacterial produce the brightly colored masses that can sometimes be seen on forest debris?
   a. bacteroides
   b. myxobacteria
   c. geosmini
   d. *Candida*

   ans. b   diff. 6   obj. 3   p. 14

31. Which of the following groups of organisms is (are) important in cycling the biologically important elements nitrogen, oxygen, carbon, sulfur, and phosphorus?
    a.   protozoa
    b.   nitrogen-fixing bacteria
    c.   soil fungi
    d.   soil bacteria

    ans. d.   diff. 7   obj. 3   p. 15

32. Which of the following types of fungi occur in symbiotic relationships with plant roots?
    a.   actinocycetes
    b.   deuteromycetes
    c.   mycorrhizae
    d.   streptomycetes

    ans. c.   diff. 8   obj. 3   p. 15

33. In barren soils, which type of symbiont is common?
    a.   animal-plant
    b.   bacterial-plant
    c.   bacterial-antibiotics
    d.   algae-fungal

    ans. d.   diff. 6   obj. 3   p. 16

34. In the upper levels of the soil, the most abundant predators of soil bacteria and algae are the
    a.   protozoa
    b.   fungi
    c.   animals
    d.   green plants

    ans. a.   diff. 5   obj. 3   p. 16

35. The density of protozoa found in most soils is about _____ organisms per gram of soil.
    a. 10^2 to 10^3
    b. 10^4 to 10^5
    c. 10^6 to $10^{6.5}$
    d. 10^7 to 10^9

    ans. b.   diff. 6   obj. 3   p. 16

36. Which of the following is not a primary environmental influence on soil?
    a.   moisture
    b.   sunlight
    c.   temperature
    d.   organic matter

    ans. b.   diff. 7   obj. 4   p. 17

37. The major effect of wetness in the soil is to reduce the _____ of the soil.
    a.   nitrogen
    b.   oxygen
    c.   hydrogen
    d.   phosphorus

    ans. b.   diff. 6   obj. 4   p. 17

38. Continued waterlogging of the soil changes the flora from primarily _____ to mostly _____ species.
    a.   parasitic, saprophytic
    b.   free living, pathogenic
    c.   aerobic, anaerobic
    d.   macroscopic, microscopic

    ans. c.   diff. 6   obj. 4   p. 17

39. Highly acidic or highly alkaline conditions of soil generally inhibit the growth of common _____.
    a.   plants
    b.   fungi
    c.   protozoa
    d.   bacteria

    ans. d.   diff. 6   obj. 4   p. 17

40. As the acidity of soil increases, the relative proportion of fungi _____.
    a.   increases
    b.   decreases
    c.   remain unchanged
    d.   essentially disappears

    ans. a.   diff. 6   obj. 4   p. 18

41. The bacteria that grow best at temperatures between 20°C and 50°C are referred to as
   a. thermophiles
   b. mesophiles
   c. psychrophiles
   d. thermoduric

   ans. b.   diff. 5   obj. 4   p. 19

42. Psychrophiles grow at temperatures below _____ °C.
   a. 20°C
   b. 10°C
   c. 5°C
   d. 0°C

   ans. a.   diff. 5   obj. 4   p. 19

43. Bacteria that grow best at temperatures above 50°C are classified as
   a. thermophiles
   b. mesophiles
   c. psychrophiles
   d. pathogenic

   ans. a.   diff. 5   obj. 4   p. 19

44. The greatest population of bacteria found in compost heaps are
   a. pathogens
   b. psychrophiles
   c. thermophiles
   d. mesophiles

   ans. c.   diff. 5   obj. 4   p. 19

45. The great majority of soil bacteria are
   a. psychrophiles
   b. photosynthetic
   c. pathogenic
   d. heterotrophic

   ans. d.   diff. 6   obj. 4   p. 19

46. The size of the bacterial population in the soil is limited primarily by
   a. depth
   b. moisture content
   c. temperature
   d. organic matter

   ans. d.   diff. 5   obj. 4   p. 19

47. The soil zone that surrounds the plant roots and contains their exudate is termed the
   a. population
   b. organic zone
   c. root ball
   d. rhizosphere

   ans. d.   diff. 5   obj. 4   p. 19

48. Plants are eaten by animals; animals die and are decomposed by microorganisms; microorganisms return the nutrients to the soil. The series of events cited here is an example of
   a. biochemical cycling
   b. life after death
   c. soil enrichment
   d. food cycling

   ans. a.   diff. 6   obj. 5   p. 20

49. The organisms that convert carbon dioxide into organic material are the
   a. producers
   b. primary consumers
   c. secondary consumers
   d. decomposers

   ans. a.   diff. 6   obj. 5   p. 20

50. The organisms that metabolize the organic matter synthesized by the producers are the
   a. decomposers
   b. predators
   c. consumers
   d. producers

   ans. c.   diff. 6   obj. 5   p. 21

51. The organisms that digest and convert dead plant and animal material into small organic molecules are the
   a. producers
   b. decomposers
   c. predators
   d. consumers

   ans. b.   diff. 6   obj. 5   p. 21

52. The conversion process of changing solar or light energy into chemical bond energy is referred to as
    a. photosynthesis
    b. glucogenesis
    c. Calvin cycle
    d. respiration

    ans. a.   diff. 6   obj. 5   p. 22

53. The process carried out by chlorophyll-containing plants and microorganisms that makes light energy available to the consumers and decomposers is
    a. respiration
    b. glucogenesis
    c. Calvin cycle
    d. photosynthesis

    ans. d.   diff. 7   obj. 5   p. 22

54. The communities (of organisms) that grow on the dark ocean floor near hot undersea geysers that spew hydrogen sulfide are supported by
    a. chemoautotrophy
    b. photosynthesis
    c. respiration
    d. conversion

    ans. a.   diff. 6   obj. 5   p. 23

55. The organisms that obtain energy from the oxidation of hydrogen sulfide and use carbon dioxide to synthesize organic compounds are the
    a. converters
    b. respirers
    c. photosynthesizers
    d. chemoautotrophs

    ans. d.   diff. 7   obj. 5   p. 23

56. Oxygen is cycled by the processes of
    a. respiration and photosynthesis
    b. consumption and decomposition
    c. predation and decomposition
    d. inhalation and exhalation

    ans. a.   diff. 5   obj. 5   p. 23

57. The current level of oxygen in our atmosphere is kept in balance by chemical reactions in the atmosphere. Which of the following reactions is not one of those involved?
    a. aerobic respiration
    b. anaerobic respiration
    c. photosynthesis
    d. photosynthesis and respiration

    ans. b.   diff. 5   obj. 5   p. 23

58. The carbon cycle revolves around
    a. green plants
    b. organic compounds
    c. carbon dioxide
    d. primary producers

    ans. c.   diff. 6   obj. 5   p. 24

59. The primary function of bacteria, fungi, protozoa, and small animals in the carbon cycle is
    a. metabolism
    b. anabolism
    c. decomposition
    d. photosynthesis

    ans. c.   diff. 6   obj. 5   p. 24

60. Which of the following is mismatched?
    a. fungi degrade plant material
    b. bacteria decompose carbon dioxide
    c. aerobic decomposition proceeds by aerobic respiration
    d. anaerobic decomposition occurs by fermentation

    ans. b.   diff. 6   obj. 5   p. 25

61. The main gas involved in the aerobic decomposition of organic matter is
    a. oxygen
    b. carbon dioxide
    c. hydrogen sulfide
    d. nitrogen

    ans. b.   diff. 6   obj. 5   p. 25

62. The anaerobic bacteria that gain energy from the oxidation of hydrogen gas and the reduction of carbon dioxide are the
    a. methanogens
    b. decomposers
    c. photosynthetics
    d. heterotrophs

    ans. a.   diff. 6   obj. 5   p. 25

63. The equation $4H_2 + CO_2 \rightarrow CH_4 + 2H_2O$ is an excellent example of
    a. chemoautotrophy
    b. aerobic respiration
    c. methanology
    d. anaerobic respiration

    ans. d.   diff. 8   obj. 5   p. 25

64. A portion of the organic matter produced by the photochemical oxidation of methane is buried under anaerobic conditions in mud or ocean sediments. Under these conditions, it will eventually form
    a. organic ocean rocks
    b. food for fish
    c. the base for ocean plants
    d. fossil fuels

    ans. d.   diff. 7   obj. 5   p. 26

65. Besides carbon and oxygen, the element that organisms require in greatest quantity is
    a. nitrogen
    b. hydrogen
    c. sulfur
    d. iron

    ans. a.   diff. 7   obj. 5   p. 26

66. When nitrogen is combined with hydrogen as in ammonium, or oxygen as in nitrate, the result is _____ compounds.
    a. fixed nitrogen
    b. organic
    c. ionic
    d. fertilizer

    ans. a.   diff. 7   obj. 5   p. 26

67. The industrial fixation of molecular nitrogen that is used to synthesize nitrogen fertilizers is known as the _____ process.
    a. synthetic
    b. fertilizing
    c. Haber-Bosch
    d. Beijerinck

    ans. c.   diff. 6   obj. 5   p. 26

68. All forms of nitrogen undergo a number of transformations. The entire series of reactions involving nitrogen comprise the _____ cycle.
    a. nitrification
    b. decay
    c. rejuvenation
    d. nitrogen

    ans. d.   diff. 5   obj. 5   p. 27

69. The process that degrades the nitrogen found in the upper surface of soil into the components amino acids, purines, and pyrimidines is called
    a. nitrification
    b. ammonification
    c. denitrification
    d. nitrogen fixation

    ans. b.   diff. 5   obj. 5   p. 27

70. The process that converts NH_4^+ to NO_3^- is called
    a. denitrification
    b. ammonification
    d. nitrification
    d. nitrogen fixing

    ans. c.   diff. 5   obj. 5   p. 28

71. The most reduced form of nitrogen is the
    a. ammonium ion
    b. nitrate ion
    c. nitrite ion
    d. nitrous oxide

    ans. a.   diff. 6   obj. 5   p. 28

72. Which of the following nitrogen compounds is toxic to both plants and animals?
    a. nitrite
    b. nitrate
    c. ammonium
    d. nitrous oxide

    ans. a.   diff. 6   obj. 5   p. 29

73. Nitrate itself is not highly toxic to humans but it can be converted to nitrite by intestinal bacteria. The nitrite is toxic because it
    a. degrades protein and causes a deficiency
    b. lowers the oxygen-carrying capacity of blood
    c. inhibits the DNA synthesis of erythrocytes
    d. destroys components of the immune system

    ans. b.   diff. 7   obj. 5   p. 30

74. Which of the following processes is not responsible for the removal of nitrogen from soil?
    a. leaching
    b. rain
    c. plants
    d. denitrification

    ans. b.   diff. 7   obj. 5   p. 30

75. Which of the following is responsible for nitrogen fixation by prokaryotes?
    a. transaminase
    b. DNA ligase
    c. protease
    d. nitrogenase

    ans. d.   diff. 6   obj. 5   p. 31

76. The best known of the free-living forms of the nitrogen fixing bacteria in the grasslands are members of the genus
    a. *Beijerinckia*
    b. *Azotobacter*
    c. *Clostridium*
    d. *Azolla*

    ans. b.   diff. 6   obj. 5   p. 32

77. In aquatic environments, the most significant nitrogen fixers are members of the genus
    a. *Anabaena*
    b. *Azotobacter*
    c. *Clostridium*
    d. *Azolla*

    ans. a.   diff. 7   obj. 5   p. 32

78. The most agriculturally important symbiotic nitrogen-fixing bacteria are members of the genus
    a. *Clostridium*
    b. *Azolla*
    c. *Rhizobium*
    d. *Bacteroides*

    ans. c.   diff. 6   obj. 5   p. 33

79. The form of phosphate that is readily used by most plants is
    a. ferric phosphate
    b. calcium phosphate
    c. orthophosphate
    d. aluminum phosphate

    ans. c.   diff. 6   obj. 5   p. 37

80. The organisms largely responsible for rendering organic phosphorus available to succeeding generations of microorganisms as well as to plants are
    a. prokaryotes and animals
    b. nitrogen-fixing bacteria
    c. protozoa and helminths
    d. bacteria and fungi

    ans. d.   diff. 7   obj. 5   p. 37

81. In many aquatic habitats, growth of algae and cyanobacteria is often limited because concentrations of _____ are too low.
    a. phosphorus
    b. calcium
    c. nitrogen
    d. sulfur

    ans. a.   diff. 6   obj. 5   p. 38

82. The sulfur cycle bears many similarities to the _____ cycles
   a. carbon
   b. nitrogen
   c. oxygen
   d. water

   ans. b.   diff. 5   obj. 5   p. 38

83. Sulfur is present in the soil chiefly as part of
   a. salts
   b. proteins
   c. hydrogen sulfide
   d. amino acids

   ans. b.   diff. 6   obj. 5   p. 39

84. The microbial flora concerned with the oxidation of H_2O to SO_4^{2-} are
   a. *Thiobacillus* and *Beggiatoa*
   b. the photosynthetic autotrophs
   c. the green sulfur bacteria
   d. purple sulfur bacteria

   ans. a.   diff. 5   obj. 5   p. 39

85. The reduction of sulfate to sulfite in soil is carried out primarily by the
   a. *Beggiatoa*
   b. *Thiobacillus*
   c. *Desulfovibrio*
   d. *Pseudomonas*

   ans. c.   diff. 5   obj. 5   p. 39

86. Chemically synthesized compounds are most likely to be biodegradable if they have a chemical composition
   a. similar to that of naturally occurring compounds
   b. that includes a halogen atom
   c. that is enzymatic specific
   d. similar to the soil in which it is found

   ans. a.   diff. 6   obj. 6   p. 40

87. Which of the following herbicides persists in the soil for the shortest period of time?
   a. 2,4,5-trichlorophenoxyacetic acid
   b. monuron
   c. Dalapin
   d. 2,4-dichlorophenoxyacetic acid

   ans. d.   diff. 7   obj. 6   p. 43

88. The organism responsible for degrading PCB's is
   a. *Agrobacterium*
   b. *Achromobacter*
   c. *Pseudomonas*
   d. *Corynebacterium*

   ans. c.   diff. 6   obj. 6   p. 43

True/False

1. The most important environment to a microorganism is its macroenvironment.
   false (microenvironment)

2. Farmers have noted that if the same selected plants are grown in the same fields over a long period the fertility of the soil is maintained.
   false (crop rotation is best)

3. The process that takes nitrogen from the air and transforms it into forms of nitrogen that the plants can use is denitrification.
   false (nitrogen fixation)

4. All the living organisms in an area are referred to as a community.
   true

5. The interaction of the living organisms with their nonliving environment is referred to as an ecosystem.
   true

6. The role that an organism plays in its particular ecosystem is referred to as its role.
   false (ecological niche)

7. The study of the relationship of organisms to each other and to their environment is ecology.
   true

8. The region of the earth inhabited by living organisms is called the ionosphere.
   false (biosphere)

9. It has been estimated that the top six inches of fertile soil may contain fewer than 100,000 bacteria and fungi per acre.
   false (more than two tons of bacteria and fungi per acre)

10. In the laboratory, colonies growing close together contain more organisms than those that are well separated.
    false (close colonies are smaller and contain fewer organisms)

11. The emergence of antibiotic-resistant organisms in hospital environments illustrates adaptation.
    true

12. The dominant organism in fresh milk that ferments the milk sugar lactose to lactic acid is *Lactobacillus casei*.
    false (*Streptococcus lactis*)

13. The bacteria that are able to grow in dilute aqueous solutions are referred to as oligotrophic bacteria.
    true

14. Humans live in an aerobic environment that contains 79 percent oxygen.
    false (nitrogen)

15. Members of the genus *Azotobacter* are able to fix atmospheric nitrogen by reducing it to ammonia, which is then available to plants.
    false (*Rhizobium*)

16. All soil microorganisms are beneficial to plants and humans in one form or another.
    false (some fungi and bacteria are pathogenic to plants and animals, including humans)

17. The organic material in soil is derived from the decomposition of plant, animal, and microbial matter. One of the most important products of this decomposition is humus.
    true

18. *Nocardia* species are producers of a number of useful antibiotics for humans animals.
    false (*Streptomycetes*)

21. The common group of soil microbes responsible for colorful masses on forest debris are the *Streptomycetes*.
    false (myxobacteria)

20. In barren soils, lichens, algal-fungal symbionts will frequently be found.
    true

21. The major cell wall component of woody plants is lignin.
    true

22. Highly alkaline or acid conditions generally inhibit the growth of common fungi.
    false (inhibit bacteria)

Chapter 31

Multiple Choice

1. In 1914, what percent of the bacteriology students at the Pasteur Institute were women?
   a. 0
   b. 10
   c. 25
   d. 38

   ans. b.   diff. 5   obj. 1   p. 1

2. Which of the following women of science made important contributions to the field of bacteriology in the mid-1900s?
   a. Susan Michelmore
   b. Marie Curie
   c. Helena Curtis
   d. Elizabeth McCoy

   ans. d.   diff. 5   obj. 1   p. 2

3. Which of the following scientists named and characterized *Clostridium acetobutylicum?*
   a. Elizabeth McCoy
   b. Louis Pasteur
   c. Robert Koch
   d. Helena Curtis

   ans. a.   diff. 6   obj. 1   p. 2

4. *Clostridium acetobutylicum* is a bacterium used in the production of
   a. neutralizing agents for butyl alcohol
   b. clostridial antitoxins
   c. acetone and butyl alcohol
   d. alcohol from organic sources

   ans. c.   diff. 5   obj. 1   p. 2

5. Photosynthetic aquatic microorganisms are generally incapable of which of the following activities?
   a. produce oxygen
   b. transformation of the sun's energy into chemical energy
   c. decomposition of waste
   d. convert carbon dioxide into carbon compounds

   ans. c.   diff. 6   obj. 1   p. 3

6. Which of the following is an important property of water?
   a. transform mechanical energy into chemical energy
   b. extremely efficient as a solvent
   c. store carbon dioxide for plant use
   d. act as a solvent for lipids

   ans. b.   diff. 5   obj. 2   p. 3

7. An important property of water in relation to the photosynthetic aquatic microbes is water's ability to
   a. absorb light of certain wavelengths
   b. act as a solvent
   c. be distilled for purity
   d. dilute toxins

   ans. a.   diff. 6   obj. 2   p. 4

8. Pure water is almost impossible to obtain. The closest form to this is _____ water.
   a. distilled
   b. rain
   c. deionized
   d. river

   ans. a.   diff. 6   obj. 2   p. 4

9. Which of the following natural sources of water has the fewest impurities?
   a. distilled water
   b. fresh lake water
   c. river water
   d. rainwater

   ans. d.   diff. 6   obj. 2   p. 4

10. Increasing world industrialization has led to a serious pollution problem known as acid rain, which has resulted in increased levels of _____ in the atmosphere.
    a. carbonic and acetic acid
    b. sulfuric and nitric acid
    c. carbon monoxide
    d. sulfur and nitrogen oxides

    ans. d.   diff. 7   obj. 2   p. 5

11. Which of the following is the major cause of acid rain?
    a. by-products of burning fossil fuels that are released into the atmosphere
    b. the change in the biomass composition
    c. change in the species of phytoplankton of the freshwater lakes
    d. destruction of the ozone layer

    ans. a.   diff. 6   obj. 2   p. 8

12. The _____ determines the pH of natural waters unaffected by acid rain?
    a. types of phytoplankton
    b. population of microbial life
    c. content of dissolved substances in the water
    d. flora and fauna on the banks of the lake or stream

    ans. c.   diff. 6   obj. 2   p. 9

13. Which of the following is not considered a freshwater environment?
    a. stagnant ponds
    b. marine marsh
    c. free-flowing rivers
    d. deep lakes

    ans. b.   diff. 5   obj. 3   p. 9

14. Much of the freshwater in the United States is present in
    a. aquifers
    b. surface lakes
    c. free-flowing rivers
    d. stagnant ponds

    ans. a.   diff. 6   obj. 3   p. 10

15. The nutrient enrichment of natural waters is called
    a. eutrophication
    b. oligotrophication
    c. pollution
    d. trophism

    ans. a.   diff. 6   obj. 3   p. 10

16. Aquatic systems high in nutrients are termed
    a. autotrophic
    b. oligotrophic
    c. eutrophic
    d. phototrophic

    ans. c.   diff. 6   obj. 3   p. 10

17. Aquatic systems low in nutrients are termed
    a. eutrophic
    b. heterotrophic
    c. oligotrophic
    d. artesian systems

    ans. c.   diff. 6   obj. 3   p. 10

18. One of the most important elements within an aquatic environment is its
    a. carbon
    b. hydrogen
    c. sulfur
    d. oxygen

    ans. d.   diff. 6   obj. 3   p. 10

19. Any substances that wash from a particular area and are carried into the water are described as
    a. effluents
    b. elements
    c. organics
    d. ions

    ans. a.   diff. 5   obj. 3   p. 10

20. When a lake is said to be stratified, it has the _____ layer at the top and the _____ layer at the bottom.
    a. anaerobic, aerobic
    b. aerobic, anaerobic
    c. soft, hard
    d. hard, soft

    ans. b.   diff. 6   obj. 3   p. 11

21. Which type of aquatic environment is most likely to support large anaerobic bacterial populations?
    a. oligotrophic
    b. eutrophic
    c. effluent
    d. aerobic

    ans. b.   diff. 6   obj. 3   p. 11

22. Swamps and bogs support abundant plant life. As the plants die, they sink to the bottom, after centuries of compression by the weight of the water, the plants turn into a substance called
   a. coal
   b. peat
   c. oil
   d. diamonds

   ans. b.   diff. 6   obj. 3   p. 11

23. What is the approximate salt content (given as a percentage) of seawater?
   a. 0.9
   b. 1.2
   c. 0.05
   d. 3.5

   ans. d.   diff. 6   obj. 3   p. 12

24. Which of the following groups of organisms is more likely to be found growing in seawater rather than freshwater?
   a. halophiles
   b. thermophiles
   c. mesophiles
   d. psychrophiles

   ans. a.   diff. 6   obj. 3   p. 12

25. In which ocean zone do the algae and cyanobacteria carry out photosynthesis?
   a. salt zone
   b. effluent zone
   c. photic zone
   d. anaerobic zone

   ans. c.   diff. 5   obj. 3   p. 13

26. The photic zone of the ocean supports the growth of the primary producers, which are the
   a. fungi and blue-green algae
   b. bacteria and protozoa
   c. algae and cyanobacteria
   d. protozoa and thermophilic bacteria

   ans. c.   diff. 6   obj. 3   p. 13

27. The transition zones of waterways, where freshwater flows into the sea, are known as
   a. estuaries
   b. photic zones
   c. gulfs
   d. salt zones

   ans. a.   diff. 6   obj. 3   p. 13

28. In which of the following environments will the obligate halophiles exhibit their best growth?
   a. Gulf of Mexico
   b. Mississippi River
   c. Great Salt Lake
   d. Indian Ocean

   ans. c.   diff. 5   obj. 3   p. 13

29. The primary producers of aqueous environments require which of these inorganic nutrients as their carbon source?
   a. acetic acid
   b. sunlight
   c. amino acids
   d. CO_2

   ans. d.   diff. 5   obj. 4   p. 14

30. In the aerobic conditions of aqueous environments, which of the following are the primary producers?
   a. protozoa
   b. cyanobacteria
   c. bacteria
   d. fungi

   ans. b   diff. 5   obj. 4   p. 14

31. Which of the following is not an organic nutrient that can be transformed into inorganic materials for use by the living population?
   a. bacteria and fungi
   b. discarded waste
   c. dead leaves
   d. animal corpses

   ans. a.   diff. 6   obj. 4   p. 15

32. The microbial population of freshwater is dictated by
    a. available nutrients
    b. surrounding soil makeup
    c. prevailing winds
    d. microbes in dust

    ans. a.   diff. 6   obj. 4   p. 15

33. Which of the following types of organisms is not among those that make up the bulk of the microbial life in freshwater?
    a. halophiles
    b. algae
    c. cyanobacteria
    d. protozoa

    ans. a.   diff. 5   obj. 4   p. 15

34. The aerobic chemoheterotrophic bacteria that can degrade organic materials are usually found at what level of the lake?
    a. the anaerobic portion of the lake
    b. where the oxygen level is highest
    c. where the light intensity is high
    d. in the oligotrophic segment of the lake

    ans. b.   diff. 6   obj. 4   p. 16

35. Cellulose from dead plants and algae is often degraded by species of bacteria that produce
    a. carbon dioxide and hydrogen
    b. protease
    c. plasmids
    d. cellulases

    ans. d.   diff. 5   obj. 4   p. 17

36. The bacteria that degrade the cellulose from dead plants and algae are members of the genus
    a. *Hyphomicrobium*
    b. *Cytophaga*
    c. *Caulobacter*
    d. *Beggatoria*

    ans. b.   diff. 7   obj. 4   p. 17

37. The genus of stalked bacteria that are able to live in water that is low in nutrients is
    a. *Caulobacter*
    b. *Cytophaga*
    c. *Hyphomicrobium*
    d. *Beggatoria*

    ans. a.   diff. 7   obj. 4   p. 17

38. The stalked bacteria are referred to as the
    a. chemoautotrophics
    b. stratifiers
    c. prosthecates
    d. sulfur oxidizers

    ans. c.   diff. 6   obj. 4   p. 17

39. The bacteria that oxidize H_2S dissolved in the water to inorganic sulfur or sulfate with the production of energy are mainly members of the genera
    a. *Caulobacter*
    b. *Hyphomicrobium*
    c. *Cytophaga*
    d. *Thiobacillus*

    ans. d.   diff. 6   obj. 4   p. 18

40. The bacteria whose metabolism is expressed by the equation $H_2S + \frac{1}{2}O_2 \rightarrow S^\circ + H_2O + energy$ is
    a. *Cytophaga*
    b. *Hyphomicrobium*
    c. *Thiobacillus*
    d. *Caulobacter*

    ans. c.   diff. 6   obj. 4   p. 18

41. The bacteria that oxidize ammonia to nitrite and then to nitrate, and derive energy from the process are referred to as
    a. reducers
    b. nitrifiers
    c. autotrophs
    d. photosynthetic

    ans. b.   diff. 6   obj. 4   p. 19

42. Which of the following genera contain the bacteria referred to as the photosynthetic purple and green sulfur bacteria?
   a. *Cytophaga*
   b. *Thiobacillus*
   c. *Chromatium*
   d. *Caulobacter*

   ans. c.   diff. 7   obj. 4   p. 19

43. The purple and green sulfur bacteria generate energy by the process of
   a. respiration
   b. chemoheterotrophy
   c. chemoautotrophy
   d. photosynthesis

   ans. d.   diff. 7   obj. 4   p. 19

44. The scientist who developed a means of isolating bacteria found in lake sediment samples was
   a. F. Iwanowski
   b. Louis Pasteur
   c. Carl von Linne
   d. Sergei Winogradsky

   ans. d.   diff. 5   obj. 4   p. 21

45. The iron-oxidizing bacteria are the
   a. *Gallionella*
   b. *Thiobacillus*
   c. *Beggiatoa*
   d. *Vibrio*

   ans. a.   diff. 6   obj. 40   p. 23

46. Which of the following is not among the major functions of freshwater bacteria?
   a. contribute to the salinity of the water
   b. decompose organic matter
   c. aid in transforming essential minerals
   d. release dissolved organic compounds into the food web

   ans. a.   diff. 5   obj. 4   p. 23

47. Which of the following statements about the seawater environment is not true?
   a. the salinity of the water is approximately 3.5 percent
   b. ninety percent of the marine environment is below 5°C
   c. the pH of sea water is high, about 8.0
   d. the number of microbial species is greater than in freshwater

   ans. d.   diff. 7   obj. 5   p. 24

48. The salinity of seawater is approximately ____ percent.
   a. 4.8
   b. 3.5
   c. 7.0
   d. 8.0

   ans. b.   diff. 6   obj. 5   p. 24

49. The pH of seawater is approximately
   a. 6.0
   b. 7.0
   c. 8.0
   d. 5.0

   ans. c.   diff. 6   obj. 5   p. 24

50. One method used in an attempt to measure the total mass of living organisms in a liter of seawater is called
   a. epifluorescence counting
   b. precipitate measure
   c. flocculation
   d. titering

   ans. a.   diff. 6   obj. 5   p. 24

51. The dye used in epifluorescence counting is
   a. acridine orange
   b. crystal violet
   c. copper sulfate
   d. eosin

   ans. a.   diff. 6   obj. 5   p. 24

52. A more accurate method used to determine marine biomass than epifluorescence counting is the
   a. Winogradsky column
   b. acridine orange method
   c. luciferin-luciferase enzyme system
   d. titering technique

   ans. c.   diff. 6   obj. 5   p. 24

53. The technique that is used to determine an estimate of the number of viable organisms in a given volume of seawater is
   a. Winogradsky column
   b. acridine orange method
   c. luciferin-luciferase enzyme system
   d. titering technique

   ans. c.   diff. 6   obj. 5   p. 25

54. Attempts to grow marine microorganisms in the laboratory are unsuccessful because the
   a. effects of ocean currents cannot be estimated
   b. exact salinity is difficult to duplicate
   c. variable pressure of the ocean depths cannot be maintained
   d. growth requirements vary so much among species

   ans. d.   diff. 5   obj. 5   p. 25

55. Water pressure in the ocean increases by about ____ atmosphere per ____ meters of depth.
   a. 5, 10
   b. 1, 10
   c. 1.5, 10
   d. 2.0, 15

   ans. b.   diff. 5   obj. 5   p. 25

56. Ocean organisms that can survive the great water pressures present deep in the sea are referred to as
   a. barophiles
   b. thermophiles
   c. psychrophiles
   d. halophiles

   ans. a.   diff. 6   obj. 5   p. 25

57. Which of the following marine microorganisms makes up a large proportion of the mass of organisms growing near the water surface?
   a. bacteria
   b. algae and protozoa of plankton
   c. fungi
   d. viruses

   ans. b.   diff. 5   obj. 5   p. 26

58. The marine bacteria found associated with organic particles function primarily as
   a. reducers of hydrogen
   b. food for larger animals
   c. mineral releasers for the growth of algae
   d. sources of carbon dioxide for algae growth

   ans. b.   diff. 6   obj. 5   p. 27

59. Most marine bacteria are classified as
   a. Gram-negative, motile aerobes or facultative anaerobes
   b. Gram-positive obligate anaerobes
   c. Gram-negative, nonmotile facultative aerobes
   d. Gram-positive, nonmotile facultative anaerobes

   ans. a.   diff. 6   obj. 5   p. 27

60. Which of the following groups of organisms are of major importance as fish pathogens?
   a. barophiles
   b. halophiles
   c. vibrios
   d. bacteroides

   ans. c.   diff. 5   obj. 5   p. 28

61. A fish ailment that is caused by *Pasteurella piscicida* is similar to the human disease
   a. tuberculosis
   b. brucellosis
   c. tularemia
   d. gastroenteritis

   ans. c.   diff. 7   obj. 5   p. 28

62. It has been estimated that the mass of phytoplankton is about ____ times that of the zooplankton of the sea.
   a. 50
   b. 5
   c. 100
   d. 10

   ans. d.   diff. 6   obj. 5   p. 29

63. A large part of the zooplankton of the sea is made up of the
   a. fungi
   b. marine worms
   c. insect larvae
   d. protozoa

   ans. d.   diff. 5   obj. 5   p. 30

64. In a microbiologic sense, what is the major problem associated with water pollution?
a. presence of pathogenic organisms
b. dead fish
c. foul-smelling lakes
d. trash-littered beaches

ans. a.   diff. 5   obj. 6   p. 30

65. An important concern for humans is that pollution of water with pathogenic organisms can lead to
a. disease epidemics
b. foul-smelling and tasting drinking water
c. death of fish
d. unsightly beaches

ans. a.   diff. 6   obj. 6   p. 30

66. The fecal indicator used to monitor the safety of drinking water in most cities is
a. *Shigella dysenteriae*
b. *Streptococcus faecolis*
c. *Staphylococcus aureus*
d. *Escherichia coli*

ans. d.   diff. 6   obj. 6   p. 31

67. The water monitoring system that uses *E. coli* as the fecal indicator is based on the fact that fecal *E. coli* is able to
a. grow on a variety of medias
b. survive in water longer than most pathogens
c. ferment lactose
d. survive freezing

ans. b.   diff. 5   obj. 6   p. 31

68. Often the source of an epidemic caused by fecal contaminants can be traced to
a. a severe drought that lowers the water level of the lake
b. heavy rainfall in the area affected
c. a break in a pipe carrying sewage
d. deliberate contamination of the water supply

ans. c.   diff. 5   obj. 6   p. 32

69. Which of the following diseases is not normally transmitted by polluted water?
a. tuberculosis
b. cholera
c. typhoid
d. dysentery

ans. a.   diff. 6   obj. 6   p. 32

70. An important microbiological aspect of water pollution is the process of nutrient enrichment from human sources known as
a. human contribution
b. accelerated eutrophication
c. sewage addition
d. agricultural addition

ans. b.   diff. 6   obj. 6   p. 33

71. An increase in the algae population of a lake frequently is due to an increase in the amount of
a. nitrogen
b. phosphate
c. sulfate
d. fish

ans. b.   diff. 5   obj. 6   p. 33

72. The most frequently added material to lakes that show algae "blooms" is
a. agricultural runoff
b. acid rain
c. treated sewage
d. dead fish

ans. c.   diff. 6   obj. 6   p. 33

73. High concentrations of nitrates and phosphates in the waters of Lake Washington (and other similar lakes) have led to the rapid and abundant growth of members of the genus
a. *Oscillatoria*
b. *Beggatoria*
c. *Volvox*
d. *Fragilagia*

ans. a.   diff. 6   obj. 6   p. 34

74. The Clean Air Act, which established the Environmental Protection Agency, was enacted in
    a. 1983
    b. 1970
    c. 1976
    d. 1981

    ans. b.   diff. 5   obj. 6   p. 35

75. The federal agency that is charged with enforcing regulations that have to do with air, water, and so forth is the
    a. CDC
    b. FDA
    c. EPA
    d. USDA

    ans. c.   diff. 5   obj. 6   p. 35

76. Experimentation has shown that rates of degradation of organics by microorganisms is ____ times _____ in the ocean depths than at the same temperature in the laboratory.
    a. 2 to 20, faster
    b. 5 to 50, faster
    c. 2 to 20, slower
    d. 10 to 100, slower

    ans. d.   diff. 6   obj. 7   p. 36

77. Which of the following is not significant when considering the influences of the deep ocean on microbial activities?
    a. lack of oxygen
    b. high pressure
    c. nutrient sources
    d. low water temperature

    ans. c.   diff. 7   obj. 7   p. 36

78. Which of the following metallic elements is often present in industrial wastes discarded into bodies of water?
    a. mercury
    b. iron
    c. chlorine
    d. gold

    ans. a.   diff. 6   obj. 7   p. 36

79. The danger to humans from the disposal of mercury into the ocean is the bacterial conversion of the salts to

    _____.
    a. synthetic food products
    b. mercuric chloride
    c. fertilizers
    d. methyl mercury

    ans. d.   diff. 6   obj. 7   p. 37

80. Mercury salts can be converted by bacteria into methyl mercury, which when consumed by humans acts as a powerful
    a. cardiac stimulant
    b. diuretic
    c. nerve toxin
    d. anti spasmodic

    ans. c.   diff. 6   obj. 7   p. 37

True/False

1. The percentage of women in bacteriology classes has increased since the early 1900s.
   true

2. One of the early female contributors to the study of bacteriology in the mid-1900s was Elizabeth McCoy.
   true

3. Elizabeth McCoy did extensive study of the role of microorganisms in salt marsh environments.
   false   (freshwater lakes)

4. The biomass of seawater decreases with increasing depth.
   true

5. Water is an extremely efficient solute and most water is a dilute solution of any number of substances.
   false   (water is a solvent)

6. The effect of acid rain on lakes and streams is to increase the pH to a level dangerous to indigenous plants.
   false   (acid rain lowers the pH)

7. Rainwater usually has an abundance of impurities.
   false   (rainwater has few impurities)

8. Eutrophication of freshwater occurs when nutrients are added to the water.
   true

9. Open oceans are nutrient rich.
   false (nutrient poor)

10. Estuaries are nutrient poor.
    false (nutrient rich)

11. The microbial population of freshwater is dictated by the surrounding soil conditions.
    false (available nutrients)

12. Swamps and bogs contain abundant plant life, which under pressure and extensive time becomes peat.
    true

13. The oceans constitute approximately 41 percent of the earth's surface.
    false (71 percent)

14. The salt content of the ocean is about 3.5 percent.
    true

15. The most abundant group of organisms found growing in seawater are the psychrophiles.
    false (halophiles)

16. The transition zones where freshwater flows into the sea are referred to as estuaries.
    true

17. Great Salt Lake in Utah is unusual in that the level of salinity of water supports the growth of several non-halophilic species.
    false (salinity is greater than seawater and supports only obligate halophiles)

18. The primary producers in aerobic conditions are the algae and cyanobacteria.
    true

19. The bulk of the microbial mass of freshwater is made up of algae, cyanobacteria, and protozoa.
    true

20. The equation $H_2 + \frac{1}{2}O_2 \rightarrow H_2O$ + energy describes the activity of the phototrophic bacteria to produce energy.
    false (hydrogen bacteria, the chemoautotrophs)

Chapter 32

Multiple Choice

1. The structures that the Romans used to deliver water long distances are called
    a. canals
    b. reservoirs
    c. aqueducts
    d. iters

    ans. c.   diff. 5   obj. 1   p. 1

2. An early proponent of the safe removal of a city's human waste was the English activist _____.
    a. Edwin Chadwick
    b. Robert Koch
    c. John Snow
    d. Joseph Lister

    ans. a.   diff. 5   obj. 1   p. 1

3. The threat of cholera stimulated the Board of Health in London to institute human waste reforms in _____.
    a. 1848
    b. 1850
    c. 1900
    d. 1823

    ans. a.   diff. 6   obj. 1   p. 2

4. The fear of which of the following diseases stimulated the Board of Health in London to institute widespread reforms in the removal of human waste from the urban environments?
    a. hepatitis
    b. smallpox
    c. typhoid
    d. cholera

    ans. d.   diff. 5   obj. 1   p. 2

5. The establishment of effective water and sewer systems in most large European and United States cities had virtually eliminated _____ as a threat by the _____ century.
    a. typhoid, middle of the 19th
    b. cholera, end of the 19th
    c. smallpox, beginning of the 19th
    d. typhoid, beginning of the 19th

    ans. b.   diff. 6   obj. 1   p. 2

6. Although he was influential in cleaning up the sewage problems of urban England, Chadwick's goal of _____ was never realized.
    a. clearing out the slums
    b. turning the waste into fertilizer
    c. food preparation sanitation
    d. a workable highway system around the city

    ans. b.   diff. 7   obj. 1   p. 2

7. Inexpensive, chemically synthesized fertilizers were introduced to the agricultural community in
    a. 1854
    b. 1843
    c. 1900
    d. the late 1800s

    ans. d.   diff. 7   obj. 1   p. 2

8. The fertilizer currently used for the growth and maintenance of Douglas fir tree groves in Seattle is
    a. sterile mulched garbage
    b. potassium phosphate
    c. euplankton from the nearby ocean
    d. treated sewage sludge

    ans. d.   diff. 6   obj. 1   p. 2

9. Major epidemic diseases of large towns and cities have frequently been traced to
   a. poor hygiene among the urban populace
   b. poor sanitation practices in food preparation
   c. contamination of water with fecal waste
   d. lack of immunity due to genetics

   ans. c.   diff. 6   obj. 1   p. 3

10. A modern city with a population of about 1 million people faces the disposal of ___ million gallons of domestic sewage daily.
    a. 120
    b. 10
    c. 1
    d. 60

    ans. a.   diff. 5   obj. 1   p. 4

11. As late as 1972, it was estimated that only _____ percent of Americans lived in areas with sewage treatment.
    a. 20 to 40
    b. 30 to 50
    c. 60 to 70
    d. 10

    ans. b.   diff. 5   obj. 1   p. 4

12. Based on the facts presented in this chapter (32), a major problem for both small towns and large cities is
    a. sewage treatment
    b. clean air
    c. clean water
    d. transportation

    ans. a.   diff. 5   obj. 1   p. 4

13. The conversion of organic materials to inorganic forms by microorganisms is known as the process of
    a. mineralization or stabilization
    b. primary treatment
    c. recycling
    d. dumping

    ans. a.   diff. 5   obj. 2   p. 5

14. The physical processes of screening and sedimentation to remove large objects and some particulate matter from sewage is the _____ treatment.
    a. sedimentation
    b. secondary
    c. tertiary
    d. primary

    ans. d.   diff. 5   obj. 2   p. 5

15. The processes of converting sewage material into odorless inorganic, reusable substances is described as the _____ treatment.
    a. secondary
    b. primary
    c. tertiary
    d. quantenary

    ans. a.   diff. 5   obj. 2   p. 5

16. The removal of phosphates and nitrogen compounds from sewage is described as the _____ treatment.
    a. primary
    b. secondary
    c. tertiary
    d. initial

    ans. c.   diff. 5   obj. 2   p. 5

17. The tertiary treatment of sewage involves the
    a. removal of large objects
    b. chemical and biological processes of converting sewage products into odorless inorganic compounds
    c. removal of phosphates and nitrogen compounds
    d. screening and sedimentation of particulate matter

    ans. c.   diff. 6   obj. 2   p. 5

18. Which of the following is not found in domestic sewage?
    a. food wastes
    b. potable water
    c. cleaning compounds
    d. human feces

    ans. b.   diff. 6   obj. 2   p. 5

19. _____ percent of sewage is made up of organic solids.
    a.   21 to 35
    b.   1 to 2
    c.   12 to 18
    d.   15 to 20

    ans. b.   diff. 6   obj. 2   p. 5

20. Industrial waste in sewage has threatened marine food sources. Examples of these industrial waste compounds are
    a.   ferrous sulfate and aluminum sulfate
    b.   food wastes and cleaning compounds
    c.   sludge and detergent residue
    d.   DDT and polychlorinated biphenyl

    ans. d.   diff. 6   obj. 2   p. 6

21. Which of the following is not on the list of toxic heavy metals found in sewage of industrial origin?
    a.   iron
    b.   mercury
    c.   copper
    d.   nickel

    ans. a.   diff. 6   obj. 2   p. 6

22. Which of the following waste products in treated sewage could inhibit its usefulness in fertilizer?
    a.   iron
    b.   waste-degrading microorganisms
    c.   cadmium
    d.   organic food waste

    ans. c.   diff. 6   obj. 2   p. 6

23. What percent of the annual potato harvest is processed to frozen and prepared potato products, thus increasing the amount of organic waste in sewage?
    a.   10 to 25 percent
    b.   30 to 50 percent
    c.   50 to 60 percent
    d.   roughly 66 percent

    ans. b.   diff. 6   obj. 2   p. 7

24. In what year was the discharge of pollutants into navigable waters stopped due to federal action?
    a.   1980
    b.   1968
    c.   1976
    d.   1972

    ans. d.   diff. 6   obj. 24   p. 7

25. What federal action eliminated the discharge of pollutants into navigable waters of the United States?
    a.   Clean Water Restoration Act
    b.   Clean Air Act
    c.   Federal Discouragement Act
    d.   Anti-pollution Act

    ans. a.   diff. 6   obj. 2   p. 7

26. The oxygen-consuming property of a waste-water sample is designated by the term
    a.   stabilization
    b.   lagooning
    c.   activated sludge disposal
    d.   biochemical oxygen demand

    ans. d.   diff. 6   obj. 3   p. 8

27. Large amounts of degradable organic materials in a sample of waste water or other material is reflected in a(n) _____ value.
    a.   aeration
    b.   low BOD
    c.   fecal sample
    d.   high BOD

    ans. d.   diff. 6   obj. 3   p. 8

28. What is the approximate BOD value per liter for unpolluted natural waters?
    a.   500 to 800 mg
    b.   2,000 to 7,000 mg
    c.   100 to 300 mg
    d.   0 to 100 mg

    ans. c.   diff. 6   obj. 3   p. 9

29. Which of the following is not an important consideration when choosing the method of treating sewage to reduce its BOD?
   a.   the BOD of the sewage
   b.   amount of sewage materials
   c.   the source of the sewage to be processed
   d.   the presence of toxic material in sewage

   ans. c.   diff. 7   obj. 3   p. 9

30. The bodies of water into which sewage treatment products are emptied are termed
   a.   affluent waters
   b.   effluent waters
   c.   receiving waters
   d.   receivable BOD

   ans. c.   diff. 5   obj. 3   p. 9

31. Primary treatment of sewage is designed to
   a.   remove materials that will settle out
   b.   aerate the sample to lower toxicity
   c.   neutralize the pathogenic toxins present
   d.   denature the organic waste present

   ans. a.   diff. 5   obj. 3   p. 9

32. The presence of secondary treatment mechanisms in a large percentage of municipal waste treatment plants in the United States are provided for by the
   a.   Clean Water Act
   b.   Clean Air Act
   c.   Federal Dispersal Act
   d.   Save Our Ocean Group

   ans. a.   diff. 5   obj. 3   p. 10

33. Secondary treatment of sewage is designed to stabilize most of the organic materials and
   a.   settle the solids for removal
   b.   reduce the BOD of the sewage
   c.   prepare the effluent for removal
   d.   convert it to inorganic materials

   ans. b.   diff. 7   obj. 3   p. 10

34. The mechanism by which populations of aerobic organisms stabilize sewage at most secondary treatment plants is called the
   a.   activated sludge method
   b.   digester
   c.   mineralizers
   d.   trickling filter method

   ans. a.   diff. 6   obj. 3   p. 10

35. Most of the biologically degradable organic material treated with aerobic organisms in the secondary treatment of sewage is converted into
   a.   the cell material of the organisms growing in the sewage
   b.   gases or oxidized products
   c.   toxic industrial products
   d.   inorganic solids

   ans. b.   diff. 6   obj. 3   p. 11

36. The major proportion of microorganisms present in sewage treatment are
   a.   fungi
   b.   protozoa
   c.   bacteria
   d.   photosynthetics

   ans. c.   diff. 5   obj. 3   p. 11

37. After the bacteria and fungi used in sewage treatment have used certain nutrients, they serve as food for
   a.   protozoa and nematodes
   b.   fungi and nematodes
   c.   algae and fungi
   d.   ciliates and protozoa

   ans. d.   diff. 5   obj. 3   p. 11

38. The anaerobic stabilization of sewage is accomplished by the
   a.   anaerobic organisms
   b.   aerobic treatment
   c.   digester
   d.   aerobic organisms

   ans. c.   diff. 6   obj. 3   p. 12

39. Aerobic activated sludge treatment of sewage usually removes about ____ percent of the BOD.
    a. 35
    b. 95
    c. 75
    d. 50

    ans. b.   diff. 7   obj. 3   p. 13

40. In which sewage treatment are pathogenic bacteria generally eliminated?
    a. primary
    b. tertiary
    c. quantenary
    d. secondary

    ans. d.   diff. 6   obj. 3   p. 13

41. How are virus particles commonly protected from destruction by chlorination of sewage effluents?
    a. within small aggregates of effluent materials
    b. by enclosure within bacterial spores
    c. within their protein envelop
    d. by producing chlorine neutralizers

    ans. a.   diff. 6   obj. 3   p. 14

42. After secondary treatment of sewage by the activated sludge method, what is added to the treated water to coagulate and precipitate phosphate-containing particles?
    a. ammonia
    b. lime
    c. silicate beads
    d. nitrogen

    ans. b.   diff. 5   obj. 3   p. 15

43. In the tertiary treatment, _____ filters remove detergents, pesticides, and other toxic materials.
    a. charcoal
    b. porcelain
    c. diatomaceous
    d. silicate

    ans. a.   diff. 7   obj. 3   p. 15

44. The final treatment of water in the tertiary treatment of sewage uses _____ to kill any remaining microorganisms.
    a. chlorine
    b. ozone
    c. fluorine
    d. iodine

    ans. a.   diff. 6   obj. 3   p. 15

45. In the tertiary treatment of sewage certain bacteria can reduce nitrates completely to N_2. The genera used are
    a. *Azobacter* and *Alcaligenes*
    b. *Pseudomonas* and *Bacillus*
    c. *Serratia* and *Beggiatoria*
    d. *Bacillus* and *Clostridium*

    ans. b.   diff. 6   obj. 3   p. 15

46. Small towns sometimes depend on a process called _____ for their waste disposal.
    a. oxygenation
    b. filtering
    c. lagooning
    d. chlorination

    ans. c.   diff. 6   obj. 4   p. 16

47. Frequently, the smaller sewage treatment plants of small communities and villages of the United States use _____ to ensure safety of waste disposal.
    a. primary treatment only
    b. septic tanks
    c. trickling filters
    d. aerobic degradation

    ans. c.   diff. 6   obj. 4   p. 16

48. Isolated dwellings or small groups of inhabitants use _____ for sewage disposal.
    a. release into a local stream or river
    b. lagooning
    c. burying
    d. septic tanks

    ans. d.   diff. 5   obj. 4   p. 16

49. The sewage treatment system recently developed for some small communities is the subdivision of the land into small lots with a common portion set aside for a series of small ponds. This method is termed
   a. artificial wetlands
   b. lagooning
   c. aerobic degradation
   d. trickling filters

   ans. a.   diff. 5   obj. 4   p. 18

50. The conversion of toxic hydrogen sulfide, formed during sewage treatment, to sulfate ions is accomplished by bacteria such as
   a. *Thiobacillus* sp.
   b. *Bacillus* sp.
   c. *Pseudomonas* sp.
   d. *Actinomyces* sp.

   ans. a.   diff. 6   obj. 4   p. 19

51. One area of waste treatment in which research and development are greatly needed is in
   a. determining the quality of receiving water
   b. the utilization of treated waste residues
   c. changing the pH of the residue
   d. determining the makeup of sewage sludge

   ans. b.   diff. 5   obj. 4   p. 19

52. Under what conditions can sewage sludge not be used as fertilizer?
   a. the presence of toxic compounds
   b. if high phosphate compounds are present
   c. when commercial products are more effective
   d. when the bacterial content is too low

   ans. a.   diff. 6   obj. 4   p. 19

53. A low-cost method of disposing of solid wastes in areas near towns and cities is to use it as
   a. fertilizer
   b. landfill
   c. dams for streams
   d. decorative areas

   ans. b.   diff. 5   obj. 5   p. 21

54. If a layer of dirt is placed over the solid waste each day, the area is called a _____.
   a. sanitary landfill
   b. decorative barking
   c. composting site
   d. synthetic dam

   ans. a.   diff. 6   obj. 5   p. 21

55. A program that has recently been implemented in many cities to reduce the amount of solid waste deposited in landfills is
   a. outlaw use of disposable products
   b. burning
   c. recycling
   d. five residents with a large volume of waste products

   ans. c.   diff. 5   obj. 5   p. 22

56. Which of the following products is (are) not currently included in recycling programs of many cities?
   a. plastic
   b. paper
   c. tin
   d. glass

   ans. c.   diff. 5   obj. 6   p. 22

57. The natural decomposition of organic solid material is referred to as
   a. lagooning
   b. filtering
   c. composting
   d. recycling

   ans. c.   diff. 5   obj. 6   p. 23

58. Which of the following organic wastes cannot be used in backyard composting?
   a. grass clippings
   b. meats and fats
   c. vegetable peelings
   d. nutrient-poor potting soil

   ans. b.   diff. 5   obj. 6   p. 23

59. The compost pile temperature at which pathogens but not thermophiles are killed is about _____ °F.
    a. 25
    b. 75–100
    c. 100
    d. 100–150

    ans. d.   diff. 5   obj. 6   p. 23

60. If a compost pile is turned frequently and other conditions are adequate for good aerobic digestion of organics, the composting can be completed in
    a. six months
    b. three months
    c. six weeks
    d. one year

    ans. c.   diff. 6   obj. 6   p. 23

True/False

1. An early pioneer in the development of waste removal from urban areas was John Snow.
   false   (Edwin Chadwick)

2. The development of an effective sewer system in England was instituted in 1854 following a cholera epidemic.
   false   (1848, due to the threat of a cholera epidemic)

3. Early settlers emptied human waste into local bodies of water.
   true

4. Today, the average American uses about 50 gallons of water daily and produces about 20 gallons of sewage daily.
   false   (uses 150 gallons of water and produces 120 gallons of sewage)

5. The primary treatment of sewage involves the removal of large objects and much of the particulate matter through the physical processes of screening and sedimentation.
   true

6. Tertiary treatment involves the chemical and biological processes of converting the remaining materials in sewage into odorless inorganic substances that can be reused.
   false   (secondary treatment)

7. Secondary treatment of sewage involves the removal from sewage of phosphates and nitrogen compounds that could cause eutrophication.
   false   (tertiary treatment)

8. The source of potentially pathogenic microorganisms in sewage is food waste and cleaning compounds.
   false   (human feces)

9. The toxic heavy metals, such as iron, zinc and gold are frequent industrial by-products of sewage treatment.
   false   (heavy metals are mercury, cadmium, copper, nickel, and zinc)

10. Aerobic oxidation of sewage compounds by microbes will degrade them to inorganic compounds.
    true

11. In the presence of hydrogen, the methanogens can convert CO_2 into methane.
    true

12. The oxygen consuming property of sewage is termed BOD.
    true

13. Low BOD values reflect large amounts of degradable organic materials in a sample of waste water or other material.
    false   (high BOD values)

14. The mechanisms by which populations of aerobic organisms stabilize sewage at most secondary treatment plants is called the lagooning method.
    false   (activated sludge method)

15. The digester treatment of sewage uses anaerobic organisms to act on the remaining solids to remove water so that a minimum of solid matter remains.
    true

16. Pathogenic bacteria are generally eliminated from sewage during its tertiary treatment.
    false   (secondary treatment)

17. Members of the genera *Bacillus* and *Pseudomonas* are used in the tertiary treatment of sewage to reduce nitrates completely to N_2.
    true

18. The channelling of sewage into shallow ponds as a means for waste disposal by cities is becoming a reasonable disposal system.
    false   (used by small towns)

19. The system of channelling sewage into shallow ponds is termed lagooning by small towns and is used as a method of waste disposal.
true

20. Lagooning is sometimes used in place of activated sludge digesters for the secondary treatment of sewage.
false (trickling filters are sometimes used)

21. Isolated dwellings or very small communities often rely on lagooning for sewage disposal.
false (septic tanks)

22. Sewage treatment that involves successive ponds for both aerobic and anaerobic stabilization and ultimately produces clean water for decorative or recreational use is described as lagooning.
false (artificial wetlands)

23. A low-cost method of disposing of the vegetetive waste material of towns and cities is by burning it.
true

24. Backyard composting is a popular way of disposing of the remains of animal and vegetable waste.
false (no animal products)

25. The heat common in an active compost pile is due to the action of thermophiles.
true

26. Aeration of a compost pile will lower its temperature and raise its decomposition rate.
false (aeration will stimulate thermophile activity and speed up decomposition)

Chapter 33

Multiple Choice

1. Most large cities have required the pasteurization of milk and milk products since
   a. 1940
   b. 1930
   c. 1875
   d. 1900

   ans. d.   diff. 5   obj. 1   p. 1

2. Two human diseases caused by milk-borne pathogens are
   a. anthrax and Q fever
   b. brucellosis and tuberculosis
   c. tuberculosis and hepatitis
   d. shigellosis and amebiasis

   ans. b.   diff. 5   obj. 1   p. 1

3. The first woman president of the Society of American Bacteriology was
   a. Alice Catherine Evans
   b. Marie Curie
   c. Margaret Mead
   d. Alice Langsfield

   ans. a.   diff. 6   obj. 1   p. 2

4. The connection between unpasteurized milk and brucellosis in humans was made in ____ by _____.
   a. 1930, Alexander Fleming
   b. 1900, Louis Pasteur
   c. 1885, Robert Koch
   d. 1917, Alice Evans

   ans. d.   diff. 7   obj. 1   p. 2

5. Dairy herds inspected and vaccinated against tuberculosis were common in the _____.
   a. 1940s
   b. 1930s
   c. 1920s
   d. 1900s

   ans. c.   diff. 6   obj. 1   p. 2

6. Dairy herds vaccinated against tuberculosis early in this century have been described as _____ herds.
   a. safe
   b. certified
   c. anti-tuberculin
   d. contemporary

   ans. b.   diff. 6   obj. 1   p. 2

7. The acceptance of the relationship between milk-borne disease and unpasteurized milk was finally made in the
   a. 1930s
   b. 1940s
   c. 1920s
   d. 1950s

   ans. a.   diff. 5   obj. 1   p. 2

8. Beer was produced as early as _____ by the Sumarians and the Babylonians.
   a. 6000 B.C.
   b. 4000 B.C.
   c. 2000 B.C.
   d. 100 A.D.

   ans. a.   diff. 6   obj. 1   p. 3

9. History records that yeast was used to make bread by the _____ in 4000 B.C.
   a. Romans
   b. Egyptians
   c. Sumarians
   d. Chinese

   ans. b.   diff. 6   obj. 1   p. 3

10. The microorganisms first used in food and beverage production were the
    a. molds
    b. yeasts
    c. protozoa
    d. bacteria

    ans. b.   diff. 5   obj. 1   p. 3

11. Food that is unpalatable or unsafe for human consumption is described as
    a. unwanted
    b. contaminated
    c. toxic
    d. spoiled

    ans. d.   diff. 6   obj. 2   p. 3

12. Microorganisms that spoil food are usually
    a. pathogenic for humans
    b. nonpathogenic for people
    c. the normal flora of humans
    d. toxic if consumed by humans

    ans. b.   diff. 5   obj. 2   p. 4

13. As compared to other foods, the average bacterial counts in ground beef are
    a. moderate
    b. low
    c. high
    d. pathogenic

    ans. c.   diff. 5   obj. 2   p. 4

14. Most human pathogens grow best at temperatures near _____ °C.
    a. 10
    b. 45
    c. 25
    d. 37

    ans. d.   diff. 4   obj. 2   p. 4

15. Which of the following organisms is most often responsible for the mold that appears on bread?
    a. *Achromobacter sp.*
    b. *Lactobacillus*
    c. *Aspergillus niger*
    d. *Rhizopus nigricans*

    ans. c.   diff. 5   obj. 2   p. 5

16. The microorganism often involved in the spoilage of a wide variety of foods is
    a. *Pseudomonas* sp.
    b. *Alcaligenes* sp.
    c. *Lactobacillus*
    d. *Flavobacterium* sp.

    ans. a.   diff. 5   obj. 2   p. 5

17. Which of the following microorganisms is able to inhibit the growth of the pathogenic *Salmonella* sp. often found on poultry and perhaps helps to prevent food poisonings?
    a. *Achromobacter* sp.
    b. *Pseudomonas* sp.
    c. *Alcaligenes* sp.
    c. *Serratia* sp.

    ans. b.   diff. 7   obj. 2   p. 5

18. The microbial genus able to transform ethyl alcohol to acetic acid, the principal acid of vinegar is
    a. *Alcaligenes*
    b. *Pseudomonas*
    c. *Salmonella*
    d. *Acetobacter*

    ans. d.   diff. 7   obj. 2   p. 6

19. Which of the following genera of bacteria is one of the principal bacteria that spoil wine?
    a. *Acetobacter*
    b. *Pseudomonas*
    c. *Salmonella*
    d. *Alcaligenes*

    ans. a.   diff. 7   obj. 2   p. 6

20. Which of the following genera of bacteria is not included with those that produce lactic acid?
    a. *Staphylococcus*
    b. *Streptococcus*
    c. *Leuconostoc*
    d. *Lactobacillus*

    ans. a.   diff. 6   obj. 2   p. 6

21. Which of the following genera of bacteria are particularly troublesome in the food industry due to their heat-resistant spores?
    a. *Streptococcus* and *Staphylococcus*
    b. *Bacillus* and *Clostridium*
    c. *Lactobacillus* and *Clostridium*
    d. *Acetobacter* and *Alcaligenes*

    ans. b.   diff. 6   obj. 2   p. 6

22. Which of the following organisms produces a powerful exotoxin that is readily destroyed by heating to 80°C for 10 minutes?
    a. *Bacillus coagulans*
    b. *Clostridium perfringens*
    c. *Clostridium botulinum*
    d. *Staphylococcus* aureus

    ans. c.   diff. 6   obj. 2   p. 6

23. Which of the following genera of fungi infects peanuts and other grains to produce aflatoxin, a carcinogen?
    a. *Mucor*
    b. *Rhizopus*
    c. *Neurospora*
    d. *Aspergillus*

    ans. d.   diff. 5   obj. 2   p. 7

24. The amount of moisture available in foods is designated by the term
    a. aqueous usability
    b. fluid available
    c. dampness quotient
    d. water activity

    ans. d.   diff. 6   obj. 3   p. 8

25. Foods containing unusually high levels of salt or sugar as compared with most fresh foods have a _____ water activity.
    a. dense
    b. high
    c. low
    d. soluble

    ans. c.   diff. 5   obj. 3   p. 8

26. Which of the following types of microorganisms can multiply in or on foods that are heavily salted or dried?
    a. bacteria
    b. fungi
    c. pathogens
    d. protozoa

    ans. b.   diff. 5   obj. 3   p. 8

27. Most acidic foods are spoiled by
    a. fungi
    b. bacteria
    c. pathogens
    d. protozoa

    ans. a.   diff. 6   obj. 3   p. 8

28. *Clostridium botulinum* will not produce its toxin under _____ conditions.
    a. aerobic
    b. alkaline
    c. acid
    d. neutral

    ans. c.   diff. 5   obj. 3   p. 8

29. Which of the following foods is rarely implicated in botulism poisoning?
    a. green beans
    b. pickles
    c. chicken
    d. eggs

    ans. b.   diff. 5   obj. 3   p. 9

30. Which of the following factors is not significant in influencing the kinds and numbers of microorganisms that will multiply in food?
    a. pH
    b. temperature
    c. osmotic pressure
    d. nutrient value

    ans. d.   diff. 6   obj. 3   p. 9

31. Which of the following naturally occurring antimicrobial substances is not found in significant amounts in eggs?
    a. allicin
    b. lysozyme
    c. conalbumin
    d. riboflavin

    ans. a.   diff. 7   obj. 4   p. 10

32. How does the wax on the skin of some fruits inhibit its spoilage?
    a. chelates iron
    b. inhibits growth of bacteria and fungi
    c. lyses cell walls
    d. retards entry

    ans. d.   diff. 7   obj. 4   p. 10

33. What is the substance in cranberries that inhibits the growth of fungi?
    a. benzoic acid
    b. allicin
    c. conalbumin
    d. lacrimatory factor

    ans. a.   diff. 7   obj. 4   p. 11

34. What is the substance in eggs that protects them from being spoiled by most bacteria?
    a. lysozyme
    b. allicin
    c. benzoic acid
    d. riboflavin

    ans. a.   diff. 7   obj. 4   p. 11

35. In 1858, Louis Pasteur reported that _____ was antibacterial.
    a. radishes
    b. onions
    c. garlic
    d. cranberries

    ans. c.   diff. 6   obj. 4   p. 11

36. In laboratory experiments, highly diluted solutions of _____ have been shown to inhibit the growth of *Staphylococcus, Streptococcus,* and several other organisms.
    a. onions
    b. garlic
    c. radishes
    d. cranberries

    ans. b.   diff. 6   obj. 4   p. 11

37. Which of the following food preservation methods is probably the oldest?
    a. low-temperature treatment
    b. high-temperature treatment
    c. drying
    d. irradiation

    ans. c.   diff. 5   obj. 5   p. 12

38. The person responsible for developing an effective means of preserving food that could be easily transported by the military was
    a. Napoleon
    b. Nicholas Appert
    c. Louis Pasteur
    d. Catherine Evans

    ans. b   diff. 6   obj. 5   p. 13

39. The stimulus for the development of an effective food preservation method that would also allow for easy transportation was the
    a. mass migration of populations to the city
    b. defeat of Napoleon's army due to starvation
    c. western migration in the United States
    d. fear of nuclear war

    ans. b.   diff. 6   obj. 5   p. 13

40. One of the primary concerns in canning foods is to be sure that sterilization
    a. maintains the food's nutrient value
    b. does not interfere with the edible value of the produce
    c. destroys endospores in the product
    d. maintains the true flavor of the food

    ans. c.   diff. 7   obj. 5   p. 13

41. The amount of time necessary to kill contaminating organisms at a given temperature is termed the
    a. thermal death time
    b. thermal death point
    c. killing time
    d. thermal time

    ans. a.   diff. 6   obj. 5   p. 14

42. When all other factors remain constant, the higher the concentration of the bacteria in the food the _____ to produce a safe product.
    a. pressure involved is inconsequential
    b. heat treatment time remains the same
    c. shorter the heat treatment required
    d. longer the heat treatment required

    ans. d.   diff. 5   obj. 5   p. 14

43. Which of the following statements about food sterilization is true?
    a. If the pH is neutral, the time must be longer but the temperature lower.
    b. If the pH is low, the time and temperature need not be so high.
    c. If the pH is high, the time and temperature are inverse to each other.
    d. The closer the food to be sterilized is to neutral, the shorter the time and the lower the necessary temperature.

    ans. b.   diff. 7   obj. 5   p. 16

44. Which of the following foods can be processed safely at a lower temperature for a shorter time than the other foods listed?
    a. baked beans
    b. tomatoes
    c. cream style corn
    d. chicken

    ans. b.   diff. 6   obj. 5   p. 16

45. The heat treatment that destroys all non-spore-forming pathogenic organisms without altering the quality of the food is
    a. freezing
    b. sterilization
    c. desiccation
    d. pasteurization

    ans. d.   diff. 6   obj. 5   p. 17

46. The minimum time and temperature requirement for the high-temperature-short-time pasteurization method is
    a. 140°C for 15 seconds
    b. 62°C for 30 minutes
    c. 72°C for 15 seconds
    d. 149°C for 0.5 seconds

    ans. c.   diff. 6   obj. 5   p. 17

47. Which of the following types of organisms are able to grow at household refrigerator temperatures?
    a. psychrophiles
    b. spore-formers
    c. mesophiles
    d. thermodurics

    ans. a.   diff. 6   obj. 4   p. 16

48. At what temperature is the average household refrigerator maintained?
    a. 0°C
    b. 10°C
    c. 4°C
    d. 25°F

    ans. c.   diff. 6   obj. 5   p. 18

49. If fruits and vegetables are to be held in cold storage for many months, the food is stored at approximately _____ °C in the dark at the appropriate humidity and oxygen concentration.
    a. 10
    b. 20
    c. 15
    d. 25

    ans. a.   diff. 6   obj. 5   p. 18

50. A form of food preservation that is safe, stops all microbial growth, and in which foods can be stored for many months with little loss of color, flavor, and other criteria of quality is
    a. drying
    b. freezing
    c. refrigeration
    d. canning

    ans. b.   diff. 5   obj. 5   p. 18

51. Drying prevents microbial growth by
    a. killing the non-spore-forming pathogens
    b. reducing the water activity of the food
    c. killing fungal spores
    d. destroying the nutrients needed by the microbes

    ans. b.   diff. 6   obj. 5   p. 19

52. Drying, salting, and smoking of food products preserves them for long periods. This method is based on the principle of
   a. facilitated diffusion
   b. diffusion
   c. active transport
   d. osmosis

   ans. d.   diff. 7   obj. 5   p. 19

53. Which of the following pathogens is able to grow under quite high salt conditions and can contaminate salted food?
   a. *Staphylococcus aureus*
   b. *Streptococcus pneumoniae*
   c. *Escherichia coli*
   d. *Pseudomonas aeruginosa*

   ans. a.   diff. 6   obj. 5   p. 21

54. The drying method of food preservation that is much better at maintaining the quality of the product is
   a. lyophilization
   b. smoking
   c. salt drying
   d. sun drying

   ans. a.   diff. 6   obj. 5   p. 19

55. In order to prevent salmonellosis from dried eggs, some states now have laws stating that
   a. only frozen eggs are safe to sell
   b. dried eggs cannot be sold in those states
   c. only pasteurized dried eggs can be sold
   d. raw eggs must be washed in detergents before sale

   ans. c.   diff. 7   obj. 5   p. 20

56. Which of the following chemical(s) has not been added to foods as preservatives?
   a. 17% sodium hypochlorite
   b. sodium benzoate
   c. sorbic acid
   d. nitrites and nitrates

   ans. a.   diff. 6   obj. 5   p. 20

57. The most important function, from a health viewpoint, of nitrites added to meat products is the
   a. inhibition of germination and toxin production of spores in cured meat products
   b. formation of a stable pink color to the meat
   c. its reduction to nitric acid
   d. the desirable appearance of the product

   ans. a.   diff. 7   obj. 5   p. 20

58. Frying converts the nitrates and nitrites in meat products to nitrosamines, which react with DNA and act as
   a. gastric irritants
   b. carcinogens
   c. dysentery stimulants
   d. flavor enhancers

   ans. b.   diff. 6   obj. 5   p. 20

59. Meat products treated with nitrates increase the storability of the product and reduce the incidence of
   a. giardiasis
   b. stomach cancer
   c. salmonellosis
   d. botulism

   ans. d.   diff. 6   obj. 5   p. 21

60. Which of the following irradiation methods used in food preservation is only useful in treating the surface of the product?
   a. X rays
   b. gamma rays
   c. ultraviolet rays
   d. microwaves

   ans. c.   diff. 6   obj. 5   p. 21

61. Microwaves are not useful for food preservation because they do not
   a. heat the product evenly
   b. penetrate the interior of the product
   c. kill microorganisms
   d. generate enough heat

   ans. c.   diff. 6   obj. 5   p. 22

62. One of the major advantages of irradiating food with gamma rays is that it
   a. is considered a food additive
   b. must be labelled as "treated with gamma radiation"
   c. is similar to canning in treatment
   d. can be carried out after the food has been packaged

   ans. d.   diff. 6   obj. 5   p. 22

63. Which of the following facts about the FDA's food irradiation approvals is inaccurate?
   a. fruits and vegetables in 1980
   b. wheat and wheat flour in 1963
   c. potatoes in 1964
   d. poultry in 1990

   ans. a.   diff. 7   obj. 5   p. 23

64. The standards for foods shipped between states or territories are set by the
   a. Federal Aviation Agency
   b. Federal Food, Drug, and Cosmetic Acts
   c. United States Department of Health
   d. United States Department of Agriculture

   ans. b.   diff. 6   obj. 6   p. 23

65. Food-borne idiosyncracies include
   a. food-borne infections
   b. food-borne intoxications
   c. chemical food poisonings
   d. food hypersensitivities

   ans. d.   diff. 5   obj. 6   p. 24

66. When food contains microorganisms that are ingested and then colonize the person, the result is described as a(n)
   a. intoxication
   b. infection
   c. hypersensitivity
   d. chemical poisoning

   ans. b.   diff. 5   obj. 6   p. 24

67. The most famous carrier of a food-borne illness was
   a. "Typhoid Mary"
   b. Annie Oakley
   c. Hepatitis Harry
   d. the flea on the infected rat

   ans. a.   diff. 5   obj. 6   p. 24

68. Which of the following diseases is not ordinarily associated with transmission via contaminated food?
   a. syphilis
   b. Q fever
   c. hepatitis
   d. listeriosis

   ans. a.   diff. 5   obj. 6   p. 25

69. When a microorganism growing in a food produces a toxin or converts the food to a toxic product the result is termed a(n)
   a. inflammation
   b. infection
   c. intoxication
   d. hypersensitivity

   ans. c.   diff. 6   obj. 6   p. 26

70. The most dramatic example of food poisoning is due to intoxication with the byproducts of
   a. *S. aureus*
   b. *C. botulinum*
   c. *Aspergillus*
   d. *Claviceps purpura*

   ans. b.   diff. 6   obj. 6   p. 26

71. Ingestion of *Claviceps purpura* in _____ can result in a condition known as ergot poisoning.
   a. grains
   b. peanuts
   c. coconuts
   d. olives

   ans. a.   diff. 6   obj. 6   p. 26

72. The commercial production of milk products are mostly due to the action of
   a. *Staphylococcus* and *Clostridium*
   b. *Lactobacillus* and *Streptococcus*
   c. *Mycobacterium* and *Haemophilus*
   d. *Aspergillus* and lactic acid

   ans. b.   diff. 5   obj. 7   p. 28

73. The first fermented milk products probably were
    a. cottage cheese and ice cream
    b. yogurts
    c. cheddar and provolone cheeses
    d. gouda and Limburger cheeses

    ans. b.   diff. 5   obj. 7   p. 29

74. The inoculation of pasteurized milk with a mixed culture of *Streptococcus thermophilus* and *Lactobacillus bulgaricus*, and then incubating it at 40°C will result in a product known as
    a. sour cream
    b. cottage cheese
    c. gouda cheese
    d. yogurt

    ans. d.   diff. 6   obj. 7   p. 29

75. The unique flavor of yogurt is due primarily to the production of
    a. lactic acid and acetaldehyde
    b. casein and alcohol
    c. rennet and coagulated protein
    d. curd and whey

    ans. a.   diff. 5   obj. 7   p. 30

76. It is thought that cheese originated in _____ over 8,000 years ago.
    a. the Americas
    b. Africa
    c. Asia
    d. Europe

    ans. c.   diff. 5   obj. 7   p. 29

77. The enzyme used in the production of cheese after the formation of lactic acid from lactose due to lactic acid bacteria is
    a. trypsin
    b. casein
    c. chymotrypsin
    d. renin

    ans. d.   diff. 6   obj. 7   p. 30

78. The _____ in the different kinds of milk used to make cheese give the cheese its characteristic flavor.
    a. sugars
    b. protein
    c. amino acids
    d. fatty acids

    ans. d.   diff. 6   obj. 7   p. 31

79. Which of the following microorganisms is used in the production of all the ripened cheeses?
    a. *Streptococcus thermophilus*
    b. *Streptococcus durans*
    c. *Penicillium candidum*
    d. *Streptococcus cremoris*

    ans. d.   diff. 8   obj. 7   p. 31

80. The soft, unripened cheese made due to the action of *Streptococcus lactis* and *Leuconostoc citrovorum* on the milk is
    a. cottage cheese
    b. Romano
    c. Brie
    d. Swiss

    ans. a.   diff. 7   obj. 7   p. 31

81. Soft cheeses contain _____ percent water.
    a. 50 to 80
    b. 20 to 40
    c. 45
    d. 35

    ans. a.   diff. 6   obj. 7   p. 32

82. The cheese ripened with *Propionibacterium* species is known as
    a. Parmesan
    b. Swiss
    c. Cheddar
    d. Brie

    ans. b.   diff. 6   obj. 7   p. 32

83. Which of the following foods is not the result of pickling due to lactic acid bacteria?
    a. olives
    b. sauerkraut
    c. pickles
    d. soy sauce

    ans. d.   diff. 5   obj. 7   p. 34

84. The addition of *Aspergillus oryzae* to a mixture of salt, soybeans, and wheat, which is then allowed to ferment eventually results in a product known as
    a. sourdough
    b. saki
    c. soy sauce
    d. sweet and sour

    ans. c.   diff. 5   obj. 7   p. 34

85. The production of leavened bread is due primarily to the action of _____ on the mixture of flour, sugar, and water.
    a. *Saccharomyces cerevisiae*
    b. *Lactobacillus sanfranciscans*
    c. *Streptococcus lactis*
    d. *Penicillium caseicolum*

    ans. a.   diff. 5   obj. 7   p. 35

86. The characteristic flavor of sourdough bread is due primarily to the action of
    a. *Penicillium caseicolum*
    b. *Lactobacillus sanfranciscans*
    c. *Lactobacillus* sp.
    d. *Streptococcus lactis*

    ans. b.   diff. 6   obj. 7   p. 35

87. The starting material for brandy is
    a. molasses
    b. grain mash
    c. rice
    d. fruit juice

    ans. d.   diff. 5   obj. 7   p. 36

88. The resulting solids and juices of grapes used to make wine are termed its
    a. germinater
    b. mash
    c. must
    d. malt

    ans. c.   diff. 6   obj. 7   p. 37

89. In the fermentation vat, _____ is often added to the must to inhibit growth of the natural microbial population of the grapes.
    a. sulfur dioxide
    b. carbon dioxide
    c. *S. cerevisiae*
    d. alcohol

    ans. a.   diff. 6   obj. 7   p. 37

90. The most prevalent cause of wine spoilage is the presence of _____ in the must.
    a. lactobacilli
    b. sulfur dioxide
    c. carbon dioxide
    d. ethyl alcohol

    ans. a.   diff. 6   obj. 7   p. 38

91. The product made by the fermentation of starch and germinated barley by species of *Saccharomyces* is commonly known as
    a. wine
    b. sake
    c. beer
    d. soy sauce

    ans. c.   diff. 6   obj. 7   p. 39

92. The germinated grain used in beer making is known as
    a. hops
    b. must
    c. wort
    d. malt

    ans. d.   diff. 5   obj. 7   p. 39

93. The highest per capita consumption of beer worldwide is in
    a. China and Japan
    b. West Germany and Australia
    c. United States and Great Britain
    d. France and Italy

    ans. b.   diff. 6   obj. 7   p. 40

94. The yeast used to make ales is *S. cerevisiae,* and is known as
    a. distillator
    b. bottom yeast
    c. top-fermenting
    d. malt

    ans. c.   diff. 7   obj. 7   p. 41

95. The distilled spirit traditionally made from the fermentation of juices of the agave plant is known as
    a. tequila
    b. scotch
    c. whiskey
    d. gin

    ans. a.   diff. 6   obj. 7   p. 41

96. Vinegar is an oxidation product of ethyl alcohol by species of acetic acid bacteria usually of the genus
    a. *Lactobacillus*
    b. *Azobacter*
    c. *Acetobacter*
    d. *Streptococcus*

    ans. c.   diff. 6   obj. 7   p. 42

True/False

1. Most large cities have required that only pasteurized milk be sold within its limits since 1900.
   true

2. The common milk-borne diseases are tuberculosis and cholera.
   false (tuberculosis and brucellosis)

3. In the 1920s, dairy herds were inspected and vaccinated for tuberculosis.
   true

4. The earliest civilizations that produced beer were the Greeks and the Romans in 4000 B.C.
   false (Sumerians and Babylonians in 6000 B.C.)

5. The microorganisms responsible for food spoilage are usually pathogenic for humans.
   false (nonpathogenic)

6. The most common food spoilage organisms are members of the genus *Staphylococcus.*
   false (*Pseudomonas*)

7. Food poisoning is often caused by *Streptococcus lactis.*
   false (*Staphylococcus aureus*)

8. Organisms that have low water activity are subject to invasion by fungi.
   true

9. Only a few species of bacteria are capable of causing spoilage of acidic foods such as tomatoes.
   false (molds or yeasts)

10. Improper home canning, especially of pickles and jellies that have an acidic pH are often the source of many cases of botulism.
    false (improper canning of vegetables and meats that have a more nearly neutral pH)

11. The antibacterial effects of garlic have been observed since the days of pyramid building in Egypt.
    true

12. Both Louis Pasteur and Albert Schweitzer accepted onions as an effective antibacterial agent.
    false (garlic)

13. The antibacterial effects of using salt in food preservation has its basis in the principles of active transport and exocytosis.
    false (osmosis and diffusion)

14. The Frenchman, Nicholas Appert, was awarded 12,000 francs for developing a method of canning food that could be easily transported.
    true

15. One of the primary concerns in canning food is to be sure that sterilization destroys the endospores of *C. perfringens.*
    false (*C. botulinum*)

16. The fact that the higher the sterilizing temperature the shorter the time needed to preserve the food reflects the thermal death point.
    false (thermal death time)

17. The time and temperature necessary for processing baked beans are much higher than those needed for tomatoes.
true

18. Pasteurization is a sterilizing treatment that destroys all spore-forming pathogenic organisms without significantly altering the quality of the food.
false (pasteurization is a heat treatment that destroys all non-spore-forming pathogenic organisms)

19. Growth of spoilage organisms is slowed or halted by high-temperature treatment.
false (low-temperature treatment)

20. Some spoilage microorganisms are mesophilic and are able to grow at household refrigerator temperatures of 32°F or 0°C.
false (psychrophiles, 40°F or 4°C)

21. Desiccation prevents microbial growth by reducing the water activity of food and therefore water is not available for metabolic reactions.
true

22. Irradiation is an effective means of preserving foods. Ultraviolet rays can penetrate the product readily and produce excellent results.
false (UV light can only treat the surface, does not penetrate)

23. Gamma radiation has the advantage that the sterilization treatment can be carried out after the food has been packaged.
true

24. Food-borne infections occur when food contains microorganisms that are ingested and then colonize the host.
true

25. The most serious example of food intoxication is due to *Staphylococcus aureus*.
false (*Clostridium botulinum*)

26. Among the first fermented dairy products were cheese and yogurt.
true

27. It is thought that cheese originated in Asia over 8,000 years ago.
true

28. The characteristic flavors and textures of different cheeses result from the temperatures at which the cheese is ripened.
false (the different kinds of milk used)

29. Soy sauce is a fermentation product of soy beans due to the action of *Propionibacterium shermanii* on the wort.
false (*Aspergillus oryzae*, the mixture of salt, soy beans, and wheat with the mold is called koji)

Matching

1.   Match the type of spoilage with the most likely organism responsible:

1.	fishy odor in fish	a)	*Aspergillus niger*
2.	sour raw milk	b)	*Streptococcus cremoris*
3.	bread mold	c)	*Lactobacillus thermophilus*
4.	slimy chicken	d)	*Pseudomonas* species
5.	sour pasteurized milk	e)	*Achromobacter* sp.

1 - e; 2 - b; 3 - a; 4 - d; 5 - c

Match the beverage with its starting material:

1.	whiskey	a)	fruit juice
2.	sake	b)	rice
3.	vodka	c)	grain mash
4.	table wine	d)	potatoes

1 - c; 2 - b; 3 - d; 4 - a